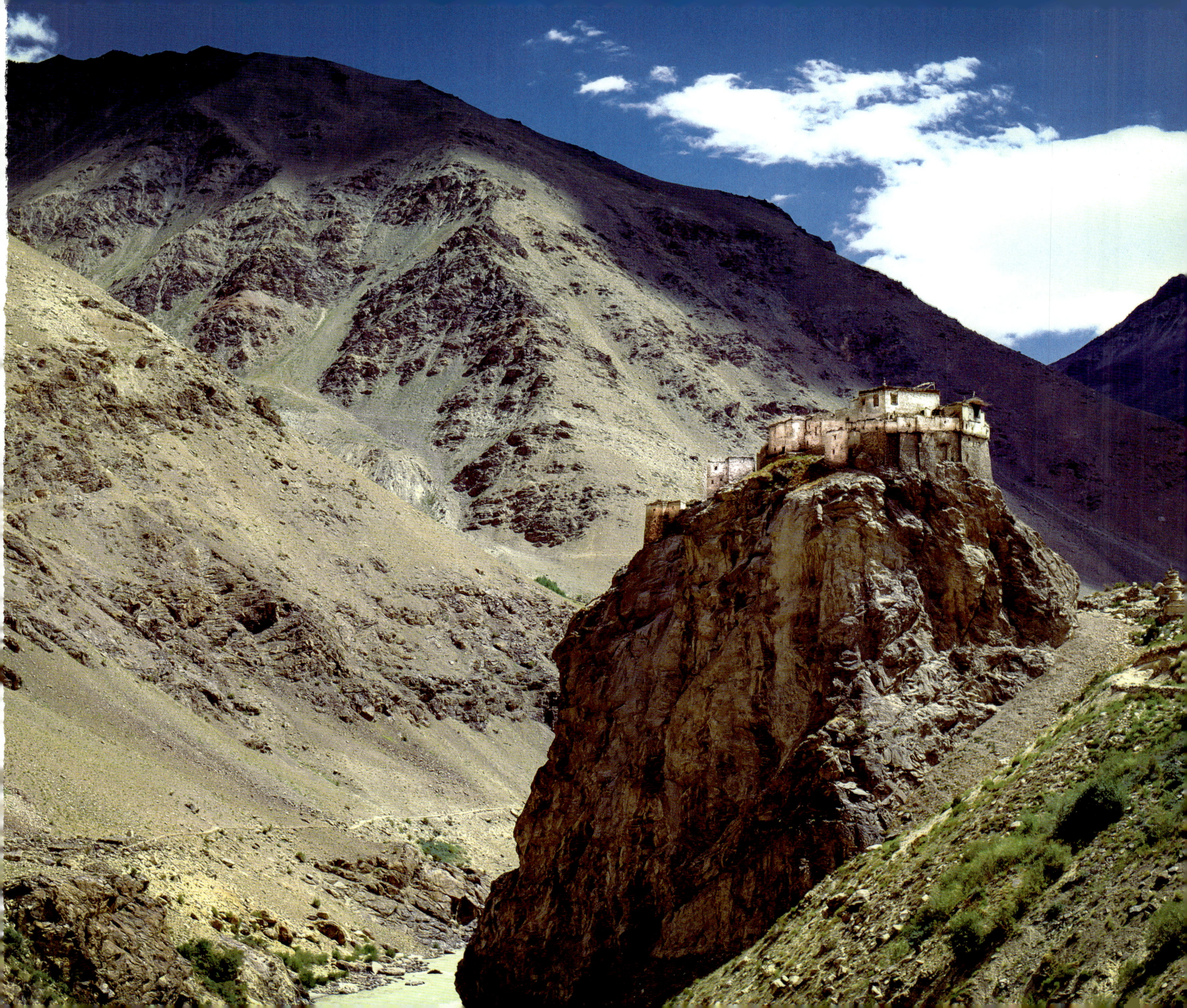

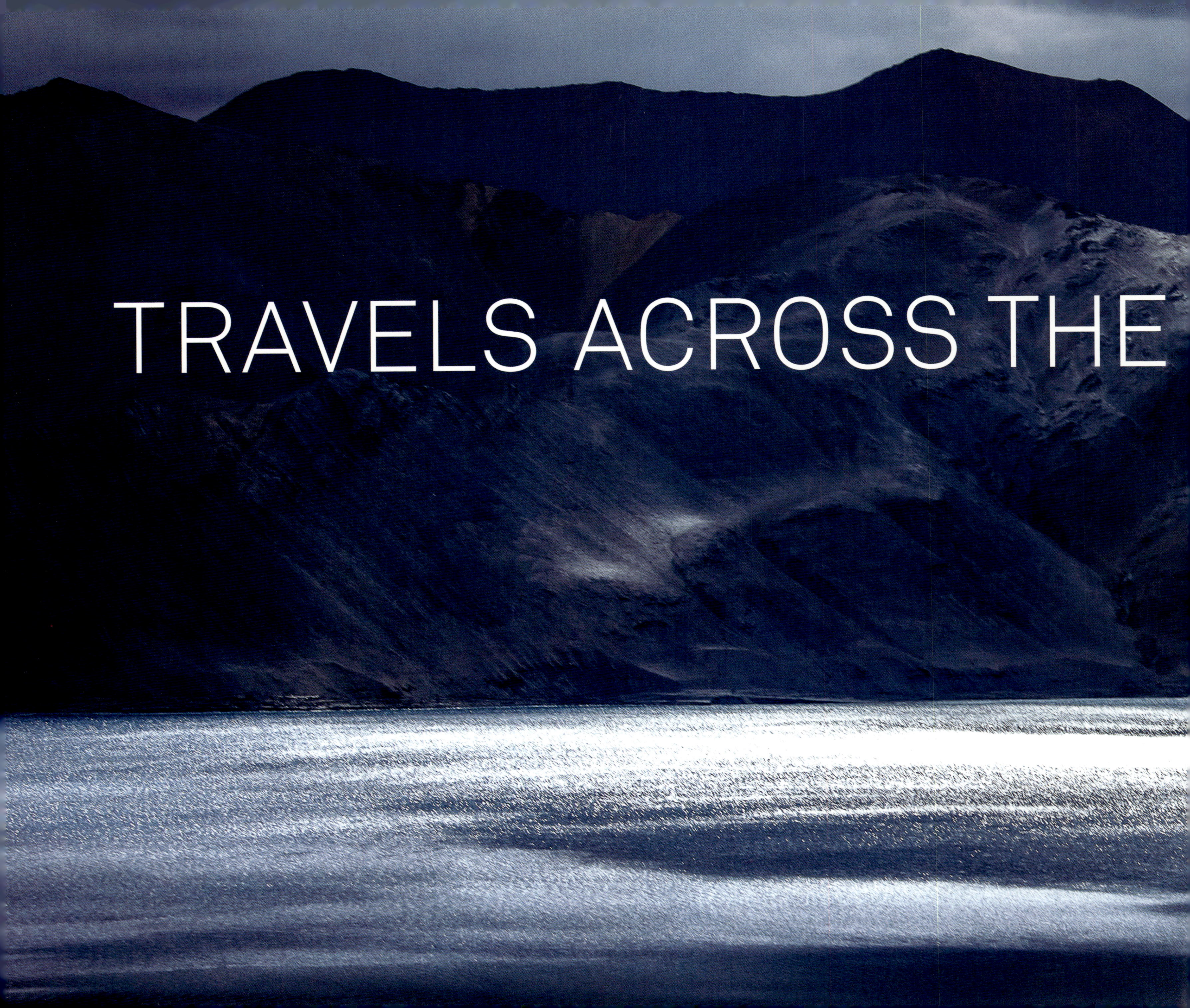
TRAVELS ACROSS THE

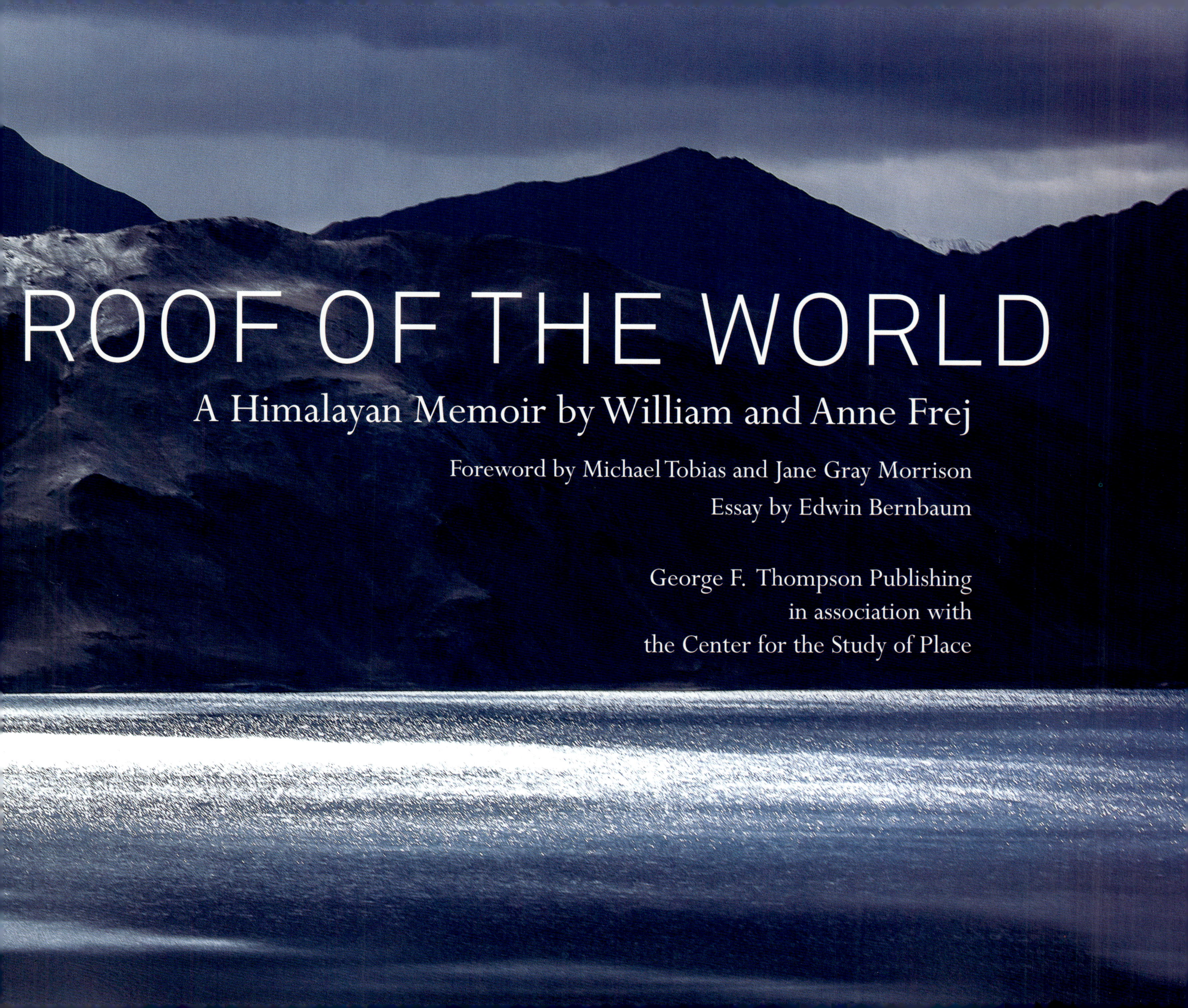

ROOF OF THE WORLD

A Himalayan Memoir by William and Anne Frej

Foreword by Michael Tobias and Jane Gray Morrison

Essay by Edwin Bernbaum

George F. Thompson Publishing
in association with
the Center for the Study of Place

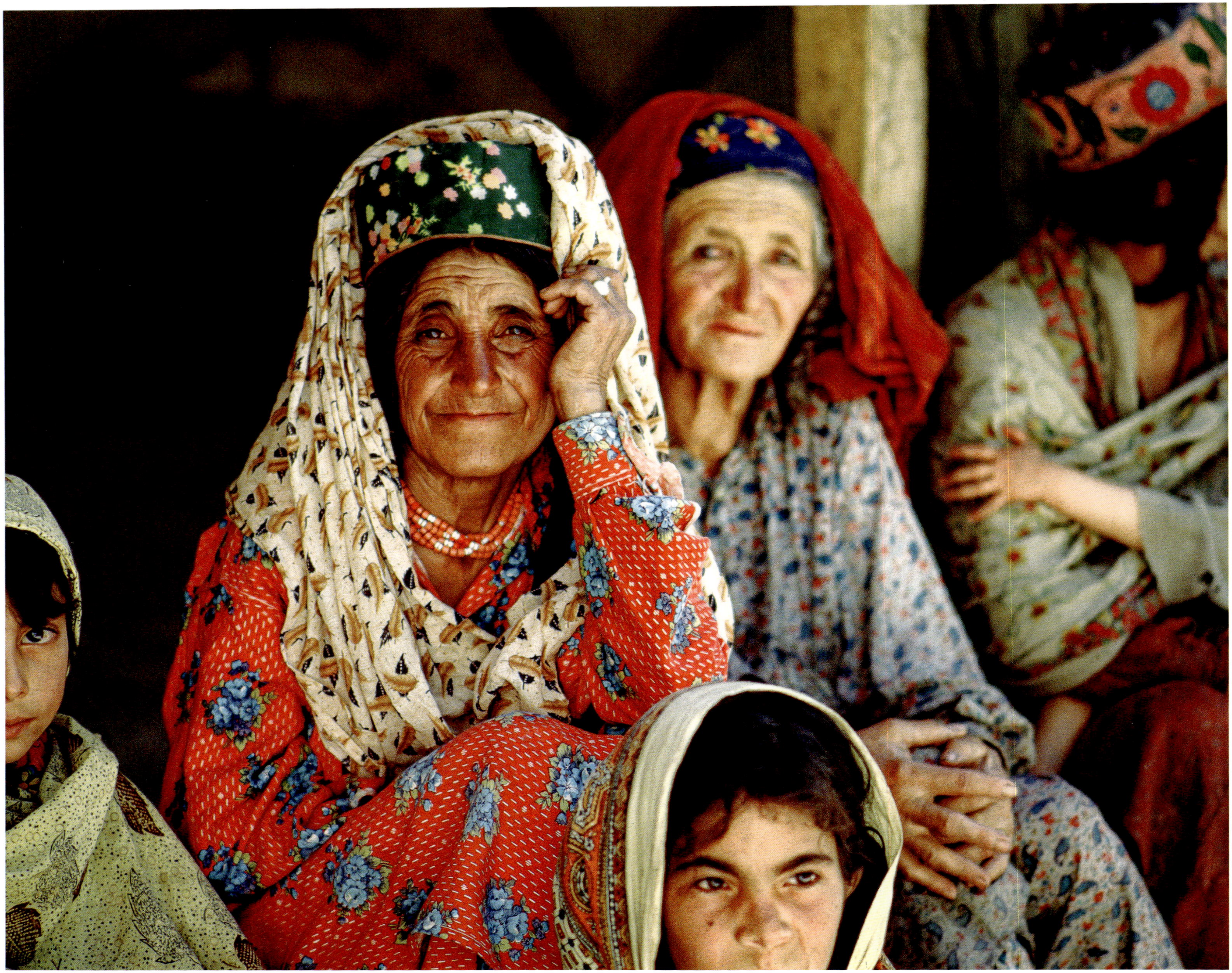

Semi-nomadic women of Yashpirt, Pakistan, 1985

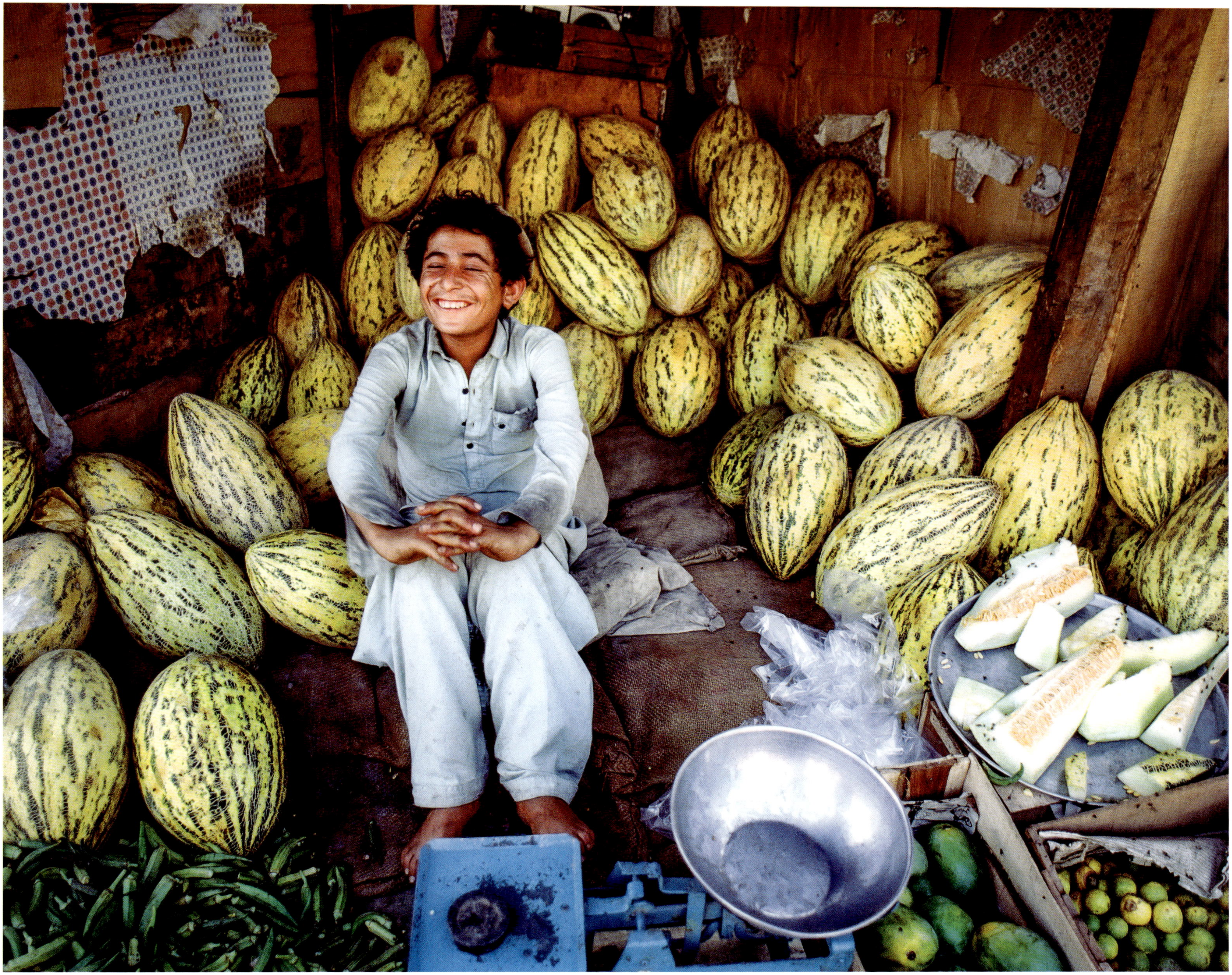

The melon seller of Gilgit, Pakistan, 1985 *Following spread: Lamayuru Monastery, Ladakh, India, 2018*

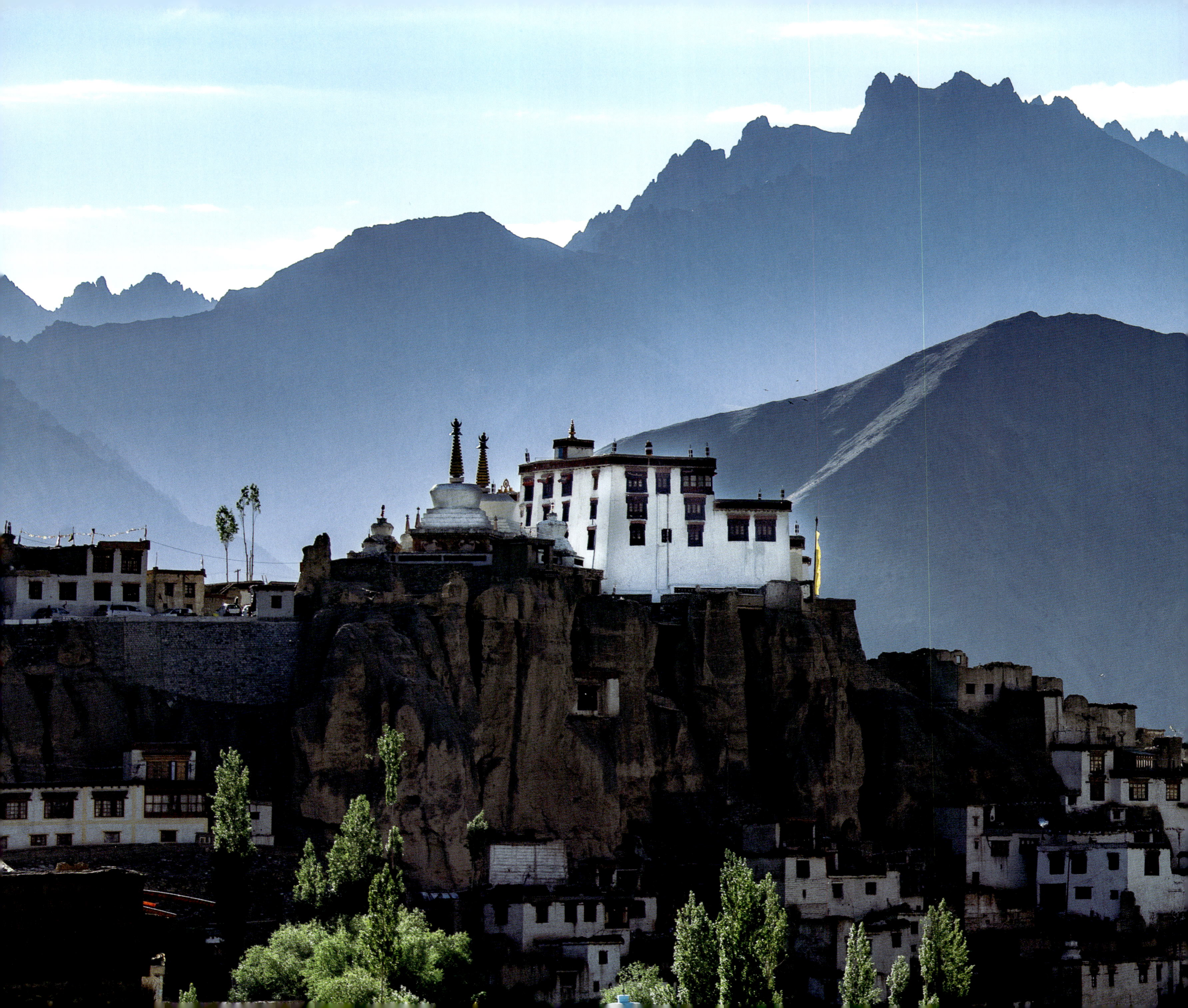

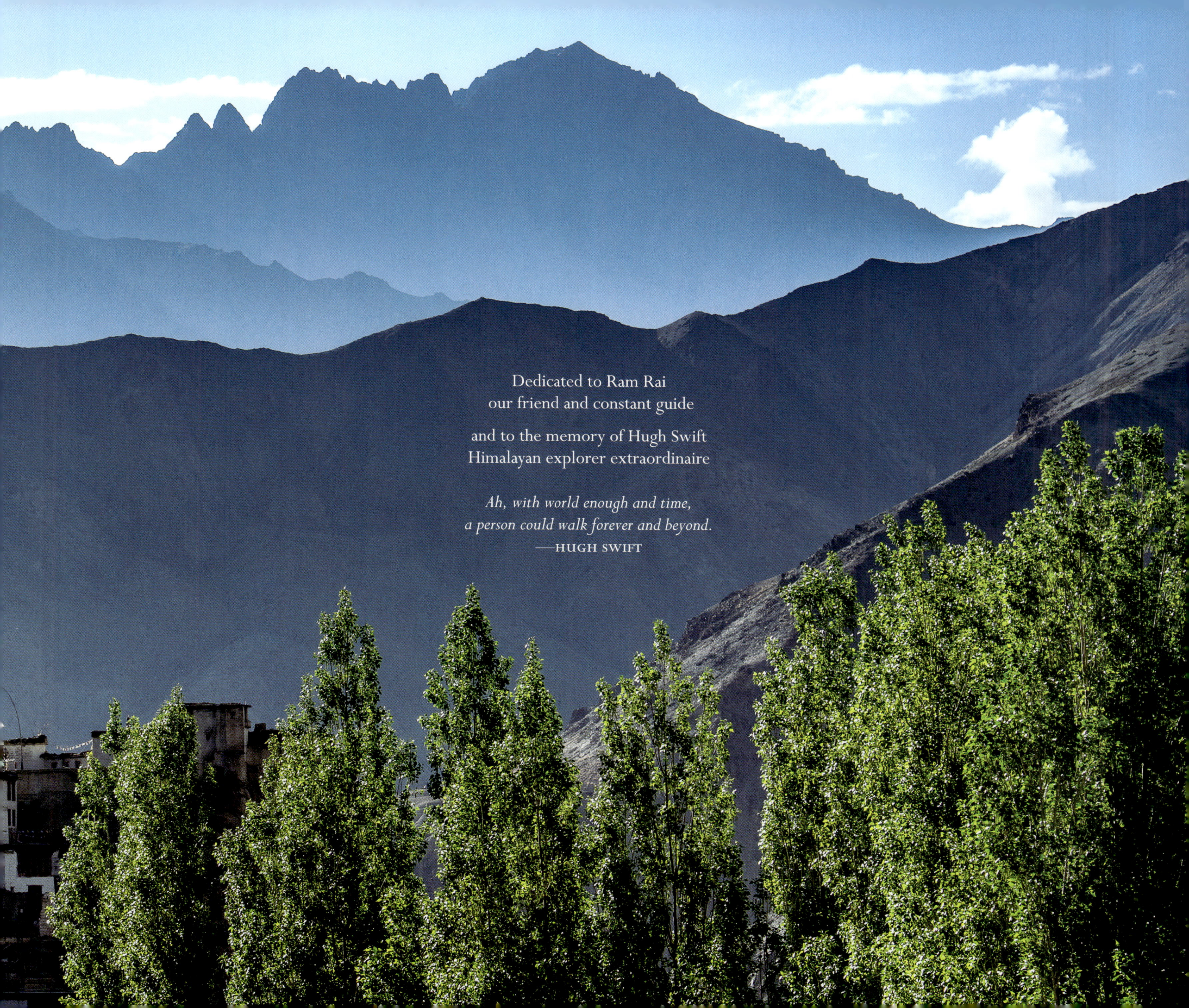

Dedicated to Ram Rai
our friend and constant guide

and to the memory of Hugh Swift
Himalayan explorer extraordinaire

Ah, with world enough and time,
a person could walk forever and beyond.
—HUGH SWIFT

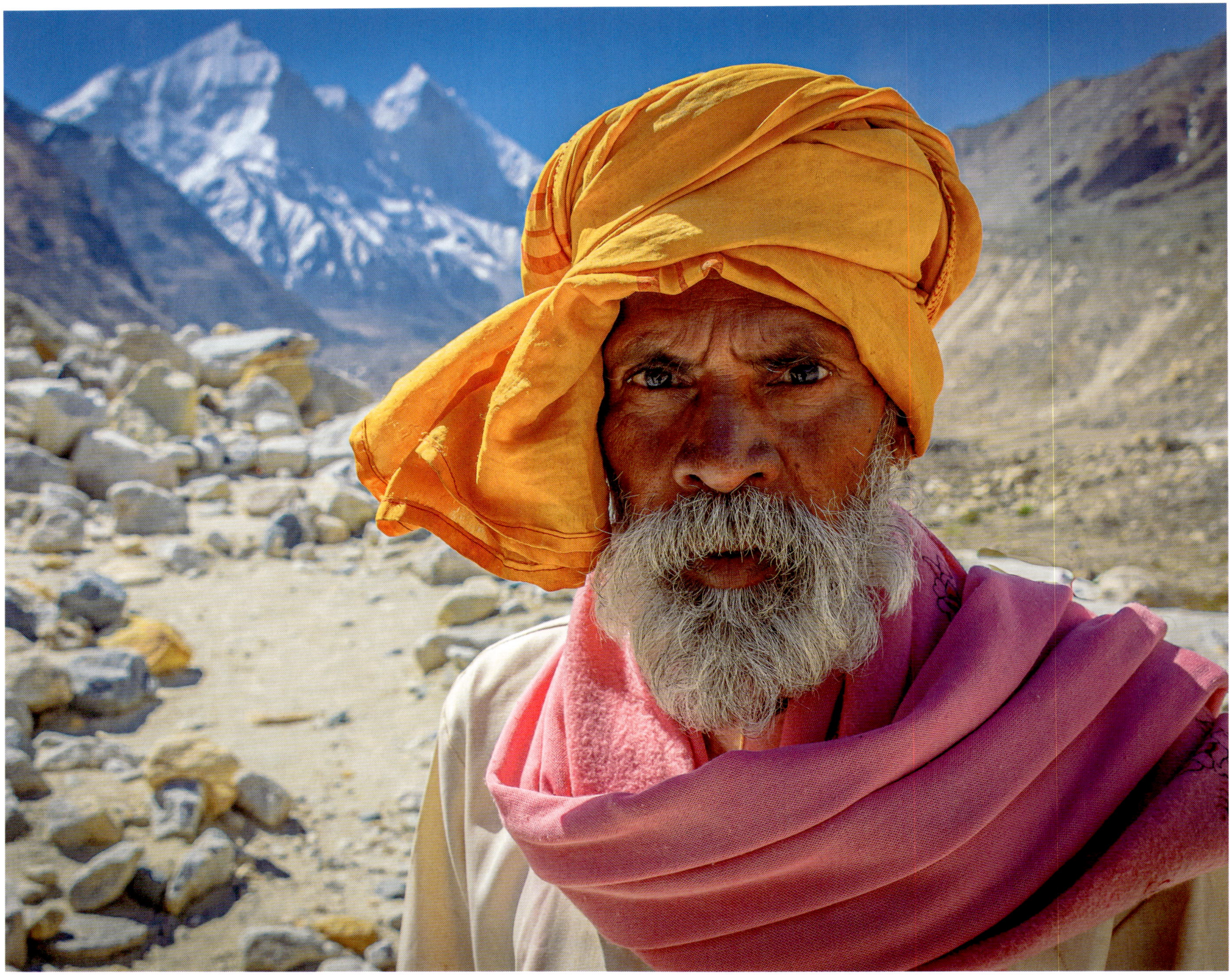

Pilgrim returning from the source of the Ganges, Uttarkhand, India, 2018

Contents

11 Foreword
MICHAEL TOBIAS AND JANE GRAY MORRISON

15 A Forty-Year Journey
ANNE AND WILLIAM FREJ

23 Sacred Mountains: Symbolism and Significance
EDWIN BERNBAUM

The Memoir
ANNE FREJ

33 Around Manaslu, Nepal
NOVEMBER 1981

43 Somewhere Beneath Mount Everest, Nepal
NOVEMBER 1982

51 Making Travel Plans, California
SPRING 1985

55 Walks Through the Northern Areas, Pakistan
JULY–AUGUST 1985

71 Treks and Travels Through Ladakh, India
AUGUST–SEPTEMBER 1985 AND JUNE 2018

99 From Kangchenjunga to Everest, Nepal
OCTOBER–DECEMBER 1985

115 Below Dhaulagiri Icefall, Nepal
APRIL 1986

121 Across Western Tibet to Mount Kailas, Tibet
JUNE 1986

145 Crossing Rivers and Glaciers to K2, Pakistan
JULY–AUGUST 1986

159 Up the Arun Valley to Makalu, Nepal
NOVEMBER 1986

169 Into the Annapurna Sanctuary, Nepal
NOVEMBER 1986

177 Through the Tsum Valley to the Ganesh Sanctuary, Nepal
MAY 1991

189 Over Kuari Pass in the Garhwal, India
JUNE 1993

195 At the Edge of the Pamir, Kyrgyzstan
AUGUST 2008

207 Around and Above the Hindu Kush, Afghanistan
2009–2010

219 The Remote Realm of Mustang, Nepal
AUGUST 2013

235 To the Sacred Mountain, Bhutan
OCTOBER 2015

245 Mani Rimdu at Chiwong Monastery, Nepal
NOVEMBER 2017

261 To the Source of the Ganges, India
JUNE 2018

269 Epilogue
271 Acknowledgments
272 Bibliography
275 Notes on Place Names and Elevations
277 A Note on the Photography
278 About the Essayist and Contributors
279 About the Authors
280 About the Book

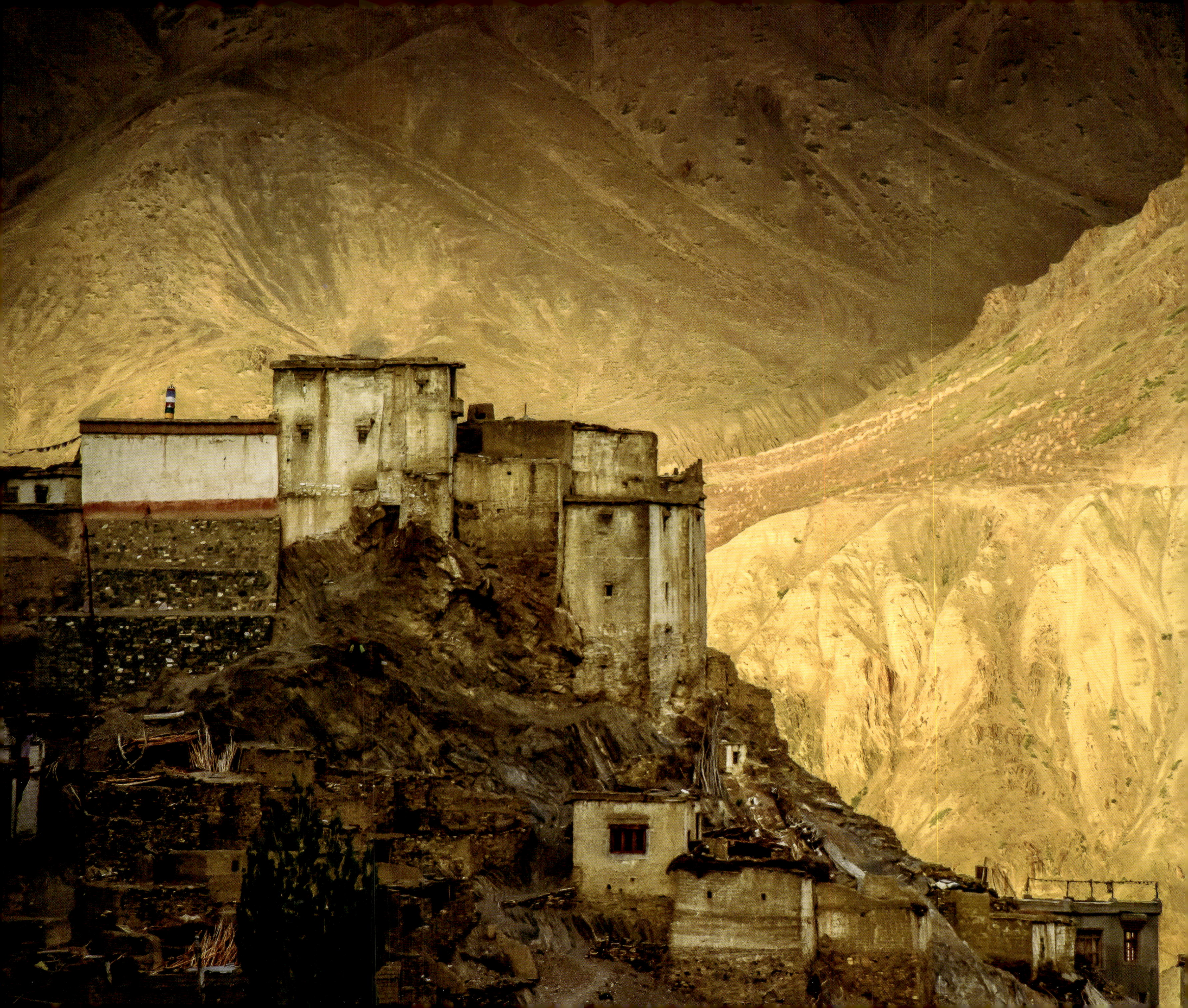

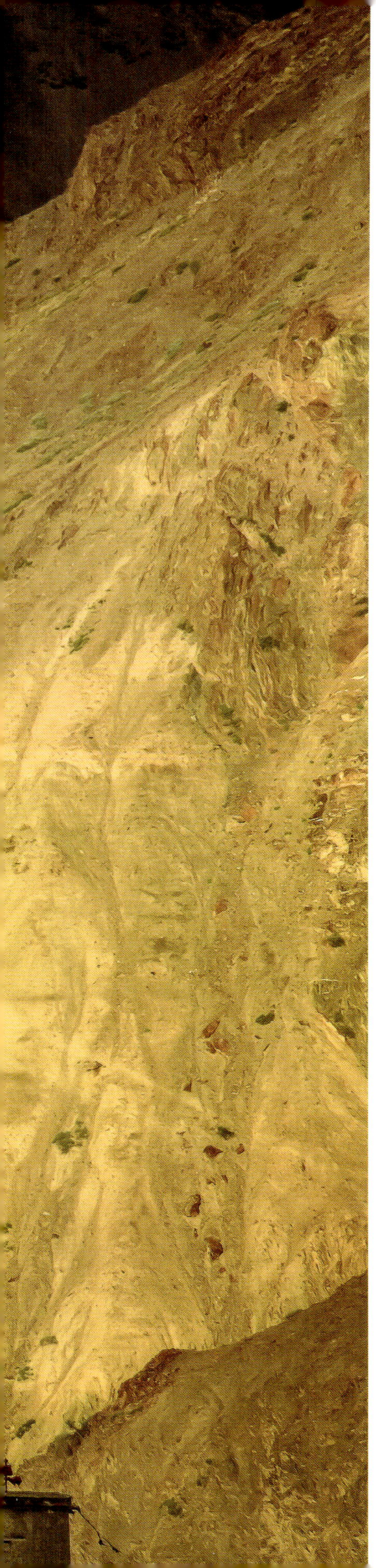

Foreword

MICHAEL TOBIAS AND JANE GRAY MORRISON

William and Anne Frej have created a magnificent book that provides intimacy, acute detail, context, history, and ethnography of a vast swath of incomparable, and ecologically critical geography—the Himalaya. With nearly 40 years of passionate exploration of this region—from one end to the other, beginning with Manaslu in 1981—they offer an engaging and deeply felt perspective on a subject more typically associated with the fever of conquest on the world's highest peaks.

Often lost in the fever pitch of high-altitude mountain-climbing endeavors and the accounts of them (at least 10,500+ expeditions have been recorded in the Himalayan Database since 1905) are the lives of more than 55,000,000 Himalayan locals and the more than 1,000,000,000 downstream recipients dependent upon Himalayan watersheds that have been increasingly denuded and/or overwhelmed by related landslides and floods from the melting of nearly every one of the more than 15,000 Himalayan glaciers. The 1,100,000,000 mountain people in the world today are utterly dependent on the ecological integrity of their environs. This co-occupancy of humans and their alpine worlds is beautifully, sensitively, and imaginatively portrayed throughout the Frejs' book.

In their Epilogue to *Travels Across the Roof of the World*, William and Anne Frej offer a clear overview of the issues pertaining to the cumulative toll of environmental impacts they have witnessed during the four decades of their Himalayan odysseys. They propound no emphatic judgment, only a warning directed to policy makers and the human conscience that what is most important, and at the source of the perennial lure that draws so many of us to this part of the world, are the people who live there and the rich biodiversity hotspots with which they must sustainably coexist. There are at least 30 major ethnic mountain communities throughout Asia, from Nuristan, Afghanistan, to India's northeastern-most Arunachal Pradesh, and tens-of-thousands of species, many listed as Vulnerable or Endangered (seventy in Bhutan alone). Given the burgeoning human populations, these are unambiguously *inhabited wildernesses*, with their own life-sustaining biophilia and interspecies imperatives. The glorious photographic and textual tributes by the Frejs to these remarkable regions and their peoples do poignant justice to an implicit remark once made by anthropologist Colin Turnbull in his classic study of the mountainous people known as the Ik, in Eastern Uganda.[1] Without the Ik, Turnbull suggested, their Mount Morungule

Lamayuru Monastery, Ladakh, India, 2018

would not be the same. And without Mount Morungule there would be no Ik. The same is certainly true in the mountain communities spread across the Himalaya.

Humanity has perennially thirsted after horizons and heights. To gain a vantage, to seek out a greener new world, if only for a glimpse of some distant, soul-replenishing moment in time. Hence, the poet Percy Bysse Shelley, gazing upon Mont Blanc during his three days in Chamonix in July 1816, wrote: "I look on high; Has some unknown omnipotence unfurled/ . . . Driven like a homeless cloud from steep to steep/That vanishes among the viewless gales? Far, far above, piercing the infinite sky, Mont Blanc appears, still, snowy, and serene. . . . "[2] For Shelley and dozens of mystics and explorers before and after him, such mountains exemplified "The secret strength of things, Which governs thought, and to the infinite dome/Of heaven is as a law, inhabits thee!"[3]

The political scientist Jack Schaar (1928-2011), a noted political theorist and Professor Emeritus at the University of California, Santa Cruz, once said to me, "I love mountains. That's why I never go there." What he prophesied was precisely what was to come. Like William and Anne Frej, he was an expert in international relations and development and saw the coming cultural and ecological consequences of say, 60,000 tourists a year to Mount Everest, the immediate path to which has over decades increasingly taken on the look of a colossal garbage dump. An even greater number annually visit Taktsang, Bhutan's most sacred monastery, with serious visual, acoustical, and ecological consequences. Ecotourism today exceeds $80,000,000,000 yearly, and most of that money does not go to the locals but to tour operators, many based in other countries. Nations like Nepal, India, Pakistan, and Bhutan have attempted to counter some of the destructive onslaught of modernity by restricting access to sacred places like India's Nanda Devi Sanctuary, which is not open to tourists, or establishing national parks that involve local communities. Bhutan made it illegal to climb any peak above 6,000 meters (19,685 feet) many years ago, happy to forego climbing revenues in favor of Buddhist conservation, as most exemplified in Bhutan, where non-violent conservation biology and animal liberation go hand-in-hand.[4]

The Frejs convey throughout their book fascinating and historic mountaineering exploits, from their own quiet, lower-elevation perches down on the trails or at various base camps of at least ten of the highest peaks in the world. Their commentary and extraordinary imagery are, ironically, nearly irresistible lures to anyone susceptible to philosophical adventure, anthropological enchantment, and aesthetic rapture. At the same time, the book is also a uniquely pragmatic guidebook, in the spirit of their late friend, Hugh Swift, as well as the work of our late mutual friend, photographer and author Galen Rowell. Both men spent much of their lives traversing the Himalayas and Transhimalayas and writing compelling, inviting commentaries on their journeys.

The Frejs point to their "unhurried pace" as a tactical but quintessential quality of mind if one is to experience the truth of the place and to truly appreciate the elegant wholeness of the landscape and its infinite components. To paraphrase the Alexandrian poet, Constantine P. Cavafy (1863–1933) in his poem "Ithaka," (1911), it is not Ithaka, per se, but the *journey* to Ithaka that is important. And it was the great Greek poet, Nikos Kazantzakis (1883–1957), who declared, "Even if I were certain I was going to heaven . . . I would pray [to] God to let me go by the longest possible route."[5] Such profound feelings infuse this deeply affecting experience, which is the Himalayan memoir, *Travels Accross the Roof of the World,* by William and Anne Frej.

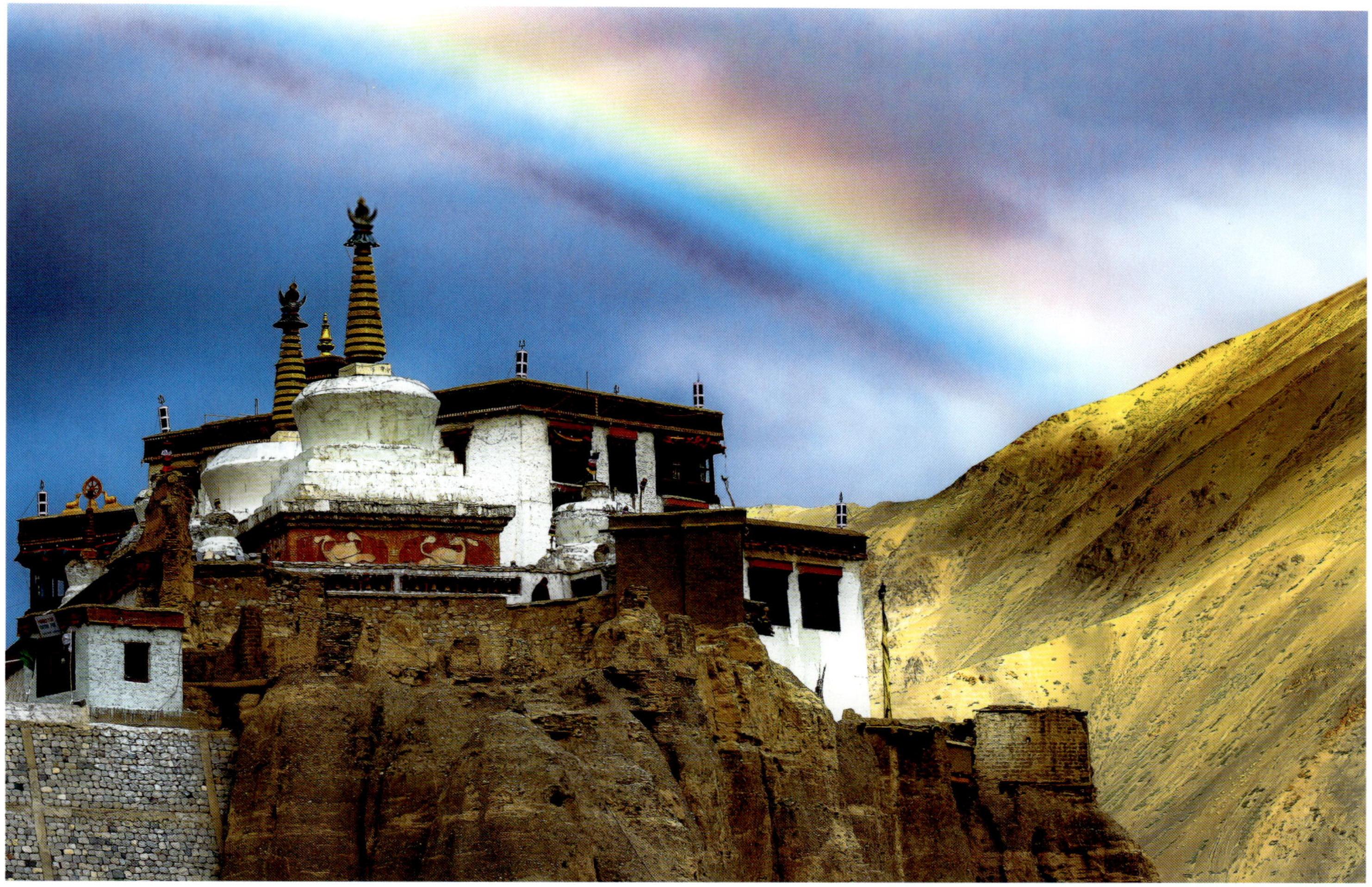

Rainbow over Lamayuru Monastery, Ladakh, India, 2018

Notes

1. Michael Charles Tobias, editor, *Mountain People*, (Norman: University of Oklahoma Press, 1986), 69–74.

2. Nathan Haskell Dole, editor, "Mont Blanc: Lines Written in the Vale of Chamouni, July 23, 1816," *The Complete Works of Percy Bysshe Shelley* (London, UK and Boston, MA: Virtue & Company Publishers, 1905), 359–367.

3. Ibid.

4. See Ugyen Tshewang, Michael Charles Tobias, and Jane Gray Morrison, *Bhutan: Conservation and Environmental Protection in the Himalayas* (New York, NY: Springer Publishers, 2021), and Ugyen Tshewang, Jane Gray Morrison, and Michael Charles Tobias, *Bionomics in the Dragon Kingdom: Ecology, Economics and Ethics in Bhutan* (New York, NY: Springer Publishers, 2018).

5. Nikos Kazantzakis, *Report to Greco*, translated by Peter A. Bien (New York, NY: Simon & Schuster, 1965), 153.

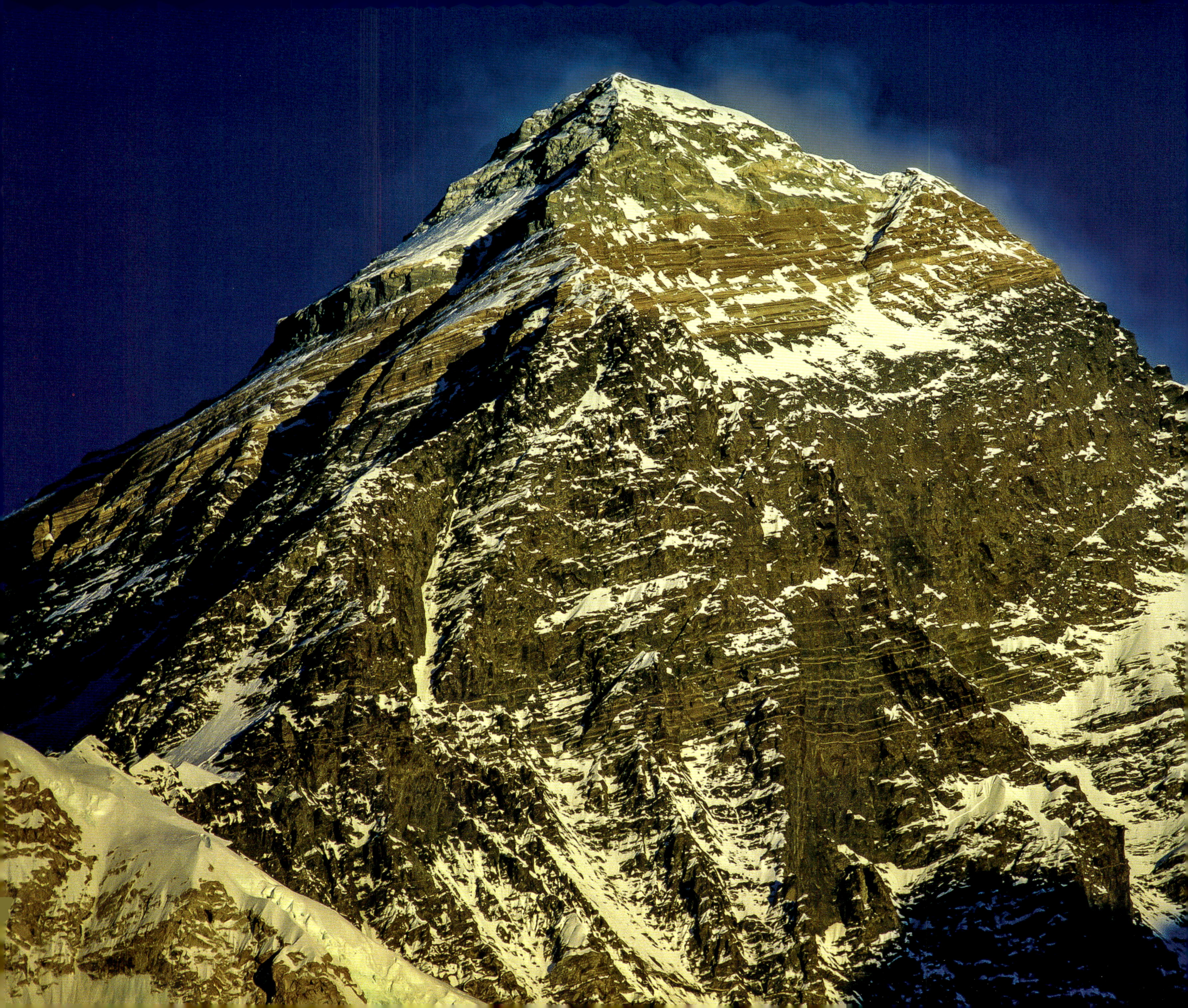

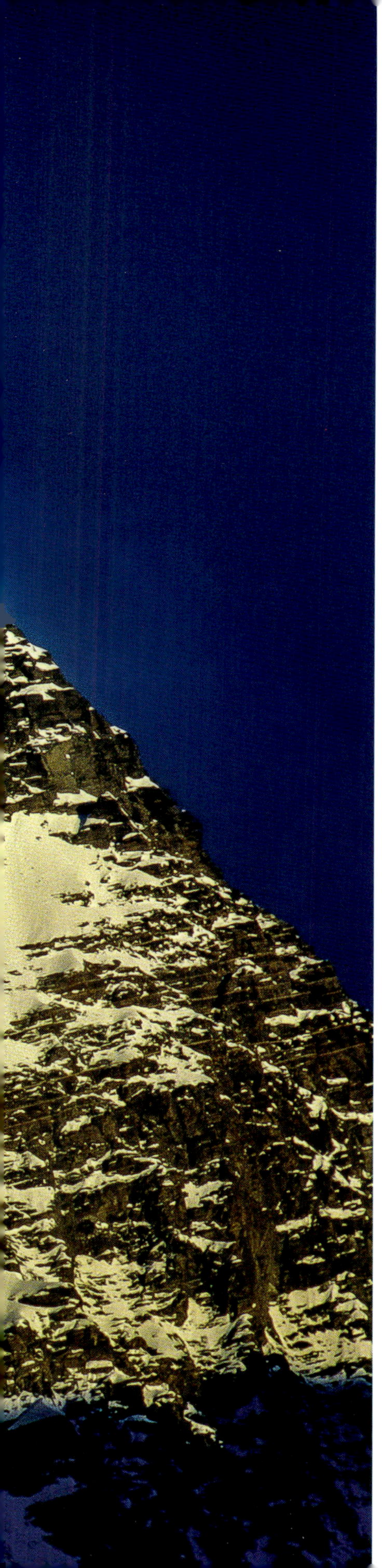

A Forty-Year Journey

ANNE AND WILLIAM FREJ

Nearly four decades after our first trip to the Himalaya in 1981, we started looking once again at photographs and journals from our travels. Thousands of Kodachrome color slides in plastic sheets lay dormant in archival boxes stacked in closets. We carried these with us from our home in California to our first overseas job posting in Jakarta, Indonesia, and then on to subsequent postings in Warsaw, Poland, Indonesia again, and Almaty, Kazakhstan, until our last posting in Kabul, Afghanistan, when the slides joined other long-forgotten personal effects in a storage unit outside Washington, D.C. From time to time we pulled out the boxes and looked briefly at a few images, and several times Bill made prints for photo exhibitions in Jakarta, Warsaw, and Almaty. For the most part, though, this storehouse of visual memories remained untouched for decades.

Memories flooded back once we started looking at the images again. There were spectacular mountain scenes as well as the familiar faces of the Nepali porters we hired to carry our bags and provisions on treks, women from faraway villages adorned in heavy gold jewelry, and playful little boys studying at monasteries who must be middle-aged monks now. We marveled that it was possible for us to remember all these people so vividly, and we wondered if they remembered us, too. We realized that what started off as hiking vacations had become a pilgrimage of sorts, leading to a life-long interest in the people and cultures of the mountain communities stretching across the roof of the world from Central Asia across Tibet, India, Pakistan, and Nepal to Bhutan.

We did two long treks in Nepal during the early 1980s before taking an extended twenty-month trekking sabbatical to the base camps of the world's ten highest peaks in 1985 and 1986. After that we traveled to Nepal and northern India and other mountain regions as often as our professional lives would allow. Our preferred mode of travel was trekking—hiking on mountain trails for extended periods and spending nights along the way, either in a tent or a simple lodge. Unlike mountain climbing, trekking does not require technical skills or equipment—a good pair of boots, a waterproof rain jacket, a warm sleeping bag and inflatable mat, plus a backpack are the essentials. Treks usually involve a destination, but the most important element is an unhurried pace with time to experience mountain scenery and local cultures.

Treks can be organized in many ways, and we have tried them all. Traveling with an adventure travel company or organization such as the Sierra Club on a group trek has clear advantages. There's no need to worry about logistics—just send in your payment, get in shape physically, pack your personal gear, and get to the arranged meeting place such as

Mount Everest, Khumbu, Nepal, 1985

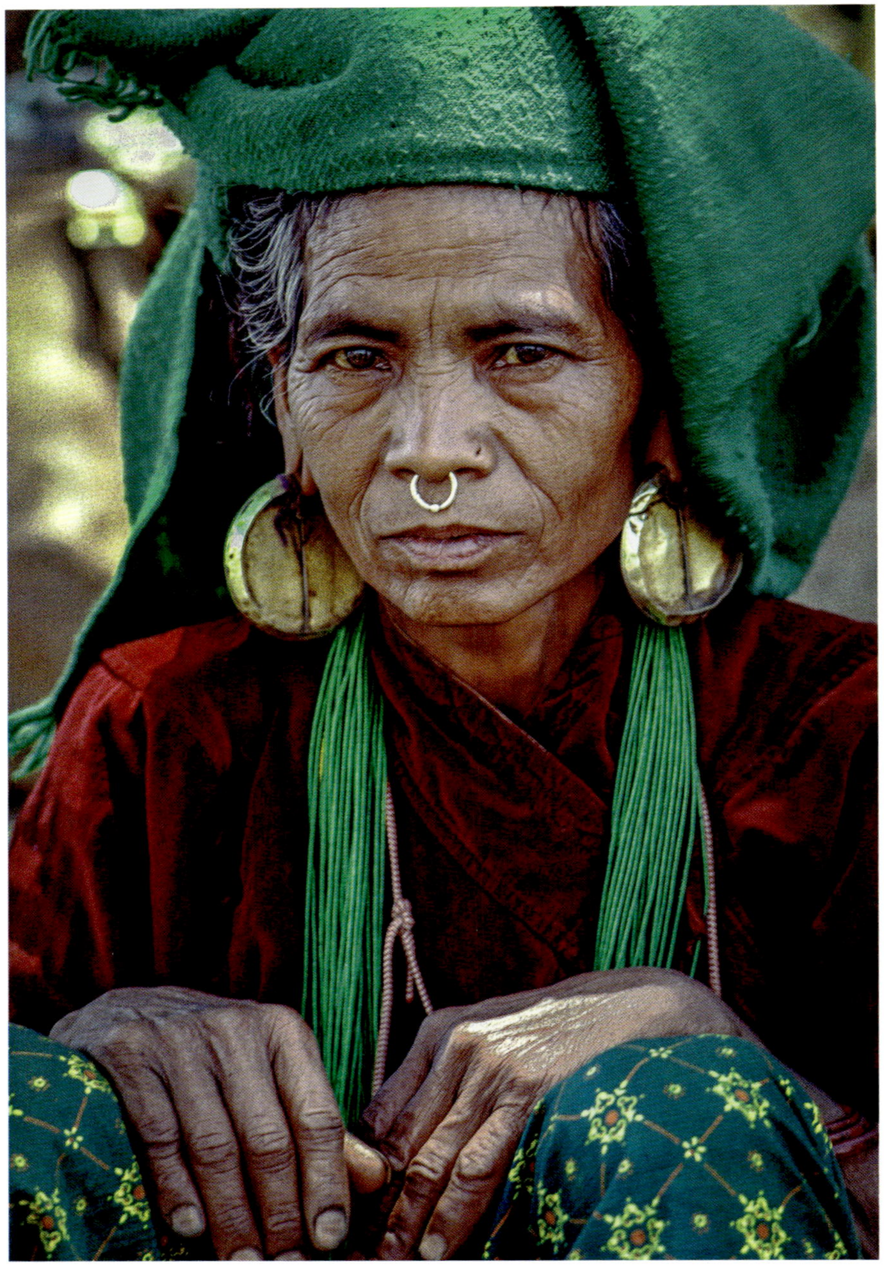

Gurung woman, on the trail to Manaslu, Nepal, 1981

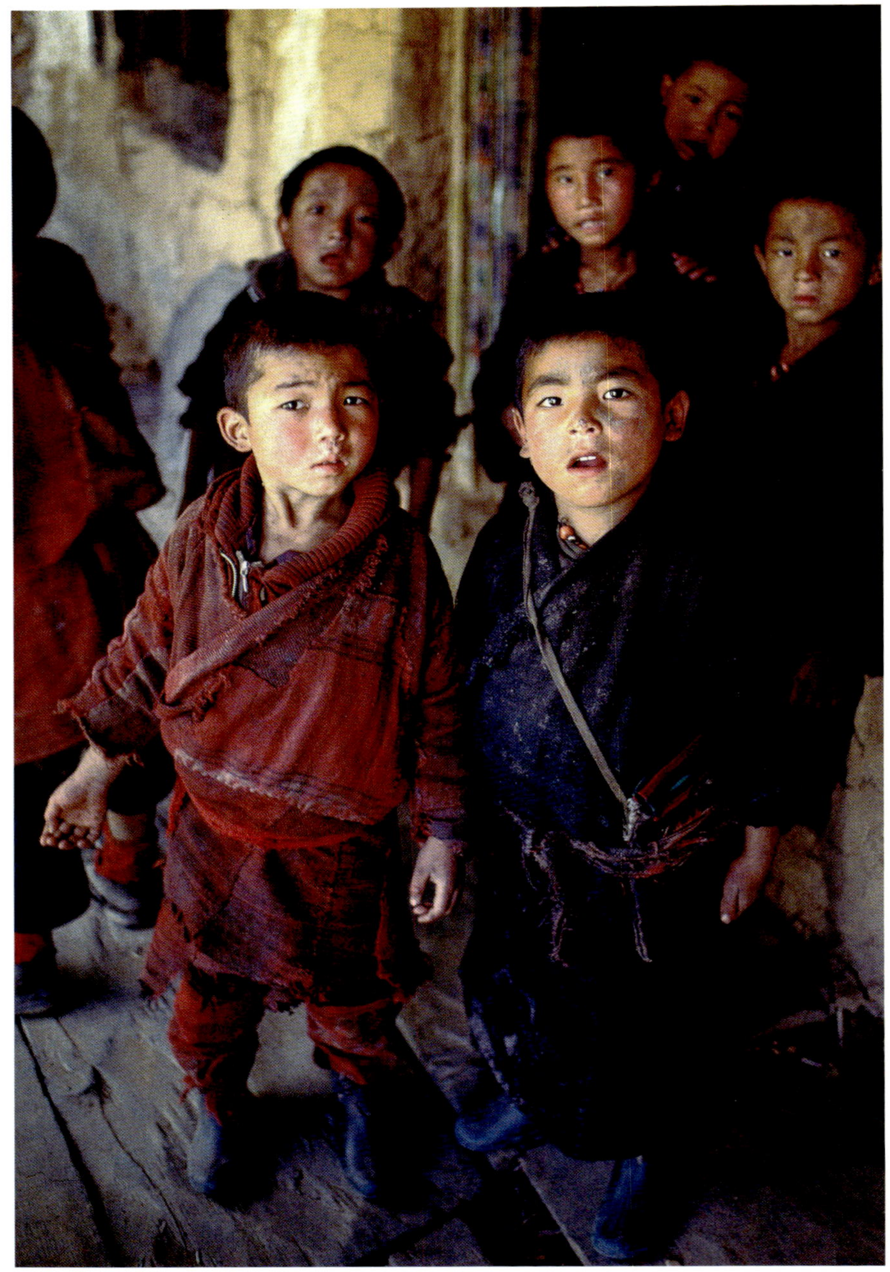

Ripchet village, Tsum Valley, Nepal, 1991

Bangkok or Kathmandu. With an organized group trip comes infrastructure like dining tents, camp chairs, toilet tents, and a large local crew for carrying loads and cooking meals. There is also the camaraderie of other hikers who generally have similar interests and are fun and interesting. The downside is that you are wedded to an itinerary and the strength, not to mention personality quirks, of your fellow trekkers. Alternatively, privately arranged treks with a small group of friends or just two people are flexible but require more advance planning—setting the route, hiring a crew, and buying food and supplies. Teahouses located along the popular trekking routes in Nepal and northern India eliminate some of these issues, since they offer a bunk and food to order, but they can be dirty and noisy, so we much prefer sleeping in our own tent. For treks to remote and infrequently visited regions, careful groundwork is essential because there are few places to restock and you must carry everything needed for weeks at a time, from clothing and gear to food and medical supplies. Thankfully, a well-established trekking culture exists in the high elevation regions of Nepal, India, and Pakistan that makes it possible to travel in less developed areas.

Since the early 1950s, when Nepal began welcoming foreign climbers and trekkers, generations of mountain people have made a profession of guiding and accommodating people like us. Their strength, tenacity, loyalty, and good cheer have made it possible for us to visit many corners of the Himalaya and beyond. This book is a memoir of the treks we have done over the years with their help, for which we are extremely grateful.

Our many journeys to remote areas on the roof of the world have inspired us, humbled us, and guided us on a path that has defined our lives. Perhaps our pilgrimage and walk around sacred Mount Kailas did absolve the accumulated sins of our lives. Perhaps the teachings of His Holiness the Dalai Lama at the Kalachakra Initiation in Bodhgaya, where Buddha attained enlightenment, encouraged us to take a different professional path and to embrace international development work in some of the world's most impoverished countries. We know that the sheer beauty of the mountains and the people we encountered on our travels were enough to keep us returning for nearly four decades.

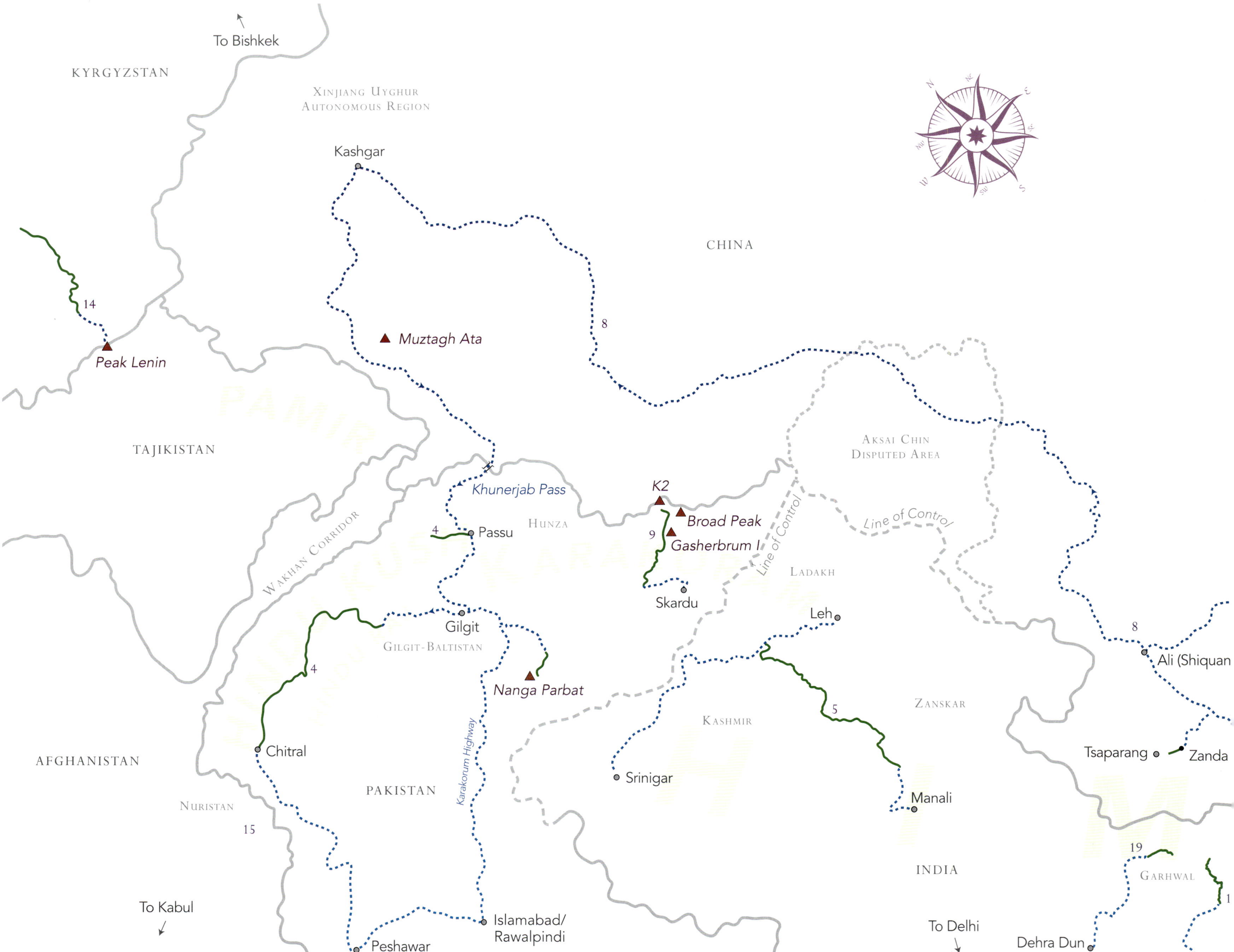

To Bishkek
KYRGYZSTAN
Xinjiang Uyghur
Autonomous Region
Kashgar
CHINA
14
Peak Lenin
Muztagh Ata
8
TAJIKISTAN
Aksai Chin
Disputed Area
Khunerjab Pass
K2
Broad Peak
Gasherbrum I
Line of Control
Line of Control
Wakhan Corridor
4
Passu
Hunza
9
Ladakh
Skardu
Leh
Gilgit
Gilgit-Baltistan
4
Nanga Parbat
8
Ali (Shiquan
Zanskar
Kashmir
5
Karakorum Highway
Chitral
Tsaparang
Zanda
AFGHANISTAN
PAKISTAN
Srinigar
Nuristan
Manali
15
19
INDIA
Garhwal
To Kabul
Islamabad/
Rawalpindi
To Delhi
Dehra Dun
Peshawar

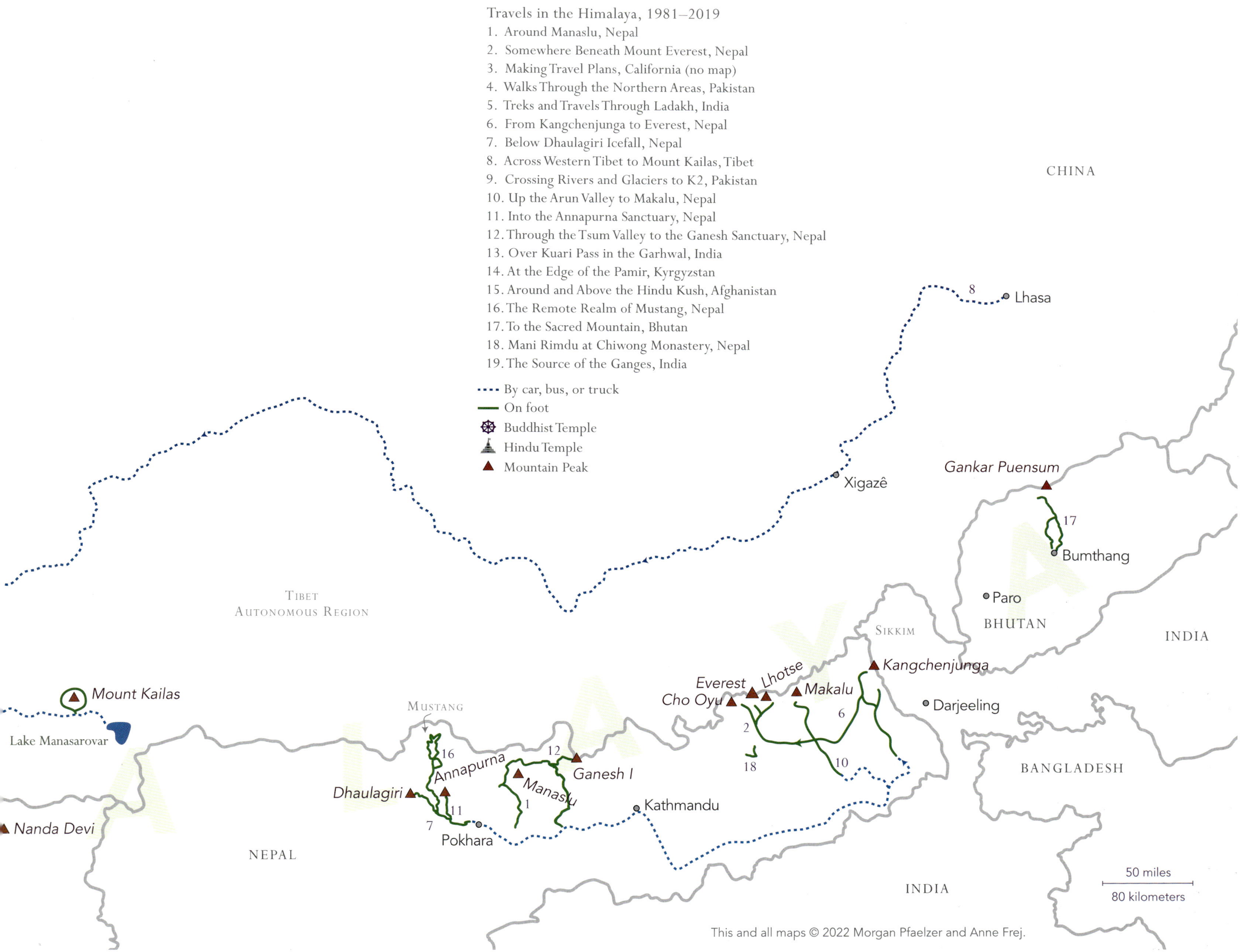
Travels in the Himalaya, 1981–2019
1. Around Manaslu, Nepal
2. Somewhere Beneath Mount Everest, Nepal
3. Making Travel Plans, California (no map)
4. Walks Through the Northern Areas, Pakistan
5. Treks and Travels Through Ladakh, India
6. From Kangchenjunga to Everest, Nepal
7. Below Dhaulagiri Icefall, Nepal
8. Across Western Tibet to Mount Kailas, Tibet
9. Crossing Rivers and Glaciers to K2, Pakistan
10. Up the Arun Valley to Makalu, Nepal
11. Into the Annapurna Sanctuary, Nepal
12. Through the Tsum Valley to the Ganesh Sanctuary, Nepal
13. Over Kuari Pass in the Garhwal, India
14. At the Edge of the Pamir, Kyrgyzstan
15. Around and Above the Hindu Kush, Afghanistan
16. The Remote Realm of Mustang, Nepal
17. To the Sacred Mountain, Bhutan
18. Mani Rimdu at Chiwong Monastery, Nepal
19. The Source of the Ganges, India
By car, bus, or truck
On foot
Buddhist Temple
Hindu Temple
Mountain Peak
CHINA
Lhasa
8
Xigazê
Gankar Puensum
17
Bumthang
Paro
BHUTAN
INDIA
SIKKIM
TIBET
AUTONOMOUS REGION
Mount Kailas
Lake Manasarovar
MUSTANG
16
Annapurna
Dhaulagiri
11
7
Pokhara
Manaslu
1
12
Ganesh I
Kathmandu
Everest
Lhotse
Cho Oyu
Makalu
Kangchenjunga
Darjeeling
2
6
18
10
Nanda Devi
NEPAL
INDIA
BANGLADESH
50 miles
80 kilometers

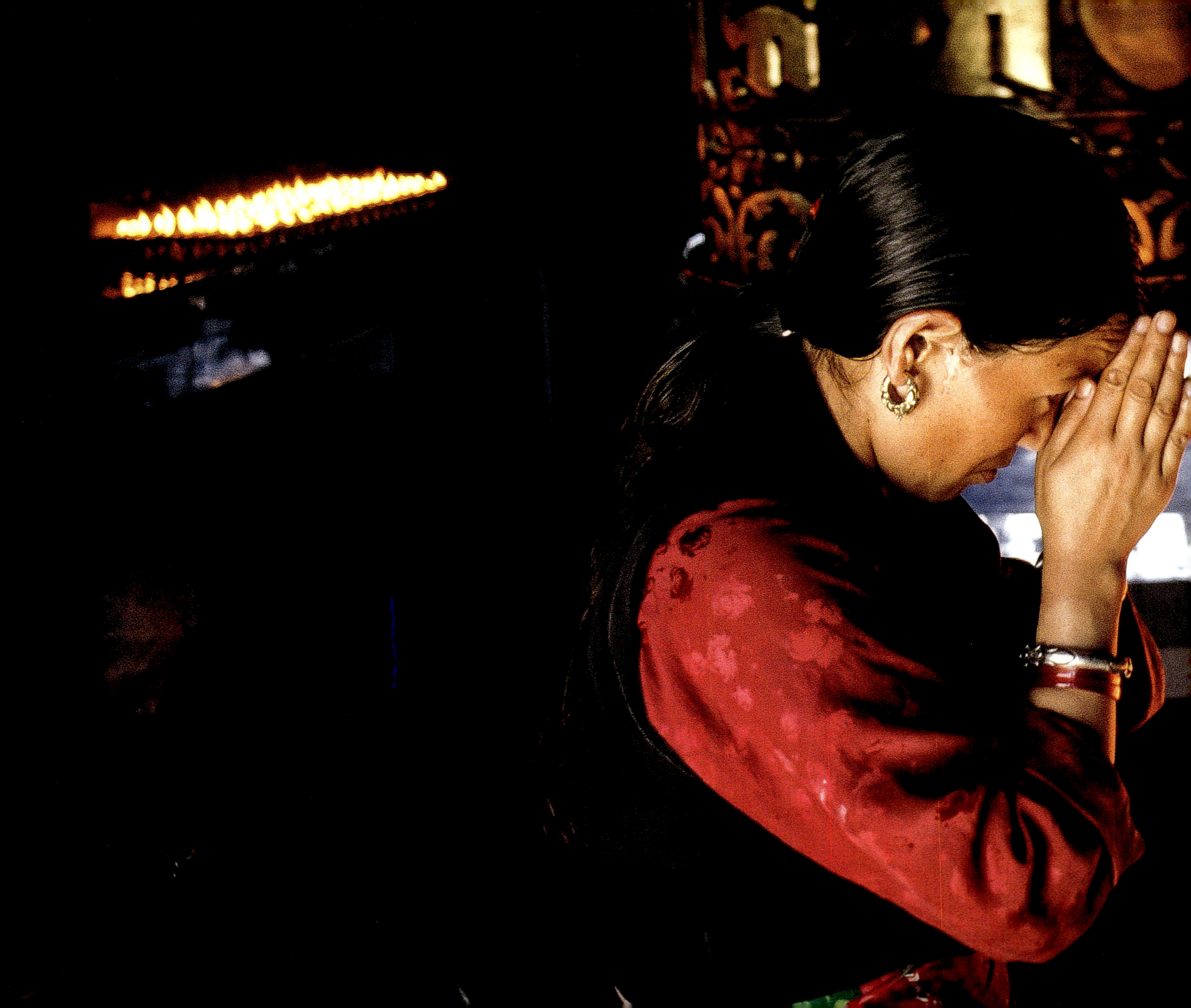

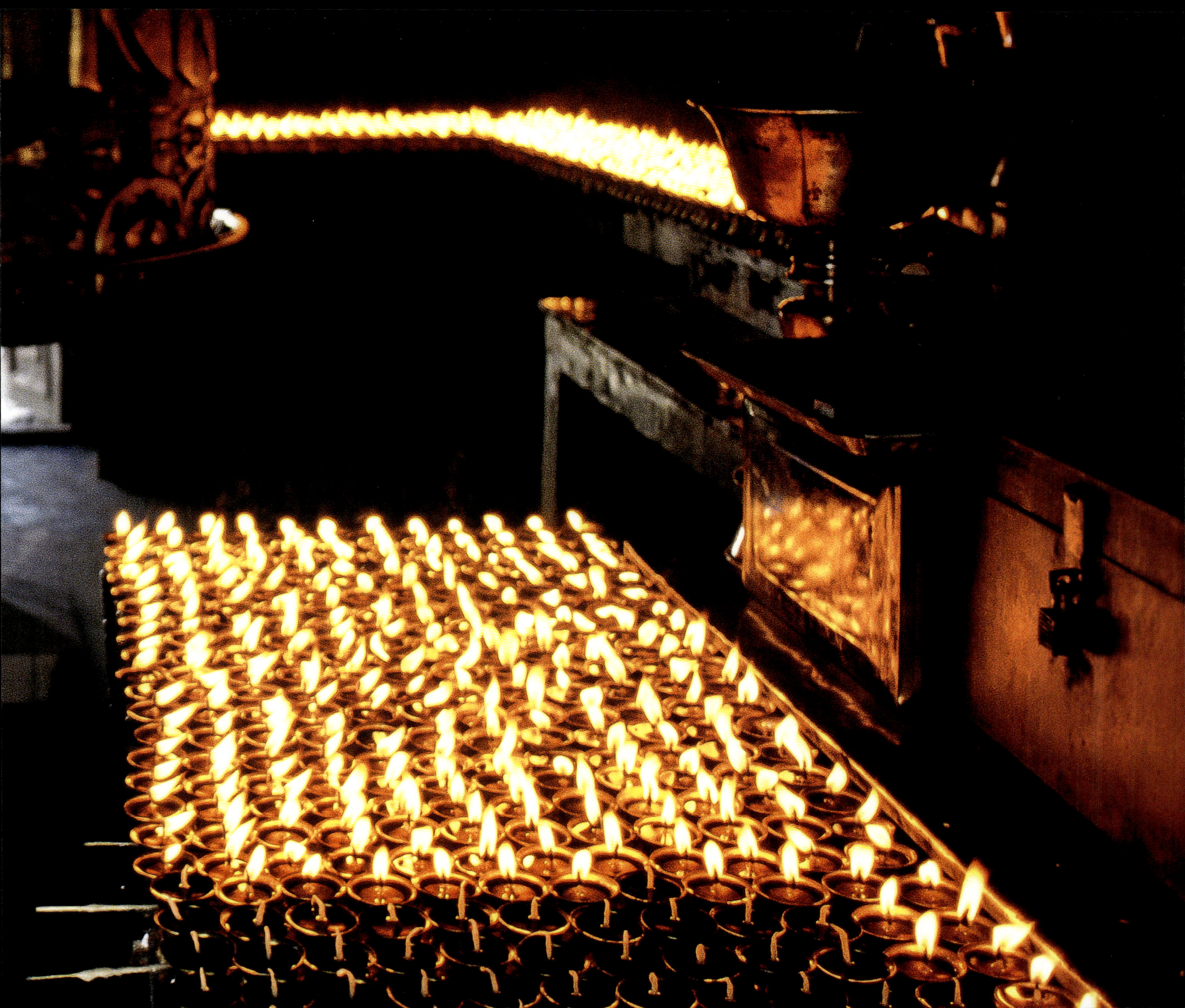

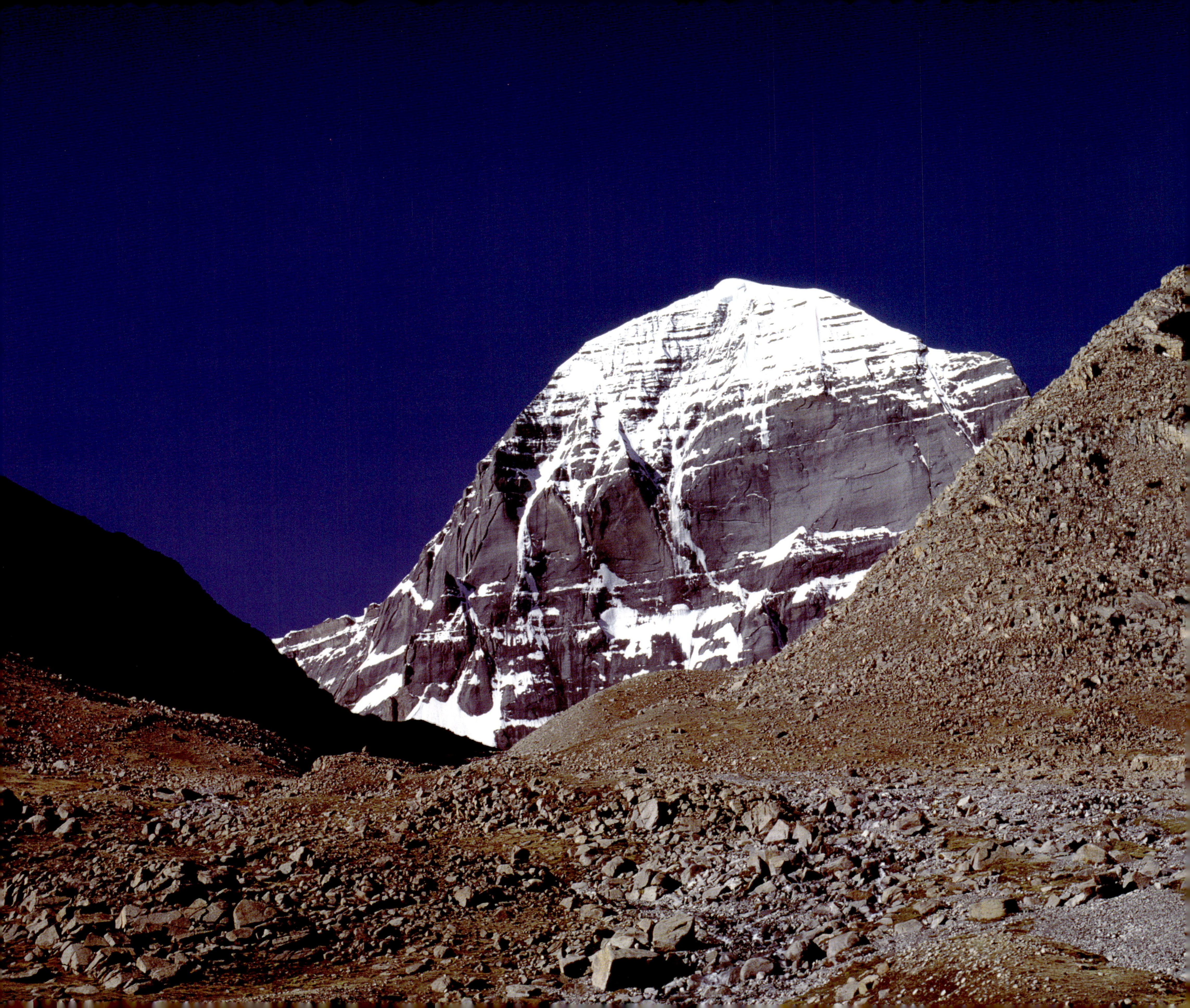

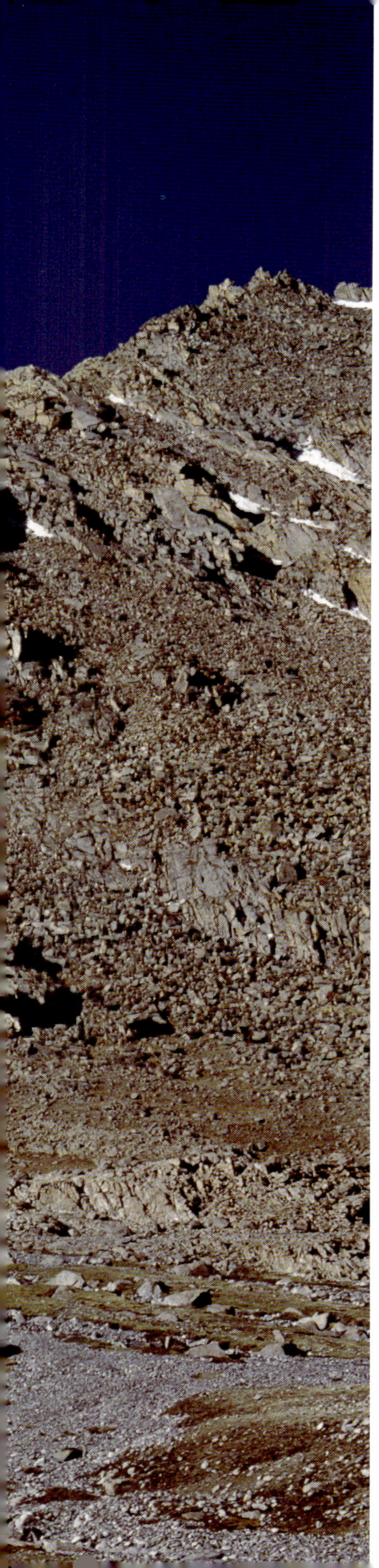

Sacred Mountains SYMBOLISM AND SIGNIFICANCE

EDWIN BERNBAUM, PH.D.

As the highest and most impressive features of the landscape, mountains have an extraordinary power to evoke a sense of the sacred. Their imposing peaks, the clouds that gather about their summits, the dramatic play of light on their slopes, the waters that issue from hidden springs, these and other characteristics combine to imbue them with an aura of mystery and sanctity. In that sanctity and mystery people of diverse backgrounds, both traditional and modern, experience a deeper reality or significance that gives meaning and vitality to their lives.

As sacred expressions of a deeper reality, however that may be construed, mountains have become associated with the deepest and most central values and beliefs of cultures and traditions throughout the world, making them cultural landscapes of the highest order. The remote Himalayan peak of Mount Kailas, rising aloof above the Tibetan Plateau, directs the minds of millions of Hindus and Buddhists toward the utmost attainments of their spiritual traditions.[1] The graceful cone of Mount Fuji in Japan has come to represent the quest for beauty and harmony that lies at the heart of Japanese culture. As the highest mountain on Earth, Mount Everest stands out, even in the modern, secular world, as an inspiring symbol of the ultimate.

The sacredness of mountains in general manifests itself in three major ways, each relevant for determining whether or not a particular mountain or mountain site qualifies for World Heritage status. First, certain peaks are singled out by particular cultures and traditions as places of special sanctity directly linked to their deepest values and aspirations. These mountains—the ones traditionally known as "sacred mountains"—have well-established networks of myths, beliefs, and religious practices such as pilgrimage, meditation, and sacrifice. A primary example is Mount Kailas in Tibet.

Second, mountains that may or may not be revered frequently contain sacred sites and objects such as temples, monasteries, hermitages, stones, springs, and groves. Great numbers of people, for example, visit pilgrimage shrines located in mountainous regions, such as the Hindu shrine of Badrinath in the Indian Himalaya. Singling out this role of mountains, a passage in the *Mahabharata*, the religious epic of ancient India, describes a Himalayan peak as a "refuge of hermits, treasury of sacred places."[2]

Third, mountains in general commonly awaken an individualized sense of wonder and awe that sets them apart as places imbued with evocative beauty and meaning. Many tourists, hikers, and climbers today go to the Himalaya to be inspired by views of the soaring peaks, symbols of transcendence rising high above the profane concerns of workaday lives and evoking their highest values and aspirations.

Mount Kailas, a sacred mountain for followers of Buddhism, Hinduism, Jainism and Bon, western Tibet, 1986

Previous spread: A Tibetan woman praying and lighting candles to venerate the light within the teachings of Buddha and to celebrate Buddha's enlightenment, Jokhang Temple, Lhasa, Tibet, 1986

People of different cultures generally experience the sacredness of mountains through the particular views they have of them, such as the mountain as center of the universe or the source of life. These views or themes differentiate the experience of the sacred and provide a basis for developing a framework for appreciating the diverse ways in which mountains have cultural and spiritual significance for different peoples and cultures. In the course of studying sacred mountains around the world, I have found nine themes to be particularly widespread in the Himalayan region. A brief description is provided for each, with an example or two to illustrate it.

Themes

Height Many people revere mountains first as high places. Mount Everest, as the highest mountain on Earth, has assumed the status of a sacred mountain in the modern world. Its summit of 29,032 feet symbolizes for many the highest goal one can strive to attain, whether one's pursuit be material or spiritual. Because of its importance in the West as the ultimate high place, the Tibetan name of Mount Everest, Jomolangma, is almost always mistranslated in Western literature as "Goddess Mother of the World." Foreigners assume that for the Tibetans and Sherpas the highest mountain must be the home of a supremely important deity. In fact, Jomolangma is short for Jomo Miyolangsangma, the name of the Tibetan goddess of Mount Everest, one of the Five Sisters of Long Life and a relatively minor deity, invoked for worldly benefits, such as longevity, food, and wealth.

Center An extremely important theme is that of the mountain as center: of the cosmos, the world, or a local region. Many Hindus and Buddhists view Mount Kailas as the physical manifestation of Mount Meru or Sumeru, which serves as a cosmic axis around which the universe is organized. For the Sherpas, Khumbhila, which stands at the center of their homeland of Khumbu, is much more sacred than Mount Everest, even though it's 10,131 feet lower at only 18,901 feet. They will climb Everest on the edge of their territory, but they don't dare climb Khumbhila for fear of incurring the wrath of the deity who dwells on its summit and watches over Khumbu.

Power Many sacred mountains are revered as places of power—political, physical, and spiritual. According to Tibetan tradition, the first king of ancient Tibet descended from heaven on a sky rope to land on Yarlha Shampo, a sacred mountain in the Valley of the Kings southeast of Lhasa. From this valley his successors spread their power and rule over the rest of the country and returned at death to be buried in huge mounds still visible today. Tibetans and Sherpas throughout the Himalaya perform rituals to appease local mountain deities, such as the country god of Khumbhila, who are portrayed in paintings as fearsome warriors who can either protect or wreak havoc, dispatching yetis to kill their yaks and setting loose devastating landslides and floods. According to the Hindu Puranas, the mere sight of the sacred Himalaya has the power to spiritually cleanse worshippers of their sins.

Deity or Abode of Deity The power of many sacred mountains derives from the presence of deities—in, on, or as the mountain itself. Indians regard the graceful peak of Nanda Devi (elevation 25,646 feet), a World Heritage site, as the abode of Nanda Devi, the Goddess of Bliss and a form of Parvati, daughter of Himalaya and consort of Shiva.
A temple at the hill station of Almora has a well-known image of the deity. Villagers in the region will perform rituals to summon the goddess from her mountaintop so she can possess a medium who can then dispense her blessings. Annapurna, the tenth-highest mountain in the world (elevation 26,545 feet), is the abode of Annapurna: "She who is filled with rice," another form of Parvati. As the consort of Shiva, the supreme

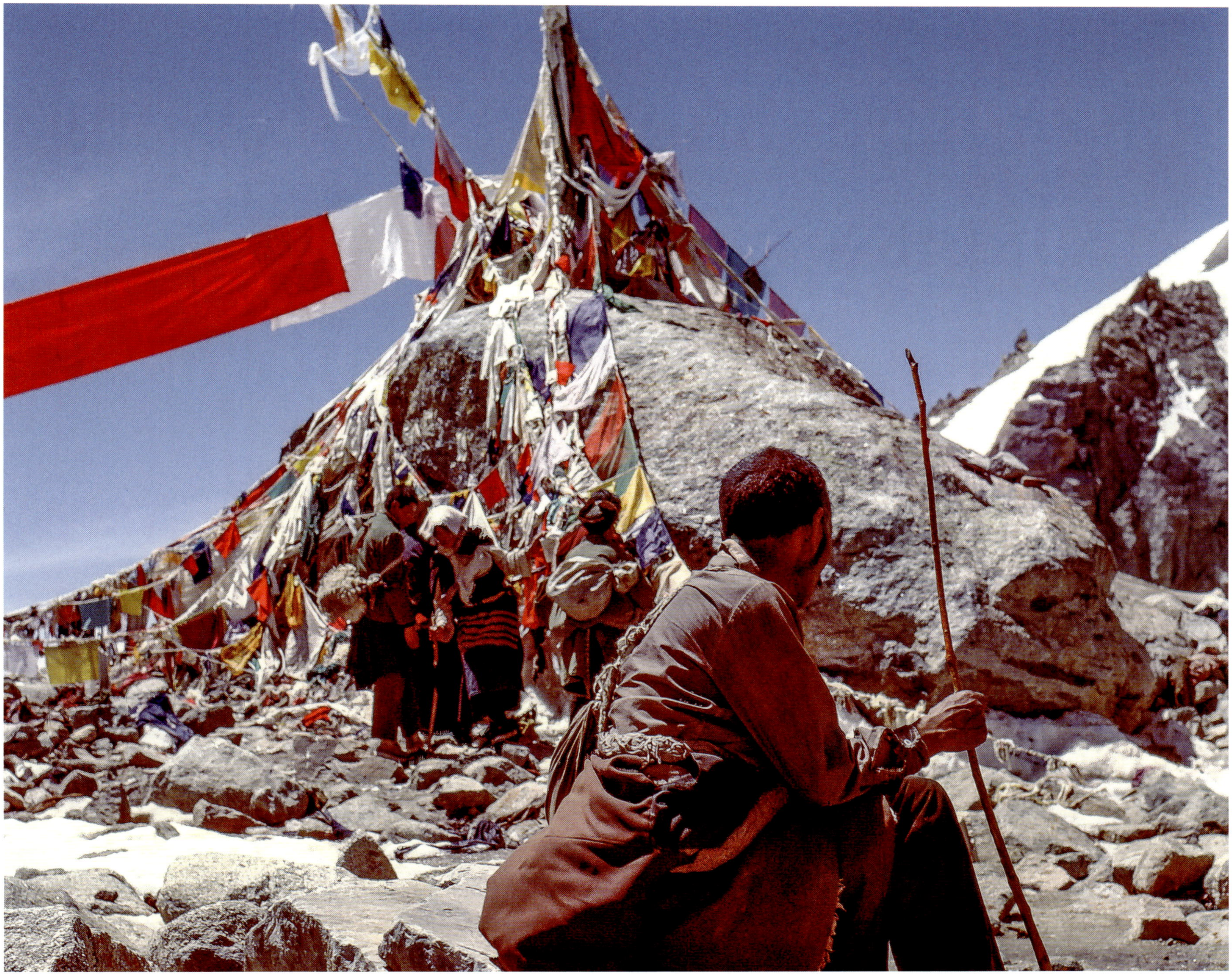

Pilgrims at Dolma La, on the sacred Kailas pilgrimage, western Tibet, 1986

deity in the form of destroyer and archetypal yogi, Annapurna is the presiding goddess of Varanasi, the most sacred city in India. She provides sustenance in life while Shiva grants *moksha* or spiritual liberation at death.

Temple or Place of Worship Many Asian traditions revere sacred peaks as temples or places of worship. Tibetan Buddhists, for example, view Mount Kailas as the pagoda palace of Demchog, the One of Supreme Bliss, a tantric deity embodying the ultimate Buddhist goal of enlightenment. A number of temples in India are modeled on the sacred peak and bear the name of Kailasanatha, Lord of Kailas—a reference to Shiva, the Hindu deity who dwells on the mountain.

Paradise or Garden Numerous cultures, both traditional and modern, view mountains as gardens and paradises: heavens on Earth. In a passage typical of many other works, the Hindu *Tantra of the Great Liberation* describes the paradise of Shiva on the summit of Mount Kailas in the following words:

> The enchanting summit of the Lord of Mountains, resplendent with all its various jewels, clad with many a tree and many a creeper, melodious with the song of many a bird, scented with the fragrance of all the season's flowers, most beautiful, fanned by soft, cool, and perfumed breezes, shadowed by the still shade of stately trees; where cool groves resound with the sweet-voiced songs of troops of celestial nymphs . . .[3]

Ancestors and the Dead Another major theme links mountains to the other world as ancestors and abodes of the dead, often involved in origin myths. The Lepchas, original inhabitants of Sikkim, revere Kangchenjunga, the third-highest mountain in the world (elevation 28,169 feet), as the place where their primordial ancestors, the first Lepcha man and woman, were created from its glaciers. According to their beliefs, when they die they go to the realm of the dead, hidden behind the great snow and ice curtain of Kangchenjunga, which they call Konglo Chu, the "Highest Veil of Ice."

Source Throughout the Himalaya people look up to mountains as sacred sources of blessings, such as water, life, fertility, and healing. The name "Himalaya" means "abode of snow." Hundreds of millions on the dry plains of India revere the snows of the highest mountain range in the world as the sources of sacred rivers on whose life-giving waters they depend for their very existence. These mountains also provide medicinal herbs and the blessings of spiritual and physical well-being. Shivling, a peak sacred to Shiva, looms over Gaumukh, a favorite haunt for yogis to meditate and the glacial source of the Ganges. Gaumukh lies up a valley from the major Hindu pilgrimage shrine of Gangotri in the Garhwal reigon of the Indian Himalaya.

Inspiration, Renewal, and Transformation Since ancient times, the Himalaya have inspired some of the most evocative passages of Indian literature. Around the fifth century CE, Kalidasa, the Indian equivalent of Shakespeare, opened the greatest of his long poems, *Kumarasambhava* or "The Birth of the Young God," with the following passage:

> There is in the north a supreme king of mountains
> named Himalaya, possessed of a divine self.
> Bathing in the eastern and western oceans, he
> stretches like a measuring rod across the earth.
>
> A source of endless jewels, snow does nothing
> to diminish his splendor:
> Just as a spot on the moon vanishes in a flood of
> moonlight, so a single blemish disappears beneath
> a flood of virtue.[4]

Inspired by such sentiments, yogis of various religious traditions have long sought out the Himalaya as the ideal place to practice meditation and realize the ultimate goal of spiritual liberation or enlighenment. In more recent times, artists, such as Nicholas Roerich, and photographers, including Bill Frej, have drawn inspiration from the evocative sight of Himalayan peaks reaching toward the sublime heights of the far blue sky. Today, many people go to see, trek, and climb in these mountains, seeking spiritual and physical renewal—and even transformation. Maurice Herzog, who made the first ascent of Annapurna in 1950 with a French expedition and lost his fingers and toes in the process, wrote of his experience:

> In my worst moments of anguish, I seemed to discover the deep significance of existence of which till then I had been unaware. I saw that it was better to be true than to be strong. The marks of the ordeal are apparent on my body. I was saved and I had won my freedom. This freedom, which I shall never lose, has given me the assurance and serenity of a man who has fulfilled himself.[5]

Each of these nine themes or views brings together different ideas, images, and associations to evoke the experience of a deeper reality. Tibetan pilgrims, for example, view the peak of Mount Kailas as the pagoda palace of Demchog. The two images fuse in their minds so that the mountain becomes the palace. The fusion or resonance of these images awakens the experience of something that imbues Mount Kailas with an aura of sanctity. The pilgrim becomes aware of a divine presence emanating from the mountain.

The process works a little like the fusion of two slightly different photographs in a stereoscopic viewer to trigger a vivid perception of the third dimension represented in each two-dimensional picture. The scene that looks flat suddenly seems to pop open with depth—in the case of a mountain like Kailas, a depth full of meaning and significance for the pilgrim who reveres it as sacred. For followers of religious traditions, the resonance or fusion of cultural and spiritual associations in a sacred mountain does not create an artificial, purely emotional effect: rather, it reveals an underlying reality present but usually hidden from awareness.

Kailas: The Most Sacred Mountain

Mount Kailas in remote western Tibet is the most sacred mountain in the world for nearly a billion people in Asia, followers of four religions, including Buddhism, Hinduism, Jainism, and Bon. Although at 21,778 feet it is 7,254 feet lower than Everest (elevation 29,032 feet), Kailas stands out from its surroundings and is a much more impressive peak, bearing the form of a Hindu temple or a Buddhist stupa. Tibetans know the mountain as Kang Rinpoche: "Precious One of Glacial Snow" in the Tibetan language.

Many Buddhists and Hindus regard Mount Kailas as the physical manifestation of Mount Meru or Sumeru, a mythical mountain at the center of the universe. It is said to rise more than 80,000 miles to the heights of heaven, and its roots go down more than 80,000 miles to the depths of hell. Mount Meru is said to lie to the north of our world, described as a continent floating in an ocean surrounding the mountain.

For Hindus, Kailas is the primary dwelling place of Lord Shiva, one of the three forms of the supreme deity. Ancient art and contemporary posters depict him in a paradise on the summit of the sacred mountain with a woman's face on top of his head and water gushing from her mouth. She is Ganga, the goddess of India's most sacred river, the Ganges. This is a reference to a very important living myth of Hinduism that has important environmental implications. According to the myth, an ancient sage asked the goddess to come down from heaven to fill the oceans. She liked it up there and didn't want to come down. So, as an excuse, she said that the weight of her waters falling from heaven would shatter the earth. Shiva offered to break her fall with the hair on his head, which is what happened.

Indian environmentalists have pointed out that the myth corresponds to what actually happens in the environment. The *Puranas*, ancient texts of Hindu mythology, say that the forests of the Himalaya can be viewed as Shiva's hair. During the summer, the Ganges does indeed fall from heaven in the form of monsoon rains, and when Shiva's hair is not there to break her fall, where Himalayan forests have been cut down, the earth does shatter under her impact in devastating landslides and floods.

Within sight of Kailas lies Lake Manasarovar, the most sacred lake in the world in Hinduism and Tibetan Buddhism. Hindu pilgrims bathe in the lake and offer its sacred waters to Shiva on the nearby summit of Mount Kailas. Shiva is known as Kailasanatha, the Lord of Kailas.

Tibetan Buddhists also revere Mount Kailas as the most sacred mountain in the world, but they see it as the pagoda palace of a different deity: Demchog, the One of Supreme Bliss. Tibetan thangka paintings show the blue figure of Demchog embracing his female consort. At a deep level, the painting symbolizes the union of the spiritual forces of masculine compassion and feminine wisdom needed to achieve the highest goal of Tibetan Buddhism: attaining enlightenment for the sake of all beings. When practitioners arouse these forces within themselves, they awaken to the true nature of reality, which is what the word "Buddha" means—one who is awake or aware—and they acquire the ability to help others to go beyond suffering, which is the meaning of nirvana.

The pilgrimage to Kailas is one of longest and most arduous in the world. In the past, pilgrims would travel by foot and horseback for weeks and months. Now, with the paved roads, it takes only a day or two from Tibet's border with Nepal. When the pilgrims finally get to Kailas, they don't climb it; rather, they walk around it, treating it with respect as they would a shrine, which it is. The circumambulation of the sacred mountain is about thirty miles long, beginning at around 15,000 feet and crossing a pass more than 18,000 feet above sea level.

One of the joys of partaking in the pilgrimage to Kailas is the people one meets along the way. They have come for the spiritual experience of a lifetime, and it shows on their faces.

As one circles around the mountain, it presents its different faces. The austere north face rises nearly 5,000 feet from base to summit. According to Tibetan tradition, Tibet's most famous yogi, Milarepa, climbed the mountain during the eleventh century CE by flying magically to the summit in an instant, certainly the most direct and elegant ascent of a mountain in mountaineering history.

Reinhold Messner, the well-known Himalayan climber, has credited Milarepa with the first ascent of Mount Kailas in an effort to discourage anyone from desecrating the sacred mountain by trying to climb it. Kailas has been designated off-limits to climbing out of respect for the many different traditions that revere the peak and to retain a sense of mystery—of the unknown—that reminds climbers of the higher values of mountaineering itself.

Milarepa sang of Kailas:

The prophecy of Buddha says most truly,
That this snow mountain is the center of the world,
A place where the snow leopards dance.
The mountain top, the crystal-like pagoda,
Is the white and glistening palace of Demchog . . .
This is the great place of accomplished yogis;
Here one attains transcendent accomplishments.
There is no place more wonderful than this,
There is no place more marvelous than here.[6]

Kangchenjunga from South Base Camp, 1985

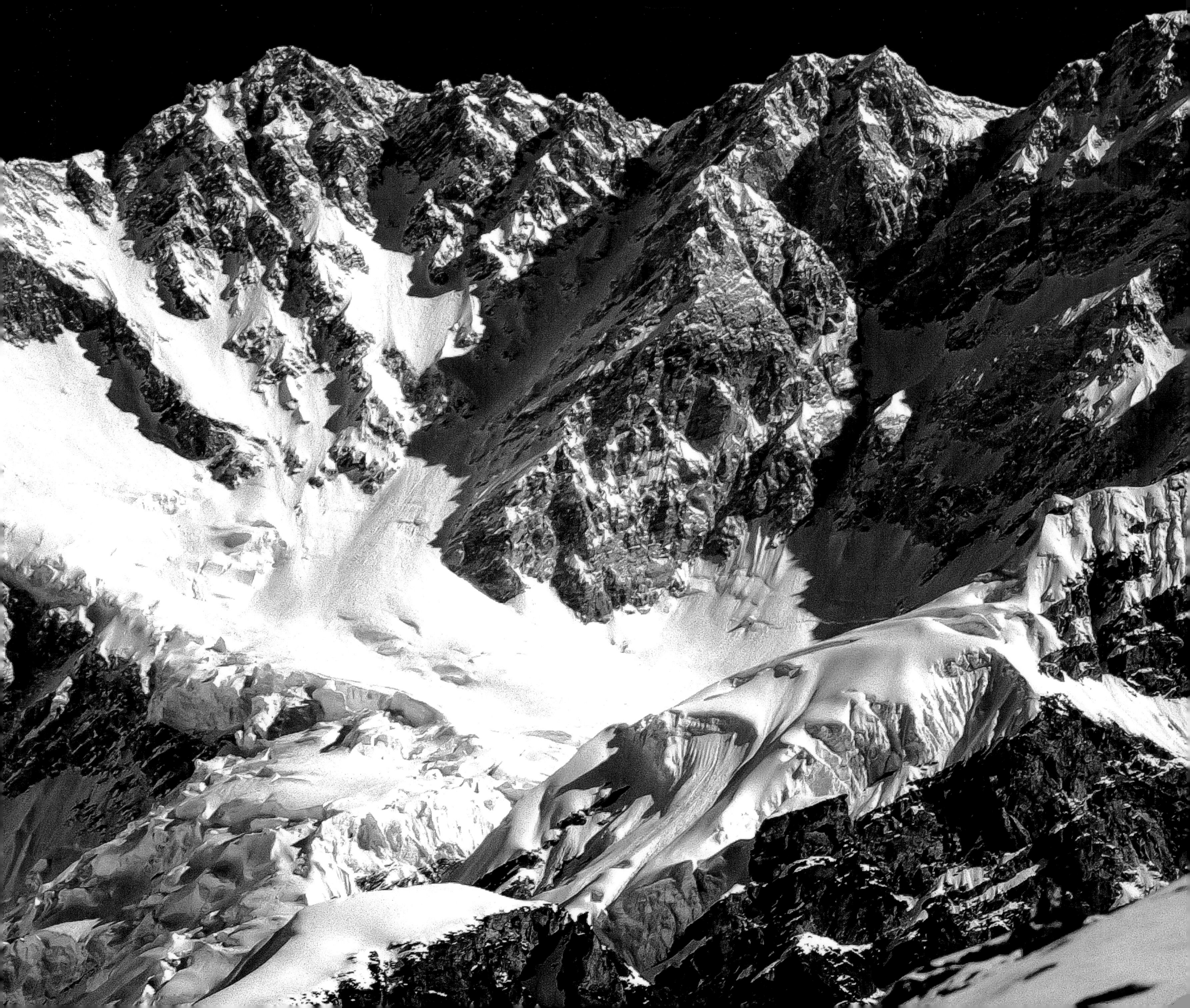

Like Milarepa, some pilgrims today walk barefoot over snowfields and icy streams, suffering no frostbite.

The high point of the pilgrimage is the Dolma La, the Pass of the Savioress (elevation 18,471 feet). Each Tibetan pilgrim leaves a part of himself, herself, or themself here, such as a lock of hair, symbolizing the fact that they have let go of attachments to the ego in order to attain enlightenment for the benefit of others.

The pilgrimage to Kailas is a journey to the center of the universe and the center of who we are. In touching that inner center, pilgrims awaken a sense of joy.

The Allure of Mountains

The many different views of Himalayan peaks such as Mount Kailas and other sacred mountains represent various ways of putting people in touch with a deeper reality that can give meaning and purpose to their lives. For some, the way involves a pilgrimage to an actual mountain, there to worship a deity, absorb blessings, or gather merit. Others use the ascent of a peak as a symbol of a spiritual path in which the climb symbolizes the efforts required to reach the goal they seek. In making this symbolic ascent, they open themselves to a deeper reality that transforms them and carries them on to their destination. The same thing may come about from the act of contemplating a mountain from the distance—or visualizing it in the mind. A well-known Hindu scripture says of Himachal, or the Himalayas:

> He who thinks of Himachal, even if he does not see it, is greater than he who accomplishes all his devotions at Benares. He who thinks of Himachal will be freed from all his sins . . .[7]

The reality revealed by a mountain can be found through other symbols and features of the landscape. But because of their natural tendency to awaken a sense of the sacred, it is often easiest to experience this reality first in mountains—or in a view of mountains—and then find it elsewhere, even in the cities and plains far from the sight of their summits.

The sense of another reality hidden within or beyond the one we know is essential. Without it we slowly die of spiritual thirst, becoming desiccated husks of ourselves. The mountains of the Himalaya stand as constant reminders of that reality and the deeper life that flows from it, like a river from its source.

1. Kailas is also written as Kailash. The original name of the mountain in the ancient Sanskrit language of India is Kailasa.

2. Mahabharata 3.43.21–25, adapted from a translation in J. A. B. van Buitenen, editor and translator, *The Mahabharata*, 3 vols. (Chicago, IL: University of Chicago Press, 1975), Vol. 2:308.

3. Arthur (Sir John Woodroffe) Avalon, translator, *Tantra of the Great Liberation* (Mahanirvana Tantra) (Madras, India: Ganesh & Co., 1913; reprint: New York, NY: Dover Publications, 1972), 1.

4. Translated from Kumarasambhava 1.1, 1.3. For another translation see Hank Heifetz, translator, *The Origin of the Young God: Kalidasa's Kumarasambhava* (Berkeley and Los Angeles: University of California Press, 1990), 21.

5. Maurice Herzog, *Annapurna*, translated by Nea Morin and Janet Adam Smith (New York, NY: E. P. Dutton, 1952), 12.

6. Garma C. C. Chang, translator, *The Hundred Thousand Songs of Milarepa*, Vol. 1 (Boston, MA: Shambhala Publications, 1977), 262.

7. From the Manasakhanda of the Skanda Purana, adapted from a translation in Edwin T. Atkinson, *Kumaun Hills: Its History, Geography, and Anthropology with Reference to Garhwal and Nepal* (Delhi, India: Cosmo Publications, 1974), 307–08.

The Memoir

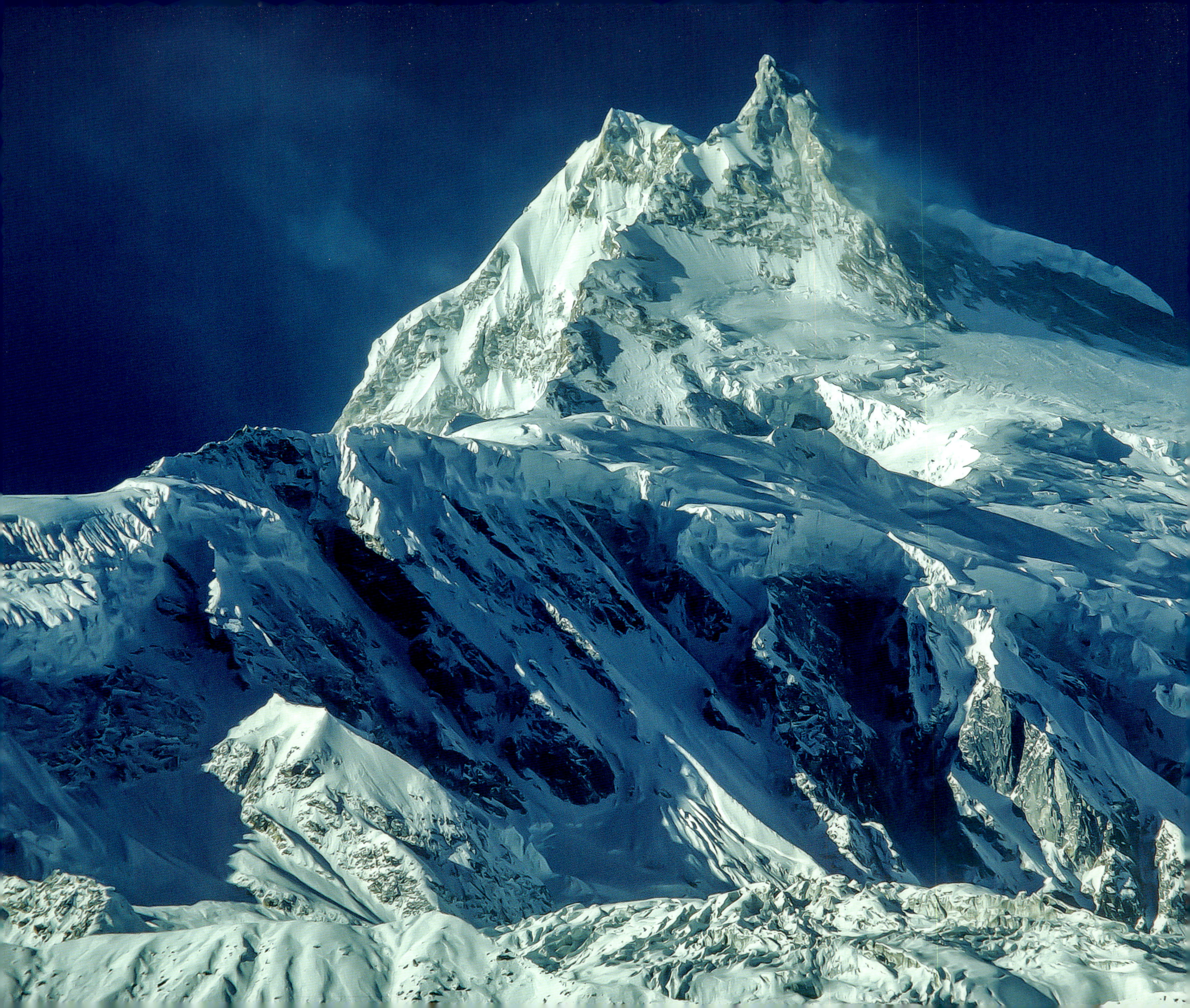

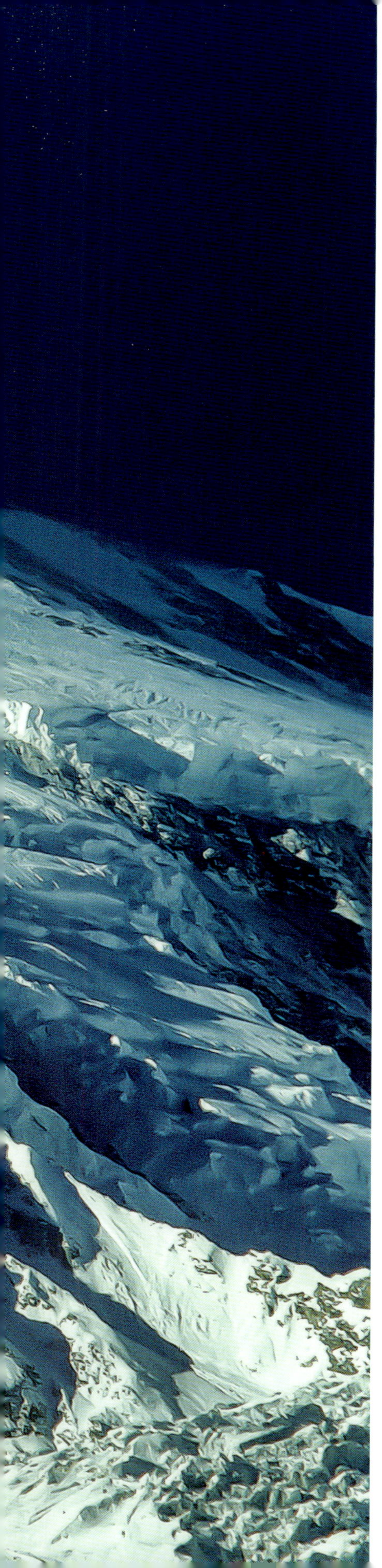

Around Manaslu, Nepal

NOVEMBER 1981

Fascination with high mountain peaks and a love of walking drew us to Nepal for the first time in November 1981. From the moment we stepped off the plane in Kathmandu, we knew that this was a magical place unlike any we had ever experienced. As we rode from the airport to our hotel in Thamel, the backpacker neighborhood, the view through the taxi windshield became clearer.

Bumping along a potholed road, we passed medieval-looking, three-story brick buildings with carved wooden latticework windows. Children played under the branches of an ancient banyan tree spreading over a courtyard. Men on bicycles and stray cows wove in and out of the traffic in front of us, competing for space with other taxis and delapidated buses. We looked to the right, and along the Bagmati River we spotted smoke rising above a row of Hindu temples with arched facades and stepped pagoda-like roofs. We learned later that this was Pashupatinath Temple, a sacred place where Hindus cremate their dead. Always in the background, towering over these scenes, were the snow-capped Himalayan peaks we had seen in the distance as we flew into Kathmandu. It was our first day in Asia and the start of a lifetime of journeys in the mountain regions of the Himalaya Range and its borderlands.

On our first trek in Nepal, we walked around Manaslu, the world's eighth-highest mountain (elevation 26,781 feet),

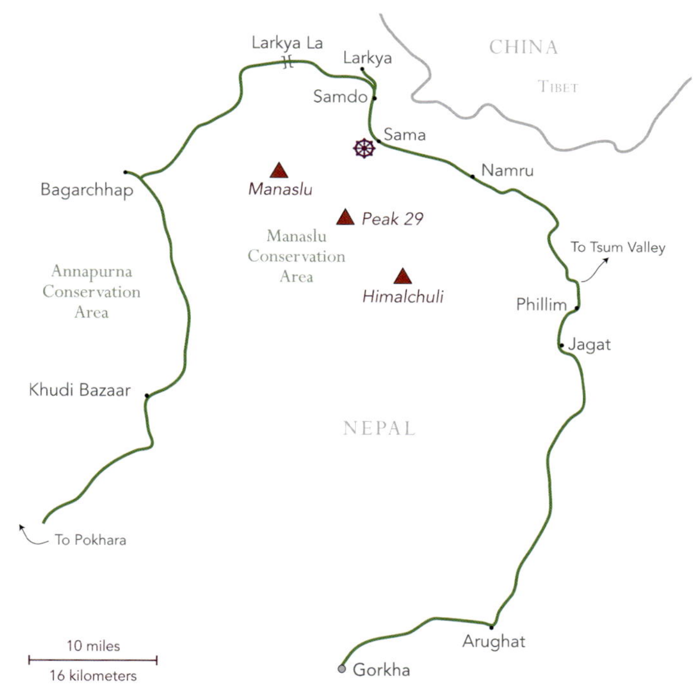

located in west-central Nepal. Known by some as the Japanese Mountain, it was first explored by Japanese climbers during the early 1950s and then successfully climbed by a Japanese expedition in 1956, in spite of an encounter a year earlier with angry villagers who believed the climbers had angered the mountain gods, resulting in an avalanche. Partly

Manaslu, 1981

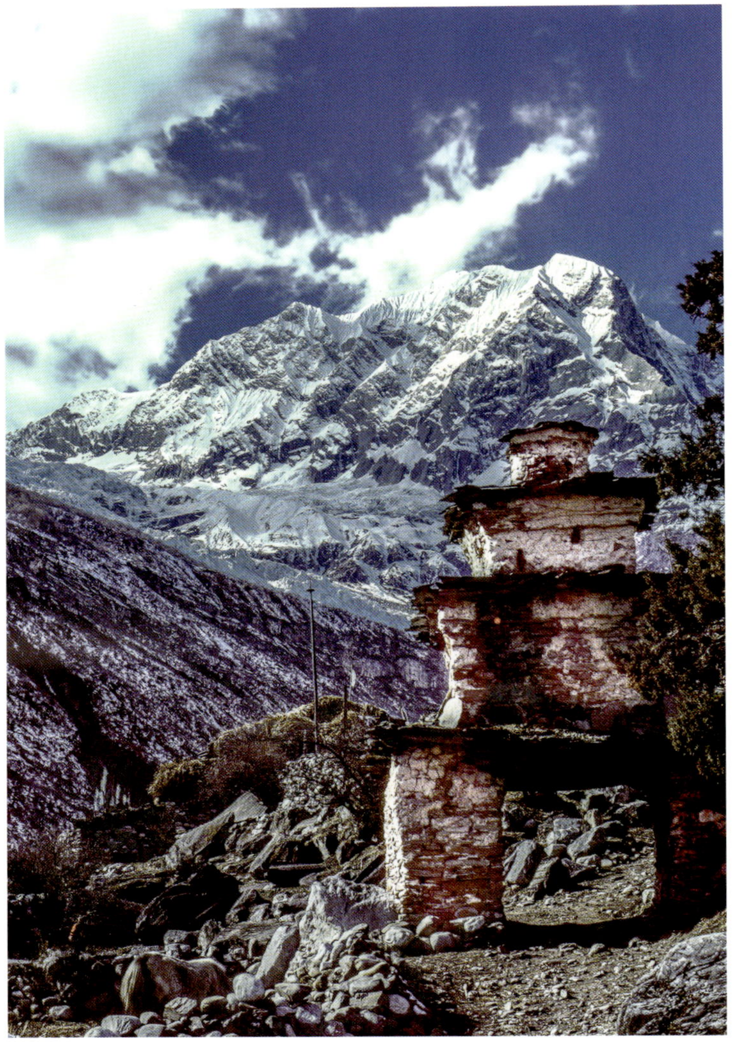

Buddhist stupa below Manaslu near Sama village, Nepal, 1981

for this reason, Manaslu was not climbed again until 1971 by another Japanese expedition. Although the peak is considered to be relatively easy to summit under ideal conditions, it has a history of deadly avalanches.

At our first group meeting at the hotel in Kathmandu, it was clear that everyone was nervous about the prospect of walking at high elevations on narrow and sometimes steep and exposed trails. Most of us in the fifteen-member Sierra Club group, with the exception of our leader, had never been on a Himalayan trek before. Reality hit me the next day as soon as we all stepped off the bus from Kathmandu to the town of Kurin Ghat and came face to face with a rickety pedestrian suspension bridge with missing wooden slats high above the Trisuli River. Swallowing my fear of heights, I started across, slowly, swaying with every step and bouncing up and down as more people entered the bridge behind me.

We soon got into the trekking pattern of waking early every morning when members of our Nepali crew thrust cups of hot lemon tea into our tents and placed a shallow pan of warm washing water outside. Before hitting the trail, we ate a breakfast of hot porridge or cornflakes and drank cups of coffee made with powdered Nescafé. We hiked all morning before lunch, which usually consisted of eggs, French fries, and warm *chapatis,* Indian-style flatbread cooked over the fire. After lunch we hiked until late afternoon, logging from twelve to fifteen miles per day before stopping at a campsite deemed suitable by Jagat Man Lama, our *sirdir* (head guide). After the Nepali porters helped us put up our tents, we spent the time before dinner writing in our journals, drinking tea, tending to blisters, and rearranging the contents of our duffel bags.

Most evenings local people from a nearby village would crowd into our dining tent, standing at the edges and marveling at us as well as the canned fruit cocktail we enjoyed after our supper of *dal bhat* (rice and lentils) and vegetables. For entertainment after dinner we took turns reading aloud from *The Ascent of Rum Doodle*, a 1956 parody of early mountain climbing "siege-style" expeditions that used large numbers of high-altitude porters, tons of equipment, supplemental oxygen, successive well-stocked camps, and fixed ropes along

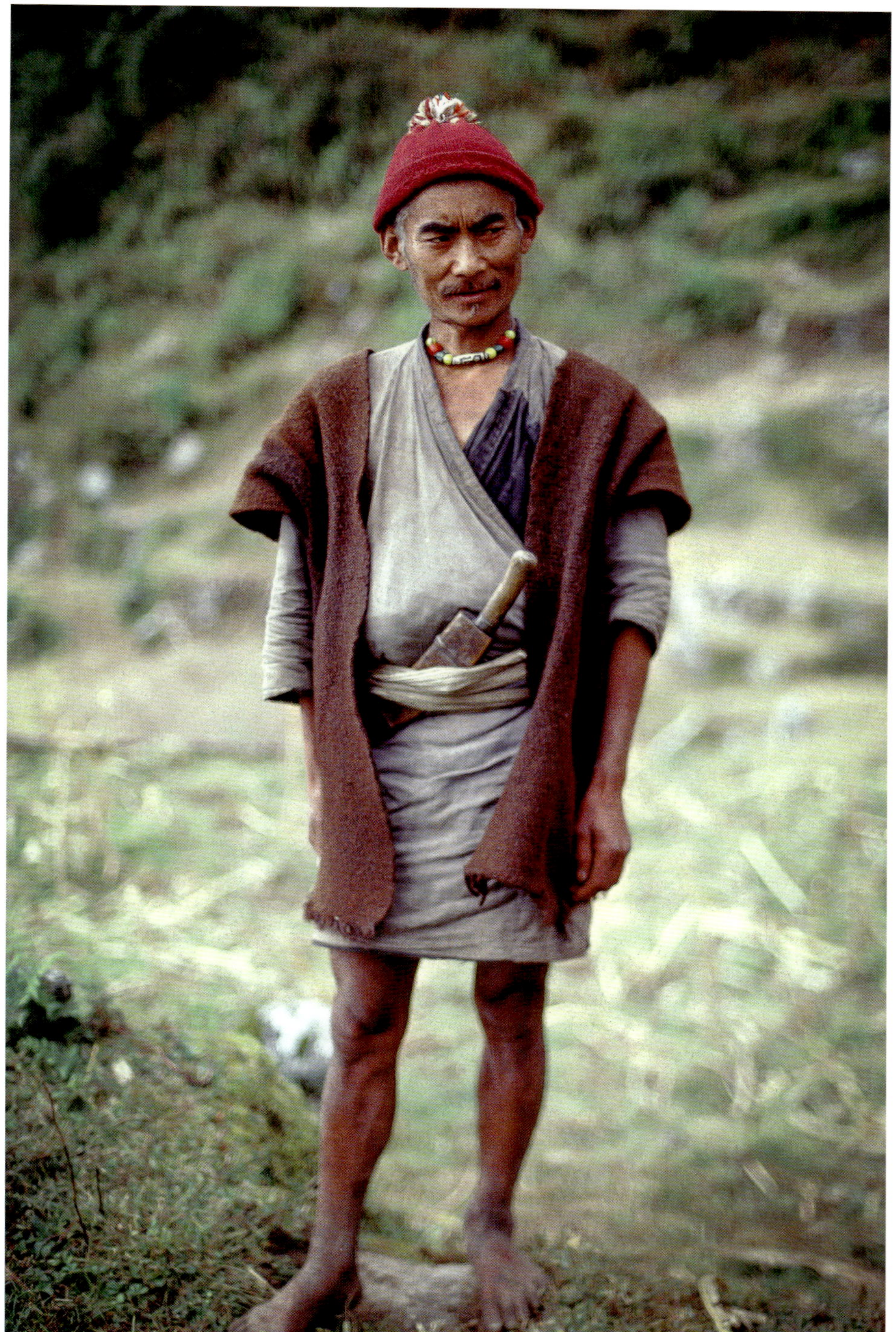

Gurung villager, Nepal, 1981

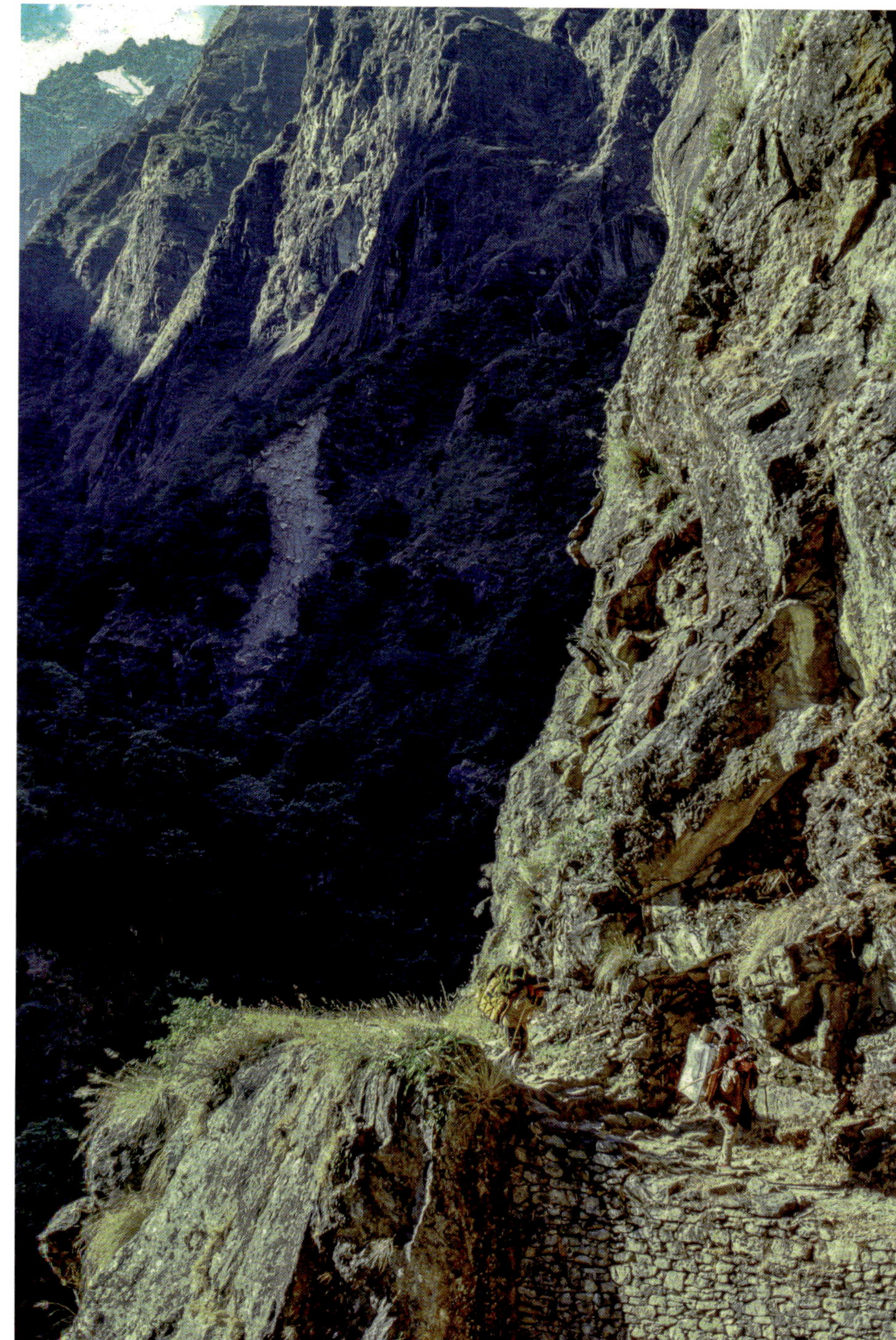

Porters on the trail to Sama, Nepal, 1981

the route to the summit. To reach the top of Rum Doodle, a mythical 40,000-foot peak located in the county of Yogistan, the fictitious climbing team employed 3,000 Yogistani porters, each carrying a load of 1,000 pounds. Our trekking group, complete with dozens of porters and a huge cook staff, seemed almost as unwieldy to me.

The first few days of our trek were hot and steamy as we hiked up and down hills in lush, low-altitude forests on our way to higher elevations. On the morning of the fourth day, we saw snow peaks for the first time: Annapurna South, Himalchuli, and the Ganesh Himal Range. The trail became steeper and more exposed, with precipitous drop-offs down to the rushing, gray-colored river below. The nights got colder, and water bottles partially froze inside the tent. On forays out of the tent in the middle of night I often saw snow-covered mountain peaks glowing luminously in the moonlight. On the ninth day, as we continued walking on the cliffs above the Buri Gandaki River, we began to encounter traders from near the border of Tibet walking alongside mules loaded with yak-skin bags filled with salt and butter and animal skins. We crossed into a restricted area at the village of Jagat and learned from authorities there that no foreign trekkers, only mountaineering teams had ever been this far before. At the next police checkpoint three days later the authorities were not so welcoming, since we were now quite close to the border with Tibet, sensitive because of the Chinese occupation since 1950 of what many consider an independent state. Somehow we were able to convince them to let us continue, arguing that we had come so far already and it would be a shame to turn around. We were less than ten miles from the Nepal–Tibet border, passing through villages where inhabitants of Tibetan descent known as the Nubri had limited contact with people outside their immediate environs. This area remained officially off-limits to foreign trekkers until the early 1990s but today is open to trekkers with a restricted area permit. Teahouses are available for trekkers.

We began seeing long *mani* walls made of hundreds of flat stones carved with inscriptions and prayers, especially the Buddhist mantra *Om Mani Padme Um* (Oh, jewel in the heart of the lotus), as well as stupas. Stupas, known as *chortens* in Tibetan, are religious monuments found all over the Buddhist world that commemorate people and events. They appear in all sizes from small, simple constructions of stone and plaster to large, two-level structures that one can walk through.

In the village of Sama, not far from the base camp of Manaslu, we were invited into a Tibetan Buddhist temple, known as a *gompa* or *gomba*, to witness a ceremony that must have been hundreds of years old. We did not know the meaning or purpose of this religious ritual but were transfixed by the sight of some twenty red-robed monks and students of all ages seated at low tables in a square configuration and chanting in unison. At the height of the ceremony a ray of sunlight struck the gold-leaf-covered statue of the Buddha on the altar at the front of the room. Soon after, strange cone-shaped offerings made of butter and *tsampa* (barley flour) were placed outside by a monk and quickly snatched up by huge black crows. This vivid introduction to Tibetan Buddhism ignited our interest and made Bill and me want to learn more about this religion and the ancient cultures associated with it.

The high point of our trek around Manaslu was Larkya La, a mountain pass (elevation 16,752 feet) that was covered in several feet of snow on the day we crossed it (*la* means mountain pass). The night before our approach to the pass we slept at 15,000 feet. Starting in the dark at 5:30 am the next day, we walked slowly up to 16,000 feet and stopped for lunch. Everyone in our group, including the porters, had a bad headache and nausea from the altitude. Several porters

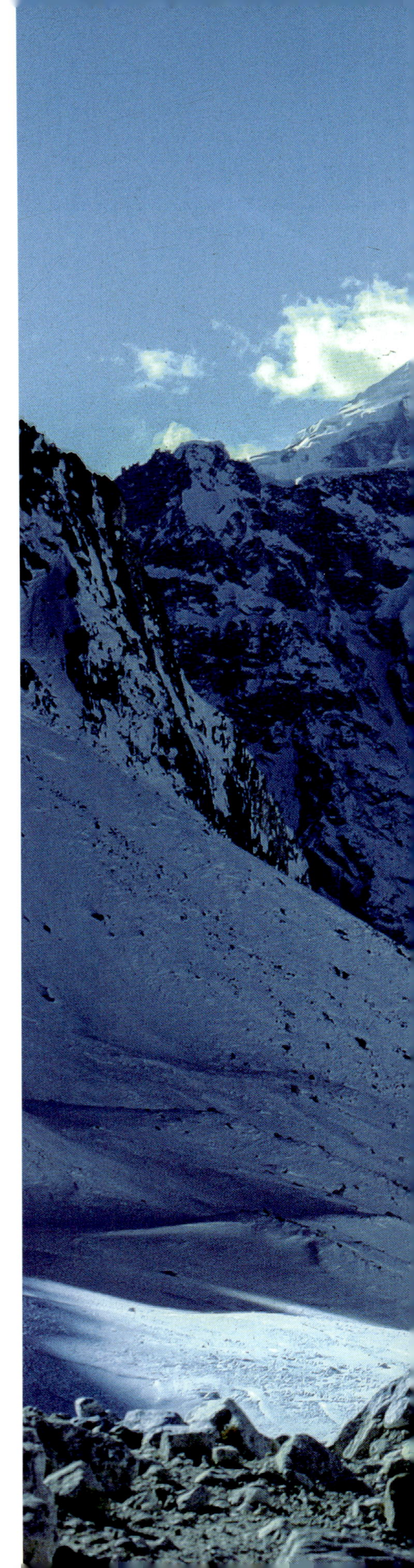

Crossing Larkya La, Nepal, 1981

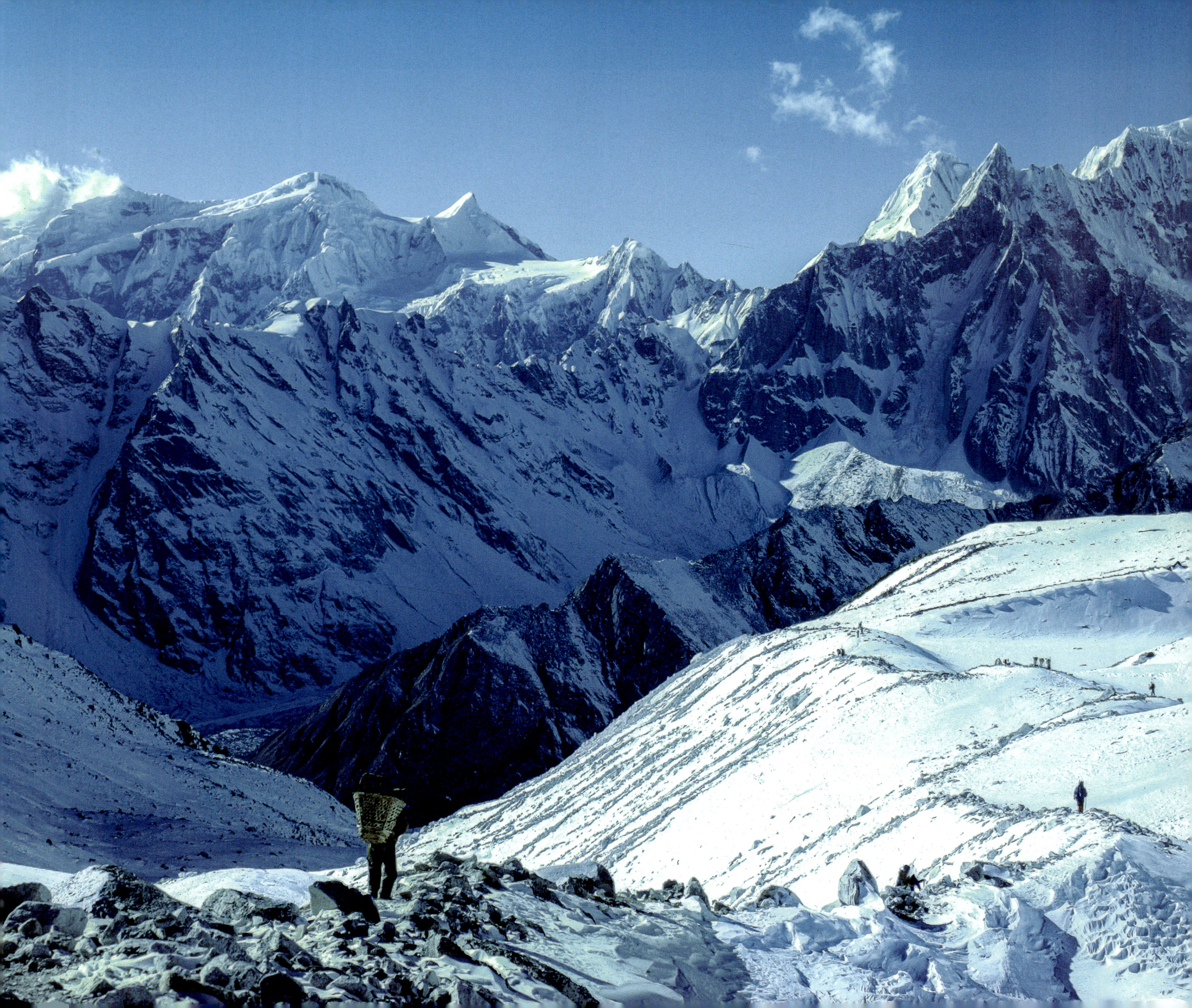

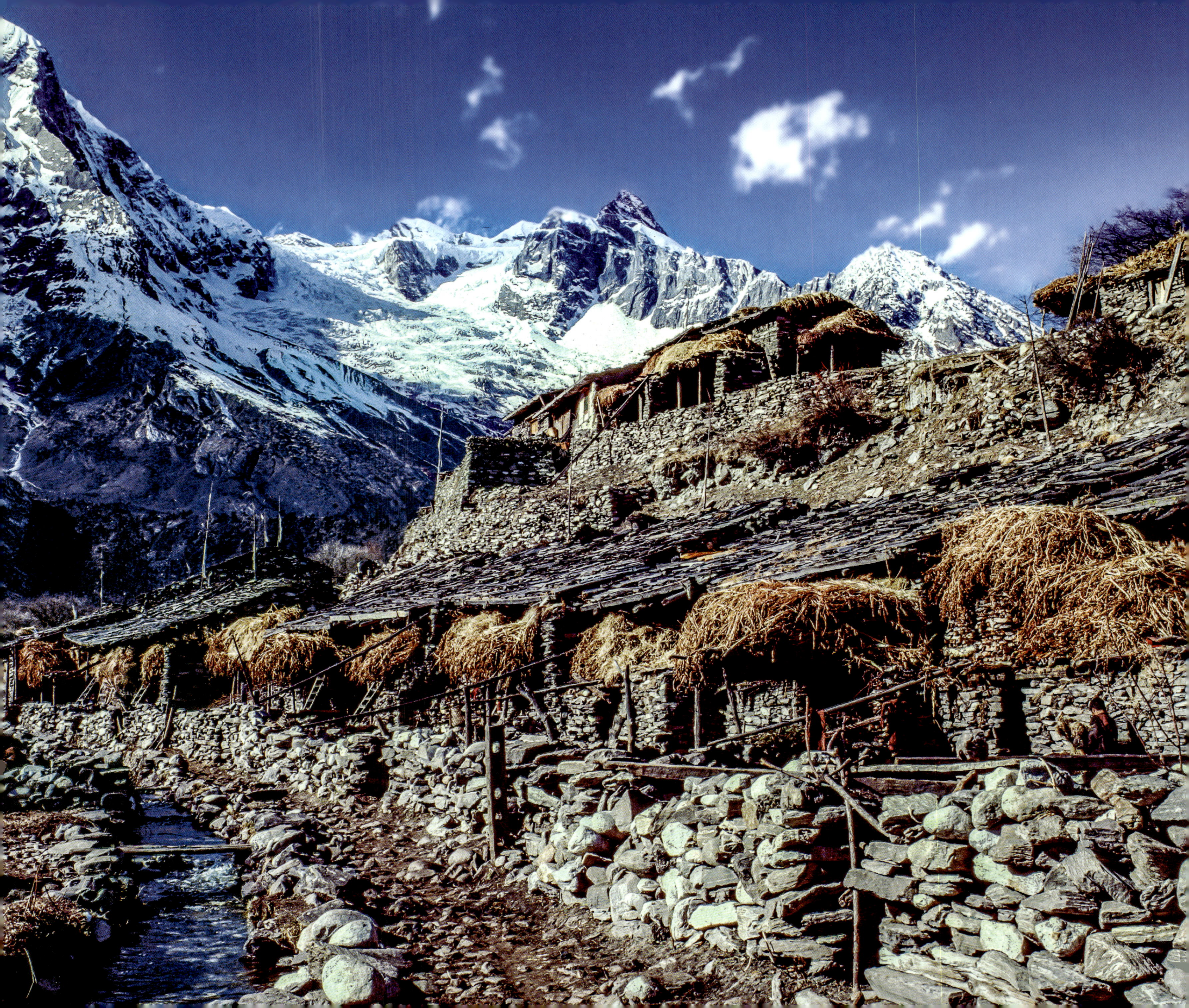

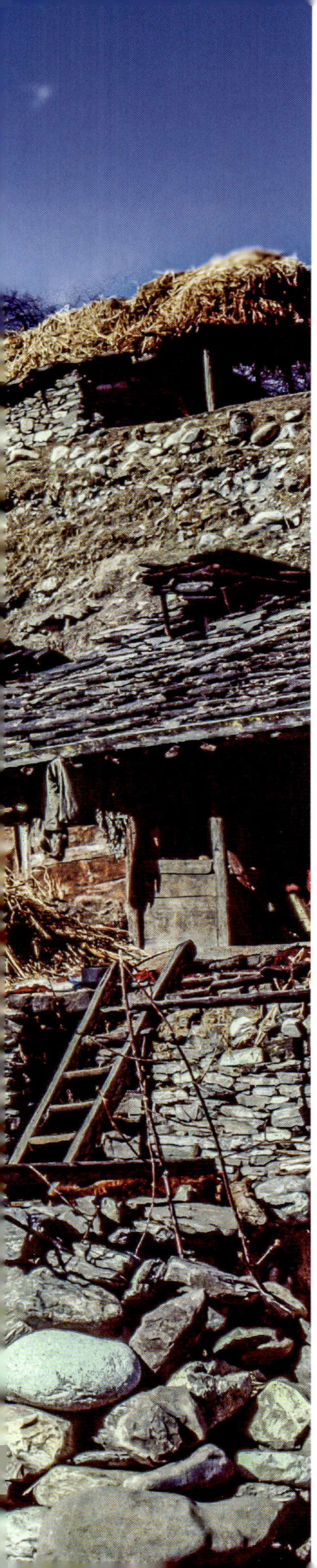

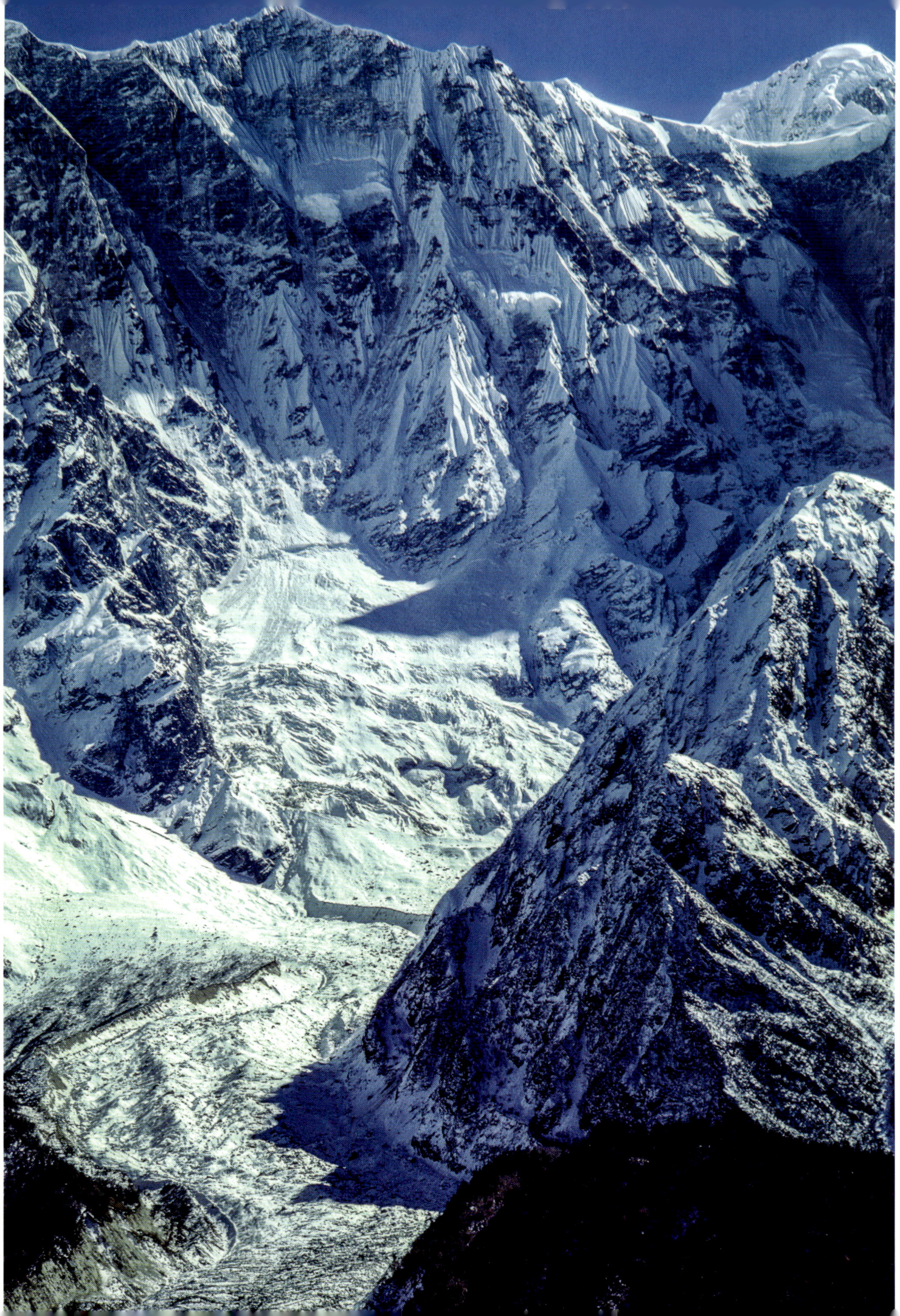

Glaciers below
Manaslu, Nepal, 1981

Far left: Sama village,
below Manaslu, Nepal, 1981

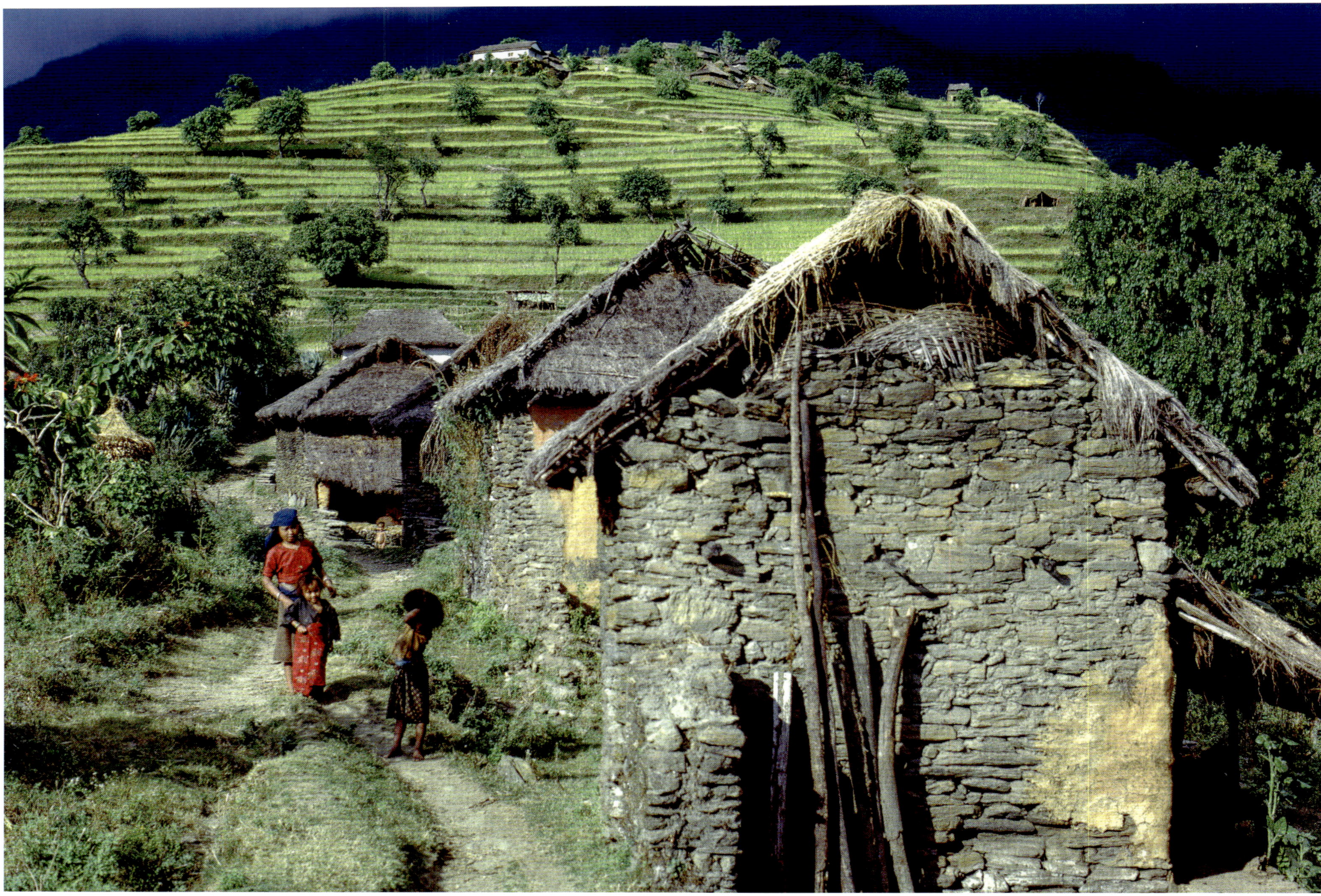

The trail to Manaslu, Nepal, 1981

were seriously sick, so some of us trekkers took extra weight from their loads and put it into our backpacks. Two hours later, we reached the top of the pass where we could see Annapurna II, the Nilgiri peaks, and many other mountains with elevations greater than 22,000 feet. Getting down the other side of the pass required stepping into deep snow on steep terrain. With lots of confusion on the way down from the pass, our small group of four people lost the route and ended up on a ridge with no trail in sight as the sun set and it got darker and darker. We turned on our headlamps and waited; finally, in the distance we saw lanterns and the flash-lights of the porters walking back to find us and lead us down the ridge to our camp.

For a week we worked our way back to the town of Pokhara along the Dudh Khola River on the western side of Manaslu. We were generally heading downward in elevation, but there were still plenty of up-and-down hikes each day. At one high point we looked back and saw a line of peaks: Manaslu, Peak 29, and Himalchuli in the foreground with Annapurna II, III, and IV, and Machapuchare in the distance.

Our food supplies began to dwindle during the last week, so our cook bought a butchered water buffalo in a small village. I spent half a day walking behind a porter carrying a basket on his back filled with furry legs and fly-covered meat that reappeared as buffalo curry that evening. Our companionable group hiked in groups of two or three, and all we talked about on the trail was the Thai food we would find in Bangkok on the way home.

Finally, on the twenty-ninth day after embarking on the trail in Kurin Ghat, our trek ended suddenly, and we boarded a bus back to Kathmandu. A mix of emotions from relief to exhilaration hit me. In Kathmandu we slept late, took showers, drank ice-cold Kingfisher beers at the Rum Doodle Bar, and ate chocolate cake at KC's Restaurant. In spite of our delight in creature comforts, we were already planning our next trek into the mountains. Bill and I realized that we could live for a while, perhaps months, without modern conveniences if it meant access to faraway, untouched mountain areas. We loved the rhythm of simply walking for hours and focusing only on the trail stretching out in front of us for miles. The most exhilarating aspect was the mountain scenery we viewed along the way. The world's highest peaks inspired us and created a reverence for their environments and the cultures surrounding them.

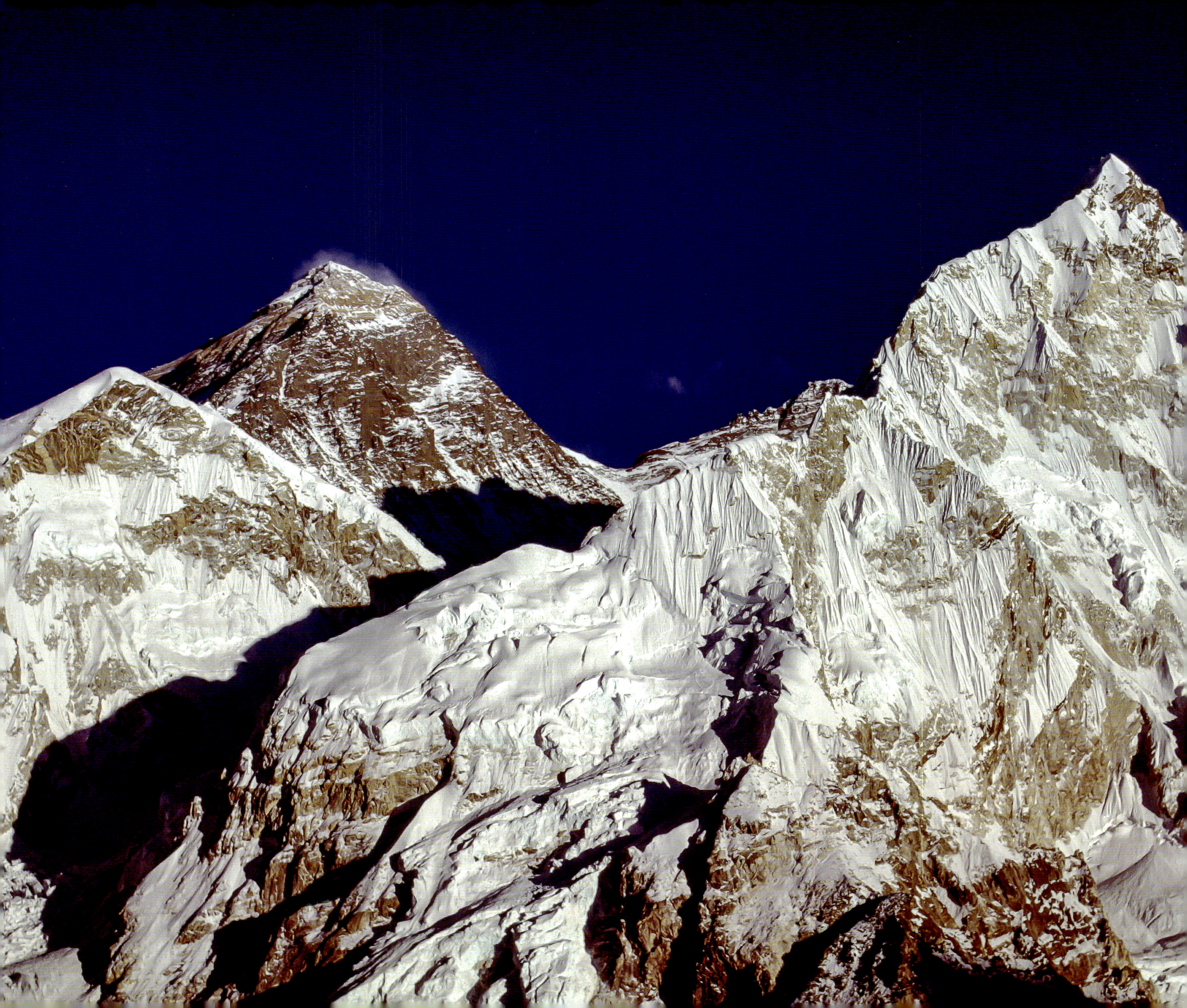

Somewhere Beneath Mount Everest, Nepal

NOVEMBER 1982

I NEED TO HIRE A YAK—and fast. We are camped at a desolate and snowy pasture below Mount Everest, and Bill is almost unconscious with signs of acute mountain sickness, including a terrible headache and loss of balance. The only remedy is to quickly get him to a lower elevation with more oxygen, and since he cannot walk unassisted, my plan is to find an animal to carry him. We are at an elevation of 15,000 feet, and the only animals around are yaks, the big, buffalo-like beasts that roam in the high elevations.

Across the field I spot a yak on a long, leather bridle held by a tiny *Sherpani* (Sherpa woman) with a baby strapped to a plank on her back. I wonder what she is doing up here in the middle of nowhere but realize she is my only option. I approach her, and through sign language we begin negotiating the yak rental. She knows I am desperate, and the negotiations are tough, but eventually we agree on the equivalent of $30 a day, a lot of money in this part of the world. I hand her a stack of Nepali Rupees, which she and her friend carefully count before leading the yak over to me. My nearly six-foot tall husband is lying in a heap on the ground, so my next task is to get him on top of the yak and seated on the tiny wooden saddle five feet off the ground.

The Sherpani and I push Bill up the side of the beast, trying to get him to straddle it, but each time he slides over to the other side. She and I run back and forth from haunch to haunch and keep pushing him back up until equilibrium is finally reached. I feel hopeful that this might work out after all, but just then the yak takes off like a rodeo steer, throwing Bill down to the frozen ground. I'm really scared now, but, several attempts later, he is in the saddle and the trip down the trail is uneventful until one hour outside the village of

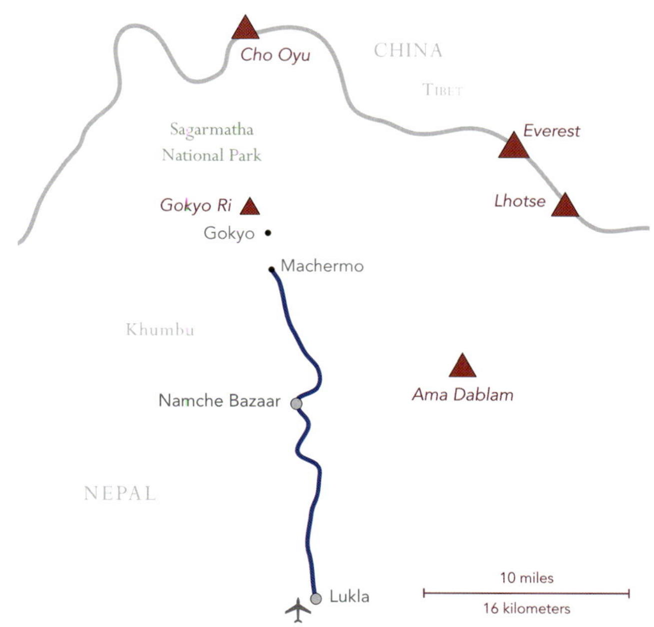

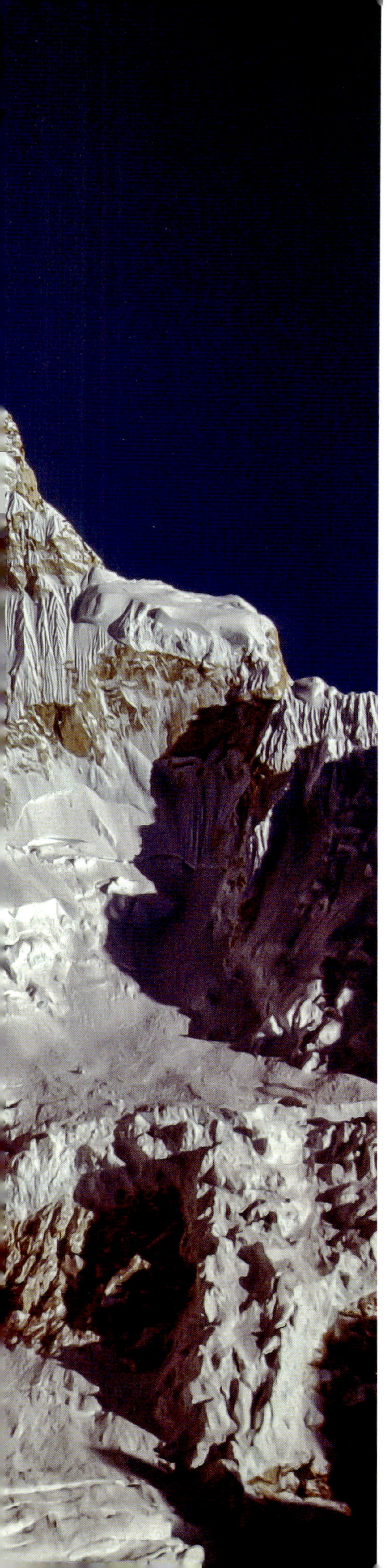

Mount Everest and Nuptse, Khumbu, Nepal, 1985

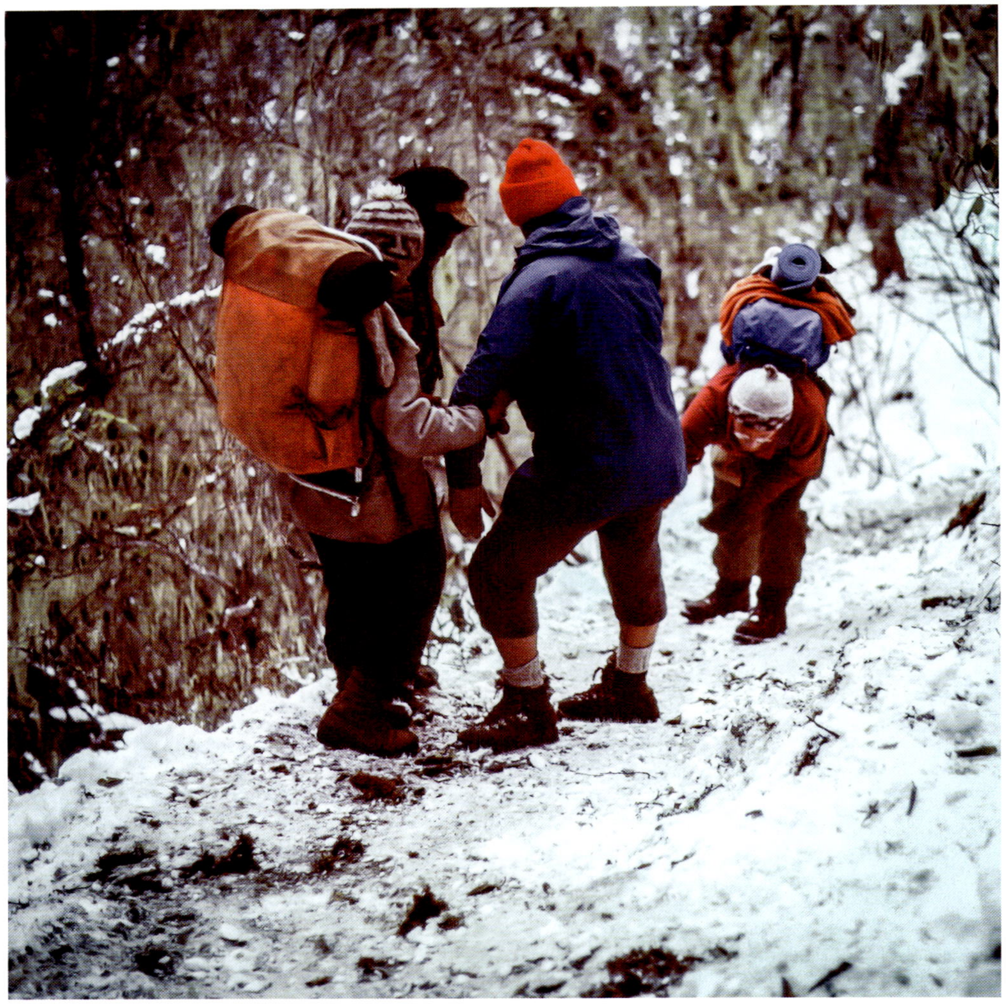

Bill being helped on the trail to Dole, Khumbu, Nepal, 1982 (Photograph by Anne Frej)

Dole when the yak abruptly bucks him off one last time. The Sherpani shrugs her shoulders and heads down the trail with the yak and her baby.

Dorje, the Nepali porter from our group who was assigned to stay behind and help me, finally shows up and lifts Bill up. We drape his arms over our shoulders and begin the walk down the narrow mountain trail. In Dole I learn that Reinhold Messner, one of mountaineering's most celebrated climbers, has set up a camp nearby in preparation for a summit attempt on Cho Oyo, the world's sixth-highest mountain (elevation 26,906 feet). Messner and Peter Habeler had climbed Mount Everest without supplemental oxygen four years earlier in 1978, something that many thought impossible. The 1982 attempt on Cho Oyo was unsuccessful, but two

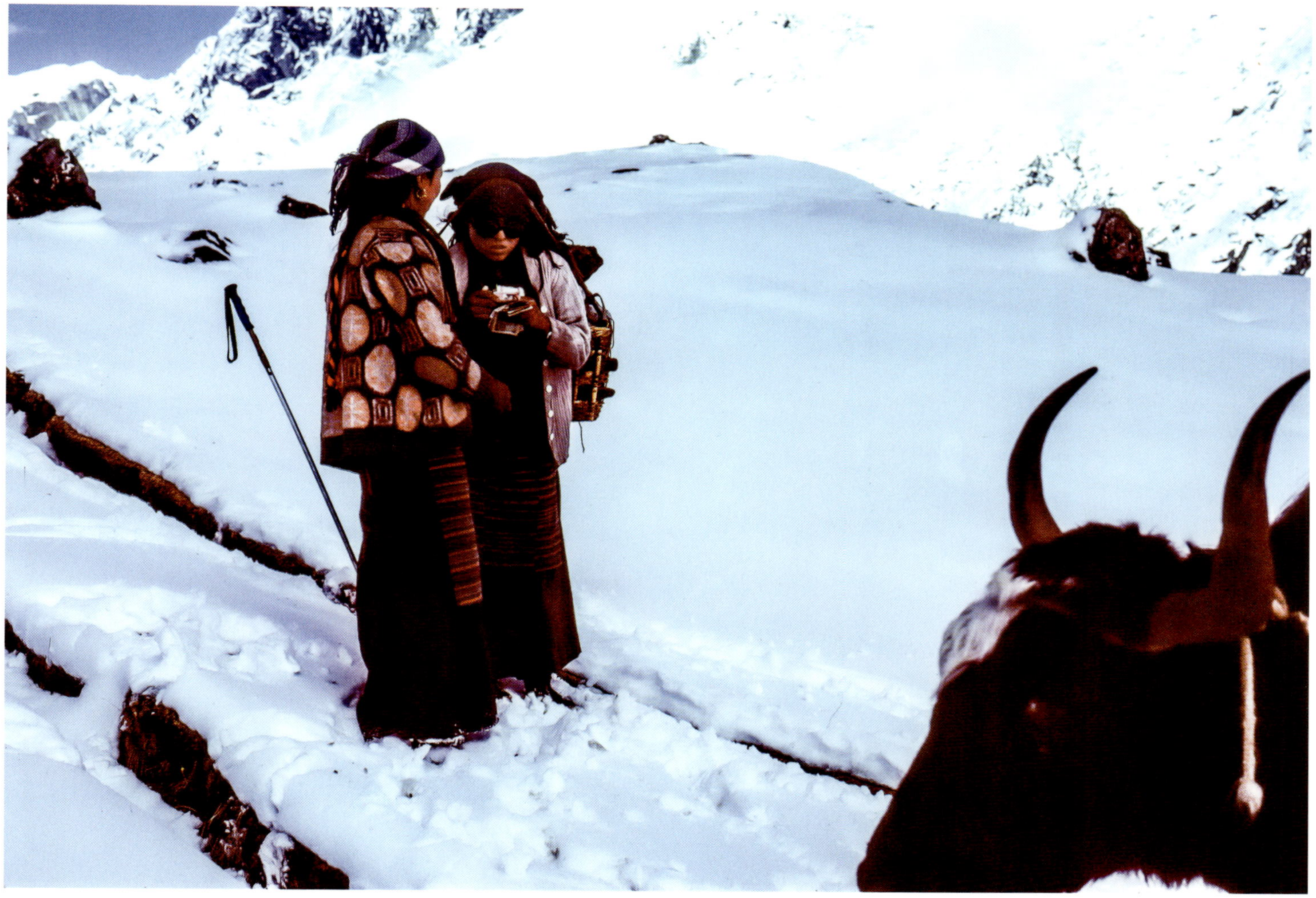

Machermo, Dudh Kosi River Valley, Khumbu, Nepal, 1982 (Photograph by Anne Frej)

years later Messner climbed Mount Everest solo and without oxygen; by 1986 he had climbed all of the world's peaks with elevations higher than 8,000 meters (26,247 feet).

I go from tent to tent until I find the climbing team's doctor, and years later I realize that I could not have found anyone more capable. Oswald Oelz, a foremost expert in mountaineering medicine and an experienced high-altitude climber, examines Bill, looks at his eyes, notes that he is unable to walk in a straight line, and tells us that he is in the early stages of high altitude cerebral edema, a serious form of mountain sickness resulting from the accumulation of fluid in the brain. Getting Bill down to a lower elevation is critical, he tells me.

We spend the night in Dole and push off the next day down the trail. It is too narrow for the yak, so Bill walks with

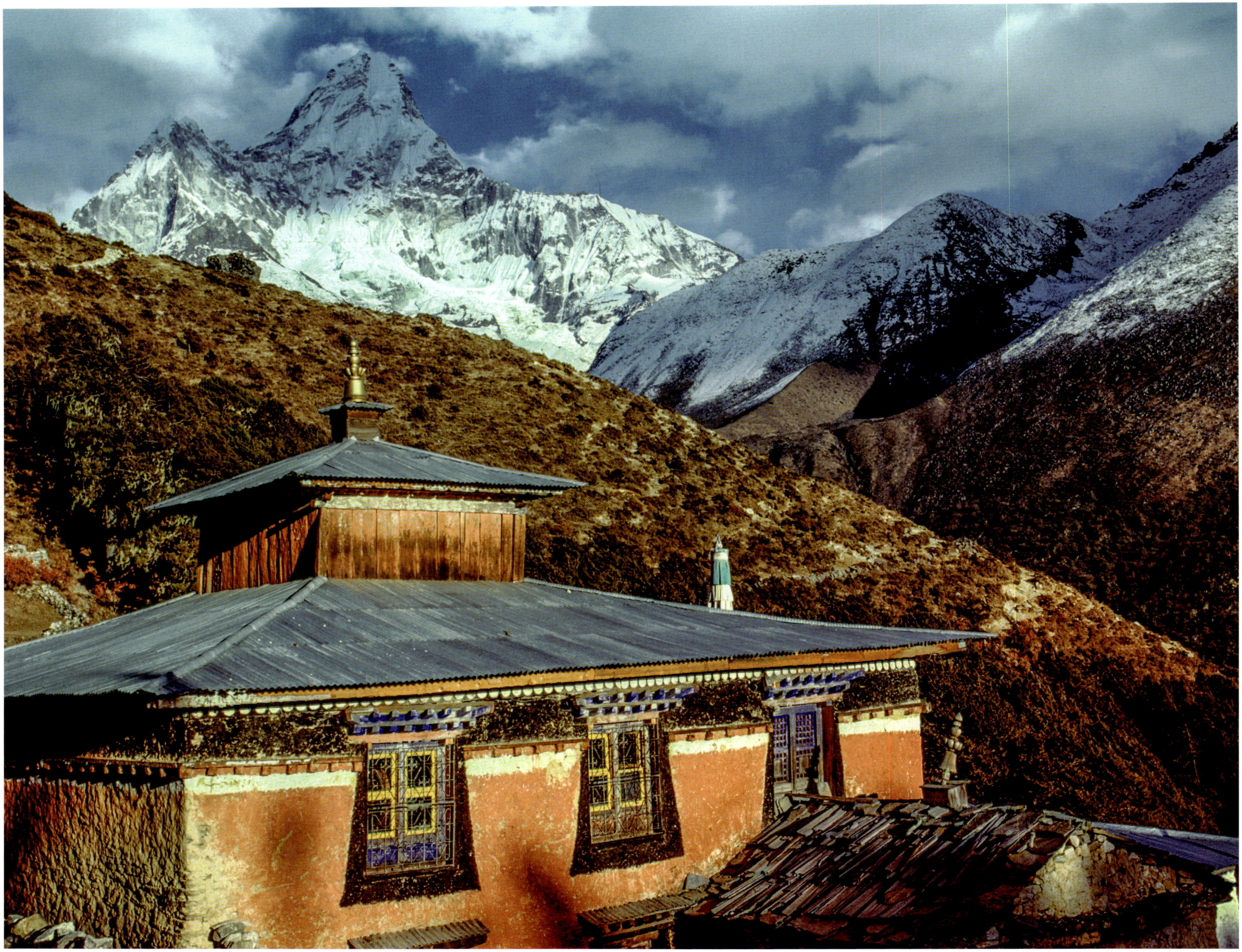

Tengboche Monastery, before the 1989 fire, Khumbu, Nepal, 1985
Opposite: Mount Ama Dablam, Khumbu, Nepal, 1982

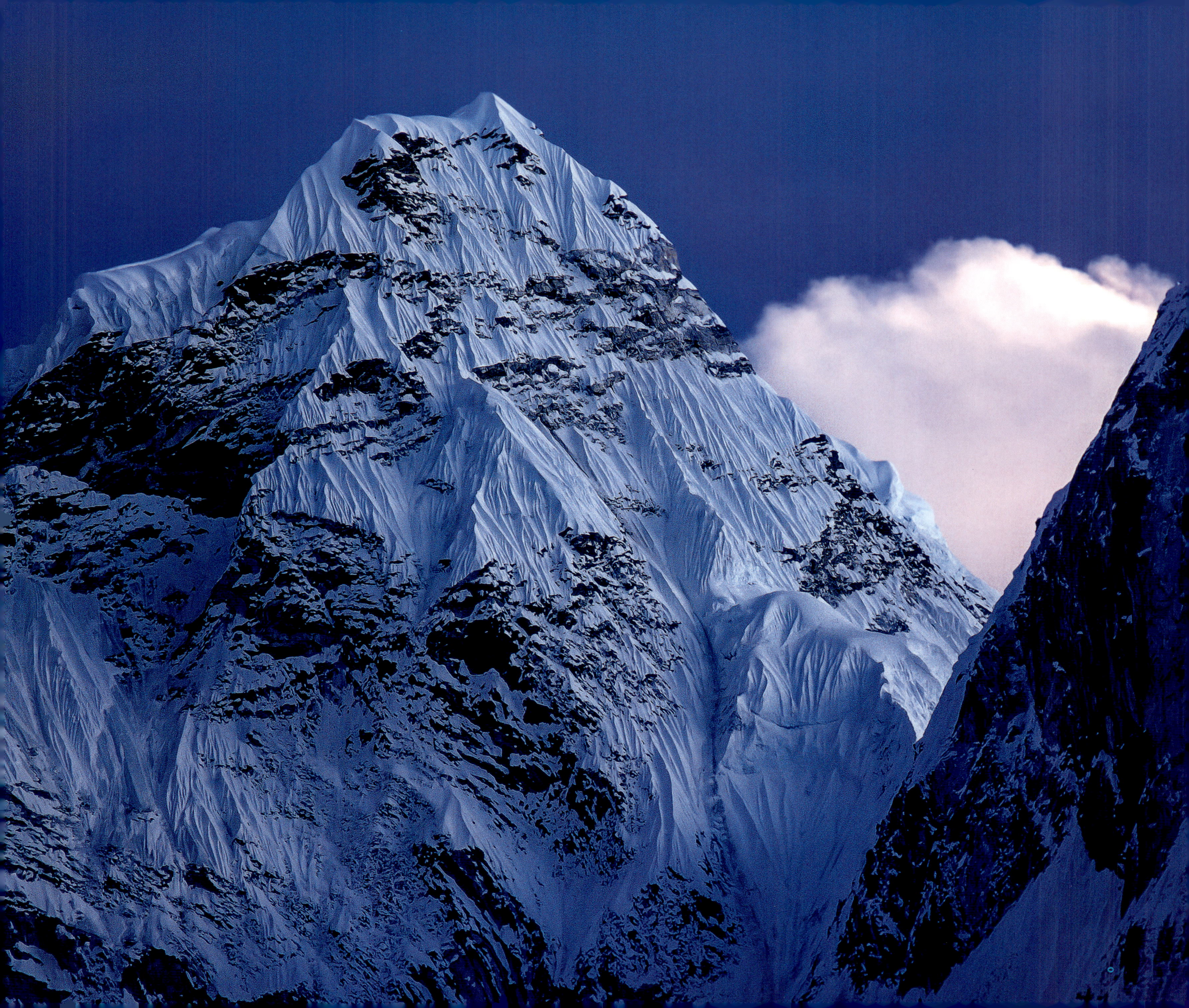

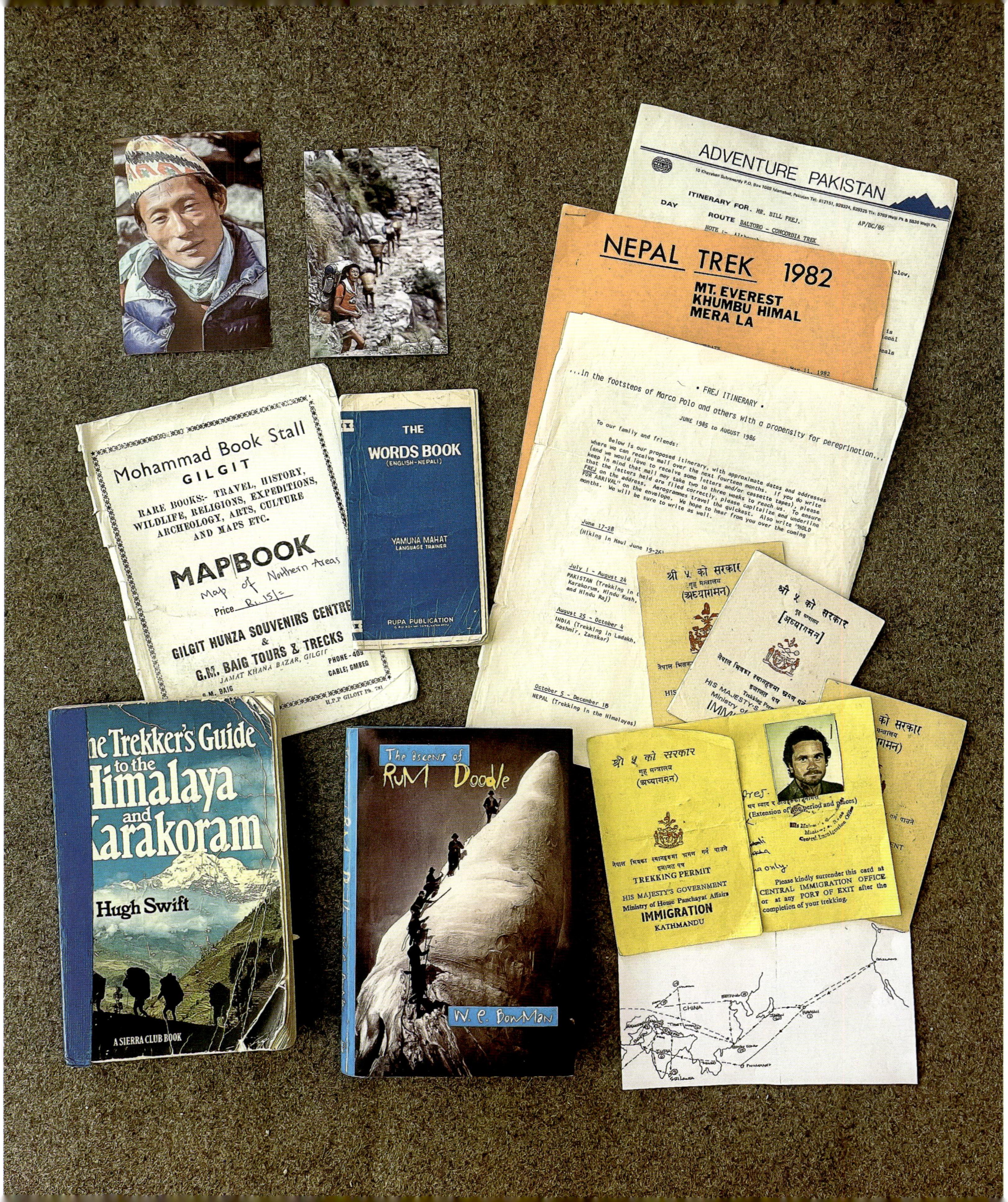

ADVENTURE PAKISTAN
ITINERARY FOR. MR. BILL FREJ
DAY
ROUTE BALTORO - CONCORDIA TREK
AP/BC/86
NEPAL TREK 1982
MT. EVEREST
KHUMBU HIMAL
MERA LA
...in the footsteps of Marco Polo and others with a propensity for peregrination...
• FREJ ITINERARY •
JUNE 1985 to AUGUST 1986
To our family and friends:
June 17-28
July 1 - August 24
PAKISTAN (Trekking in Karakorum, Hindu Kush, and Hindu Raj)
August 25 - October 4
INDIA (Trekking in Ladakh, Kashmir, Zanskar)
October 5 - December 18
NEPAL (Trekking in the Himalayas)
Mohammad Book Stall
GILGIT
RARE BOOKS:- TRAVEL, HISTORY, WILDLIFE, RELIGIONS, EXPEDITIONS, ARCHEOLOGY, ARTS, CULTURE AND MAPS ETC.
MAPBOOK
Map of Northern Areas
Price Rs. 15/=
GILGIT HUNZA SOUVENIRS CENTRE
&
G.M. BAIG TOURS & TRECKS
JAMAT KHANA BAZAR, GILGIT
THE
WORDS BOOK
(ENGLISH-NEPALI)
YAMUNA MAHAT
LANGUAGE TRAINER
RUPA PUBLICATION
श्री ५ को सरकार
(अध्यागमन)
TREKKING PERMIT
HIS MAJESTY'S GOVERNMENT
Ministry of Home Panchayat Affairs
IMMIGRATION
KATHMANDU
Please kindly surrender this card at CENTRAL IMMIGRATION OFFICE or at any PORT OF EXIT after the completion of your trekking.
The Trekker's Guide to the Himalaya and Karakoram
Hugh Swift
A SIERRA CLUB BOOK
The Ascent of RUM Doodle
W. E. BowMan
CHINA

Making Travel Plans, California

SPRING 1985

In the spring of 1985, with two treks in Nepal under our belts, we decided to take a long Himalayan "sabbatical." The idea of hiking around Asia for an extended period came to us after attending a slide show of the all-women's climbing expedition to Annapurna in Nepal led by Arlene Blum in 1978. It was the heady days of Himalayan expeditions and treks, especially in Berkeley, California, where we then lived. Attending Sierra Club slide shows by well-known climbers, perusing richly illustrated travel catalogs, and listening to firsthand stories of walks through untamed and rarely visited mountain areas plus memories of two earlier trips convinced us to do it.

Our strongest advocate was Hugh Swift, a veteran trekker who walked throughout every corner of the Himalaya and wrote three excellent guidebooks: *The Trekker's Guide to the Himalayas and Karakoram* (1982), *Trekking in Nepal, West Tibet, and Bhutan* (1989), and *Trekking in Pakistan and India* (1990). Hugh was always ready to share his insights and knowledge of trails and places, even mythical hidden valleys he had heard about and thought might be discovered one day. Hugh died in an accident in Berkeley in 1991, falling and hitting his head after a minor surgery. We think of him often when we are in the mountains and like to imagine him hiking through the hidden valleys scattered throughout the Himalaya reputed to be paradises of eternal happiness.

Once we made the decision to follow Hugh's advice to take an extended walk across the Himalaya, it was surprisingly easy to make all the necessary arrangements. Bill was then the regional director of the Neighborhood Reinvestment Corporation, and I was working as a city planner at an architectural firm in San Francisco. Twice before we had gathered the nerve to ask for four weeks of vacation time for treks instead of the usual two weeks, but now we knew we needed at least a year, and maybe more, to reach remote corners of the Himalaya on foot. This time we took the plunge and quit our jobs, rented out our house, and bought one-way airline tickets to Asia.

Bill sold his car, a new red Porsche 944, and we pocketed the cash, hoping it would last at least a year. Our families and friends were nervous and unclear about our motivations, which, it's true, we found difficult to explain. We typed up an itinerary to give them an idea of where we hoped to be on certain dates. The Internet and cell phones did not exist, so we depended on American Express offices along the way to receive letters and to cash traveler's checks. Most communication was through letters mailed from the post offices in Kathmandu or Delhi or Rawalpindi, where we learned to watch from the counter to make sure the stamps were not stolen off the envelope before it was tossed onto the airmail stack.

opposite and page 53:
Himalaya travel memorabilia

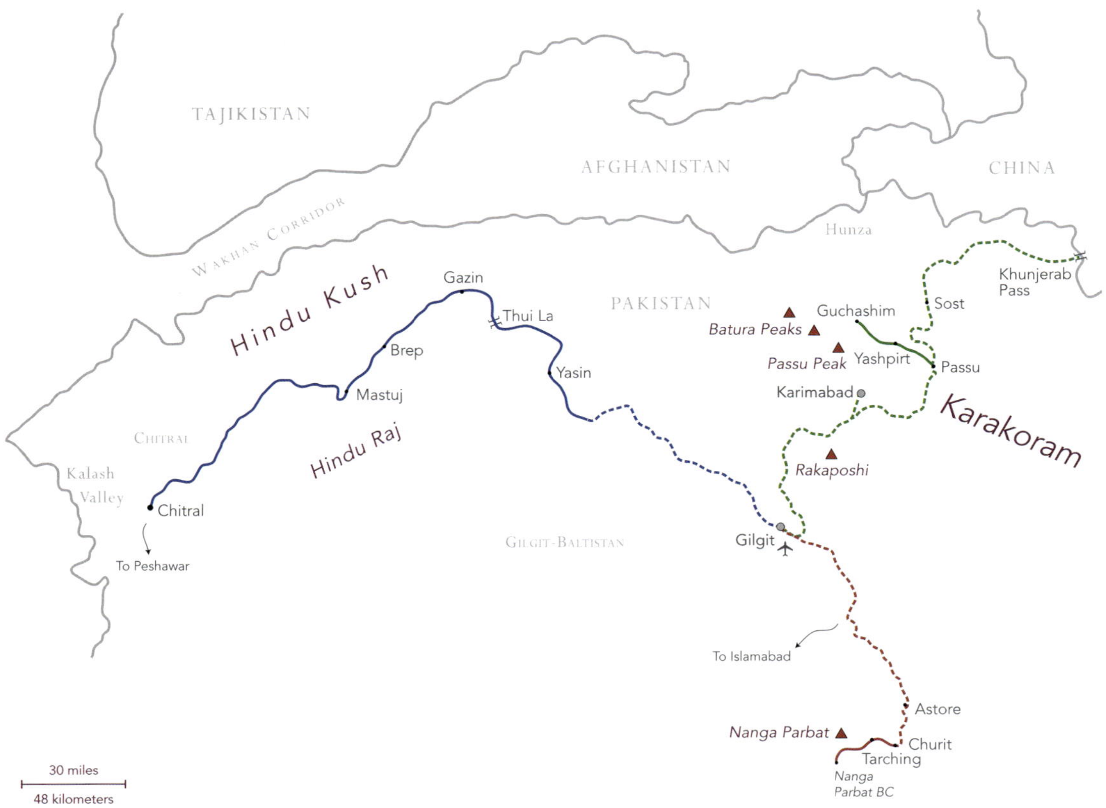

animals in the pastures high above the village, but the women coyly called us over to share chapatis and delicious fresh yak yogurt that had a thick layer of yellow cream on top. Dressed in flowered nightgown-style dresses and embroidered pillbox hats, they had a playful, feminine countenance that belied the hard life they lived.

Kukhel, our next camp, was smaller but even more glorious with the addition of juniper trees and a stream rushing with clear mountain runoff. The farthest point of our walk was Guchashim, two-thirds of the way up the thirty-mile-long Batura Glacier. From a ridge above Guchashim we could look down and see the blue-tinged glacier winding its way through huge piles of gray stone rubble. We turned around in Guchashim and walked down the way we had come back to Passu and then to Gilgit to restock our provisions for the next leg of our trek to the Yarkhun Valley in the Chitral District to the west.

Our walk to Chitral from Gilgit took the east-to-west route through the Yarkhun Valley. We rode in jeeps from Gilgit to Yasin and then to Haraf before entering a restricted area the next day. For two days we hiked through small, friendly villages until reaching our camp below Aghost Bar Glacier at 13,500 feet where we could see the peaks of Thui I and Thui II. We walked across the glacier the next day and

Baltit Fort, Hunza Valley, Pakistan, 1985

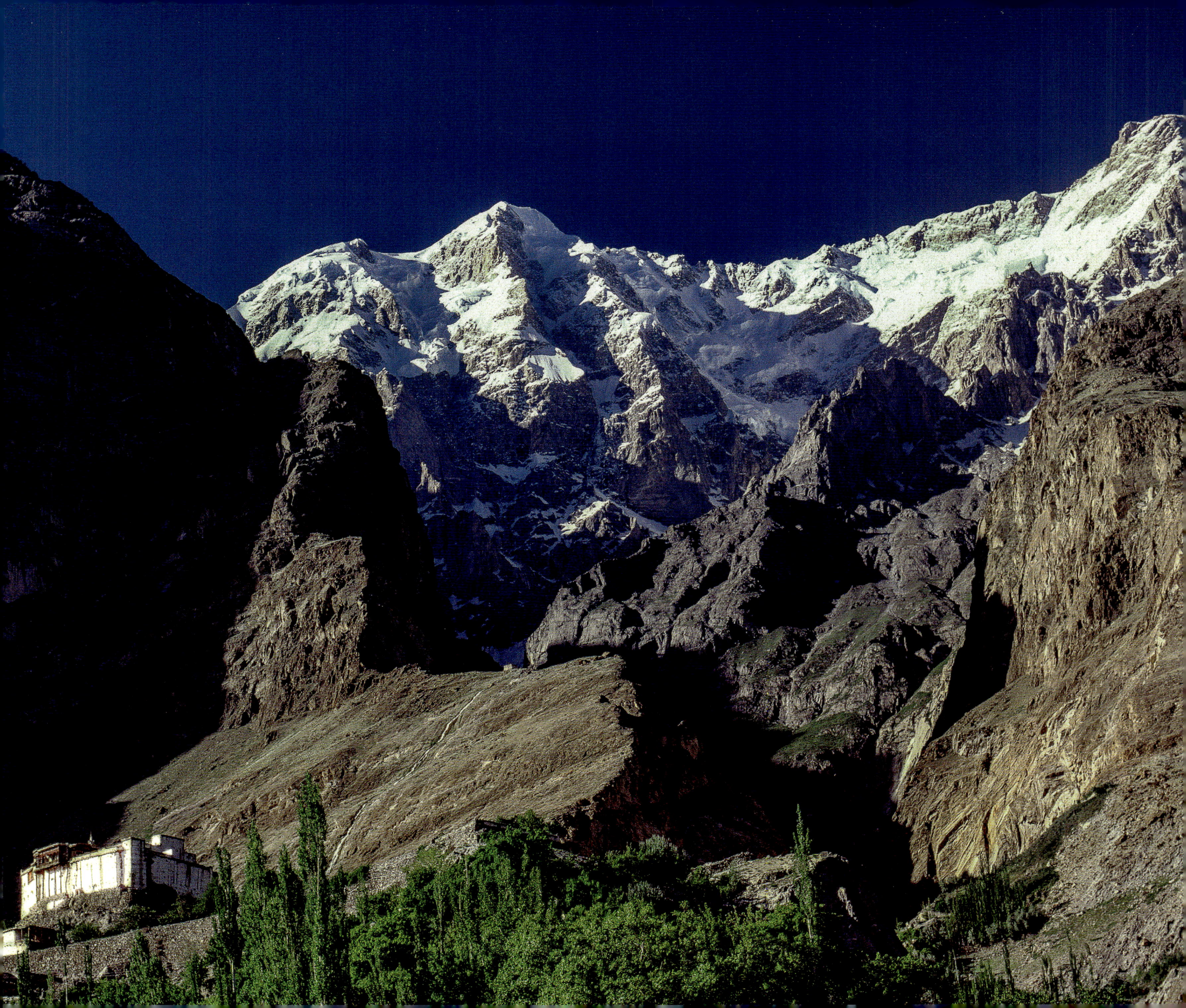

pocket. It was rainy and foggy that night and the entire next day, but the following night at Rupal Peak Base Camp the clouds finally lifted and we saw the summit of Nanga Parbat and most of its face for a few brief moments. Its south side, known as the Rupal Face, is the world's tallest mountain wall, nearly 15,000 feet from the base to the summit.

In 1970, Reinhold Messner and his brother Günther climbed the Rupal Face and then crossed the summit to the Diamir Face to the west to get down the mountain. Günther died on the descent, perhaps due to an avalanche, but Reinhold survived. Eight years later he completed a solo climb of the Diamir Face. In 2005, Americans Vince Anderson and Steve House climbed the face alpine style in six days, a remarkable achievement.

We walked back to Churit passing through wildflower-filled meadows. We set up our camp in a schoolyard, but soon a group of boys surrounded our tent and pelted us with rocks. With only an ice axe for protection and nowhere to run, we were trapped until our porters, whom we had just said good-bye to, came to the rescue and drove the boys away. It might have been a harbinger of outbreaks of terrorist activities in the region. Twenty-eight years later, in 2013, ten foreign climbers and one guide were killed by members of the local Taliban at Fairy Meadows Camp on the north side of Nanga Parbat. The situation has improved in recent years, and the spectacular mountain region of Gilgit-Baltistan continues to attract trekkers and climbers, although even the most popular attractions such as K2 Base Camp are still relatively undeveloped and require a guide.

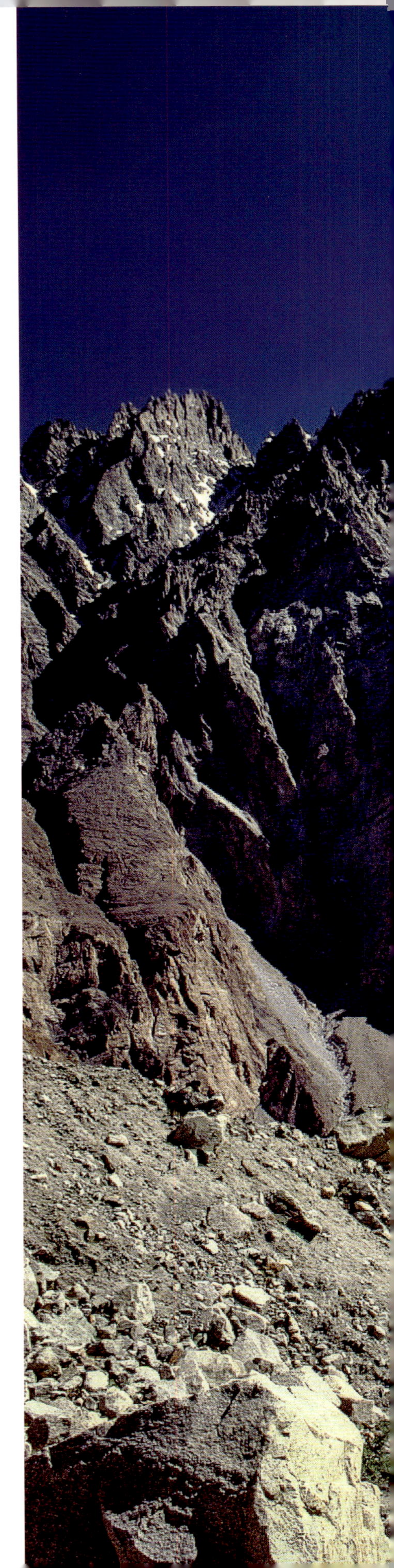

The trail to the Batura Glacier, Pakistan, 1985

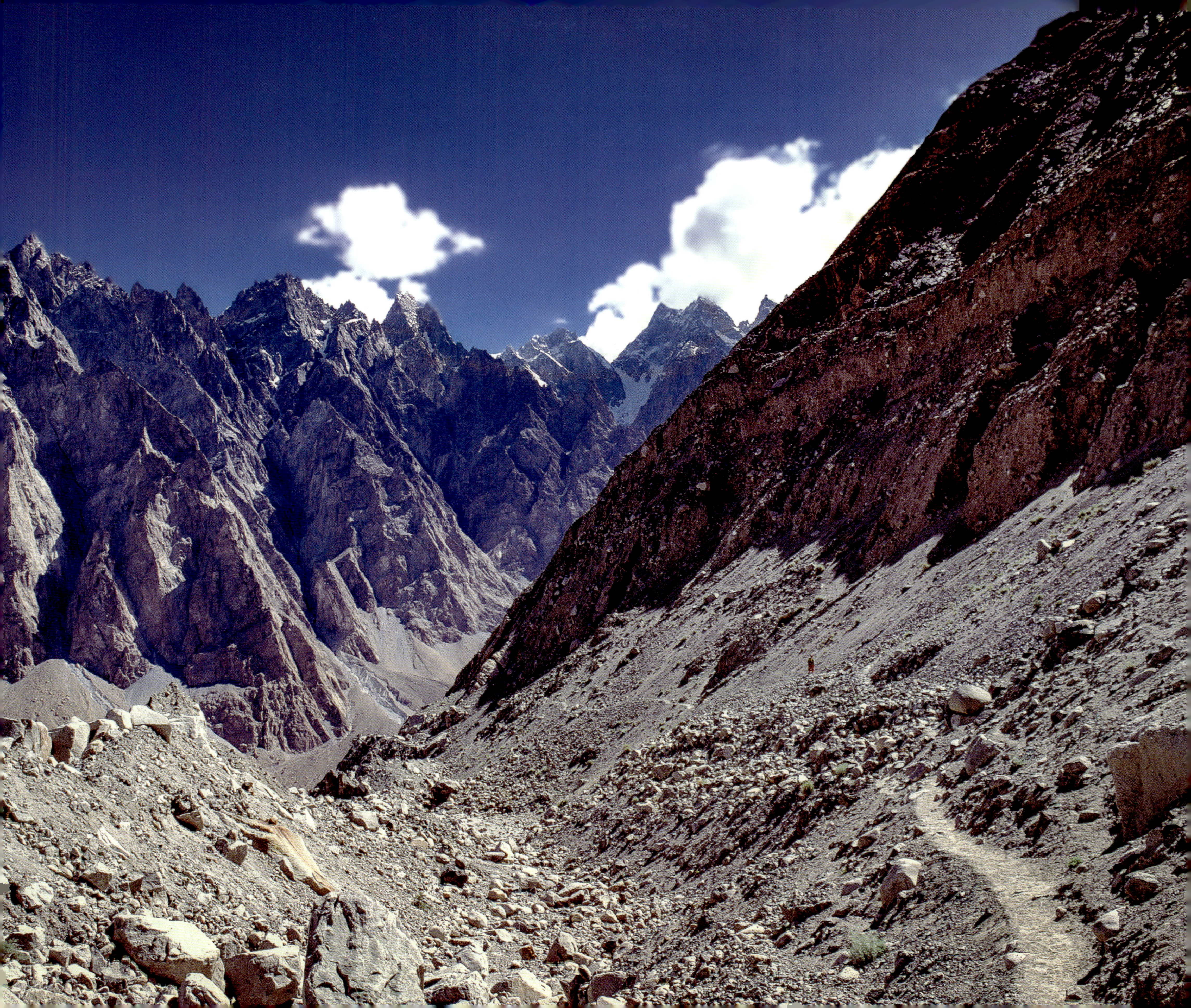

Batura Peak, Pakistan, 1985

Opposite: Passu Cones, Pakistan, 1985

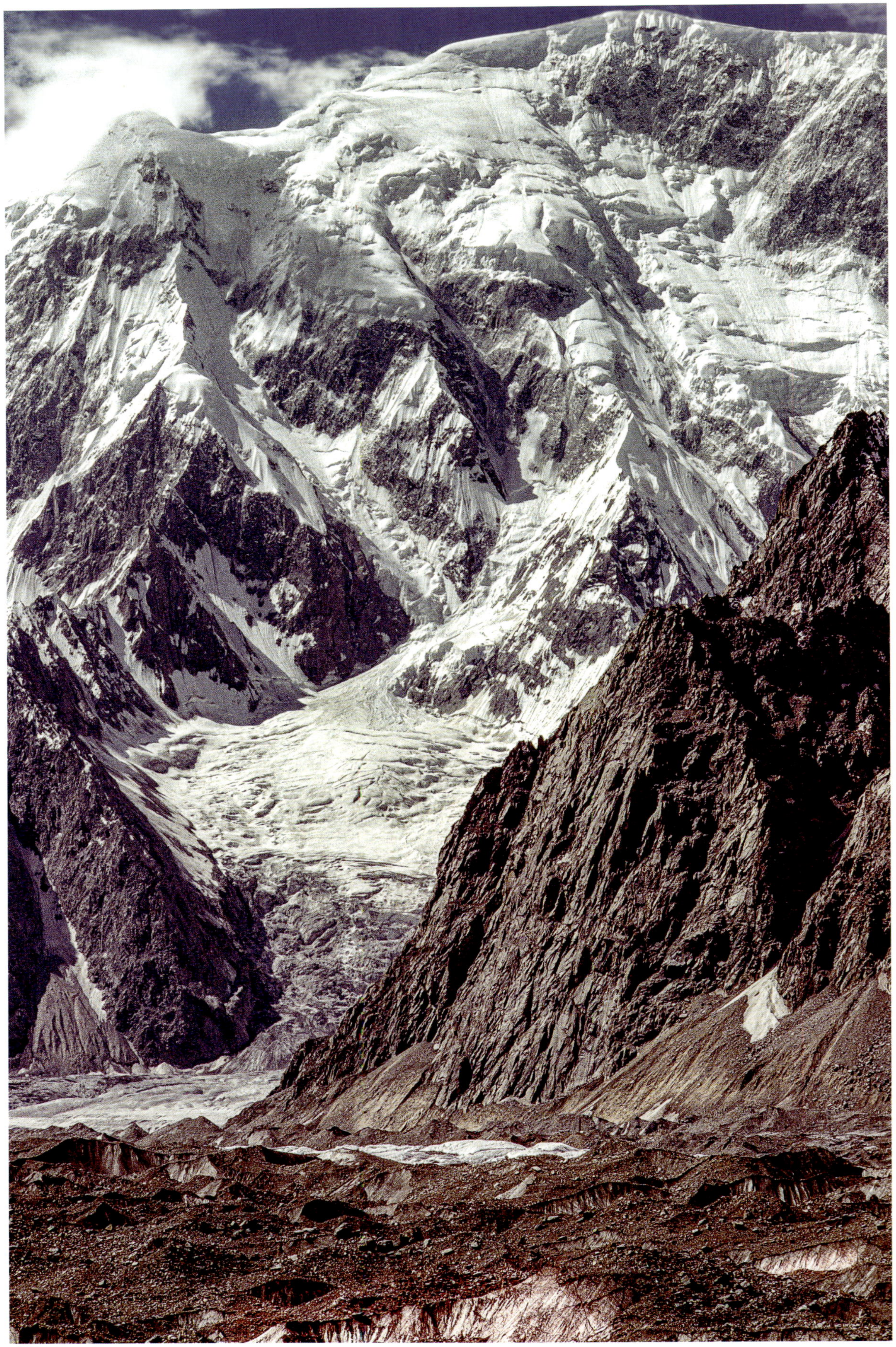

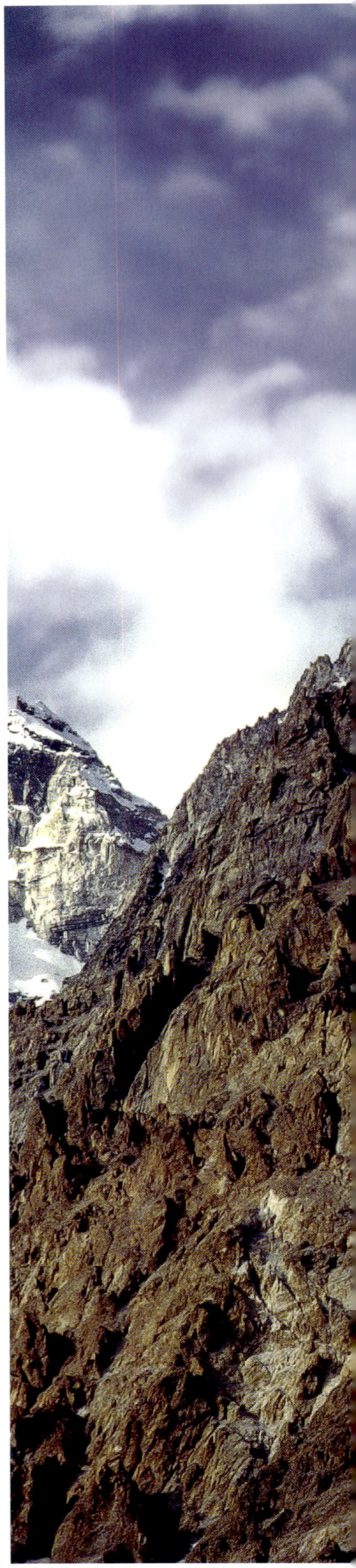

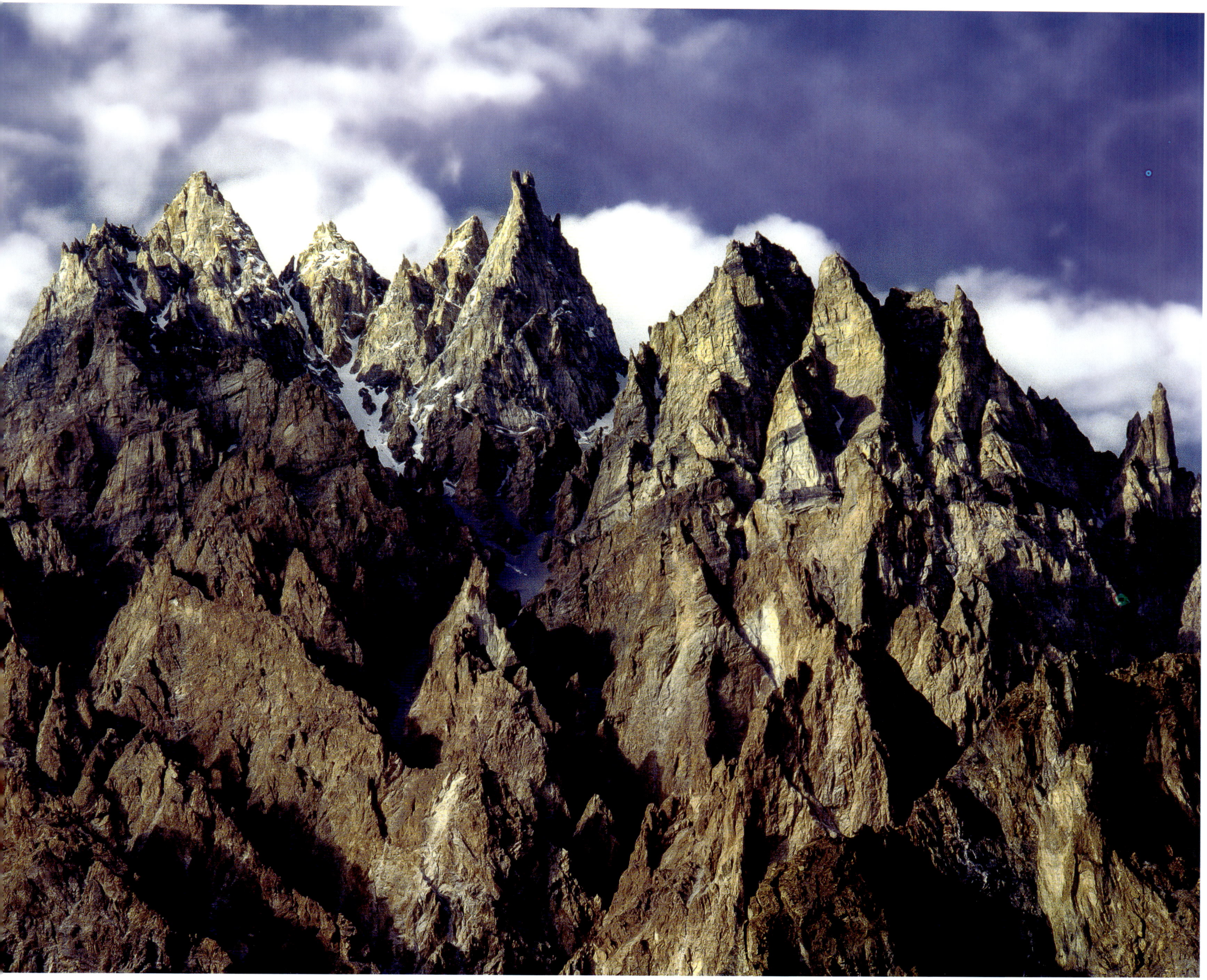

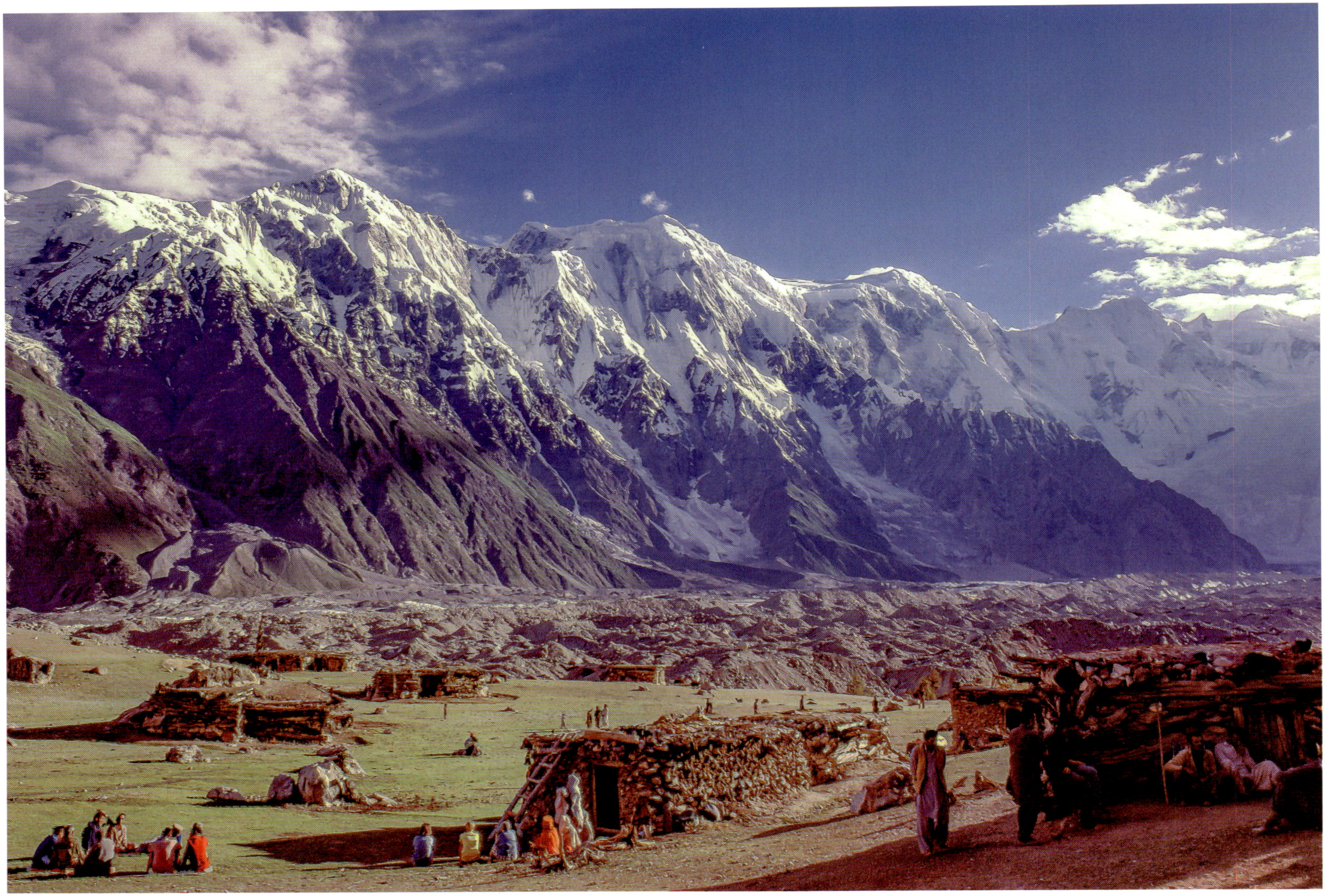

Below Batura Ridge, Yashpirt, Pakistan, 1985

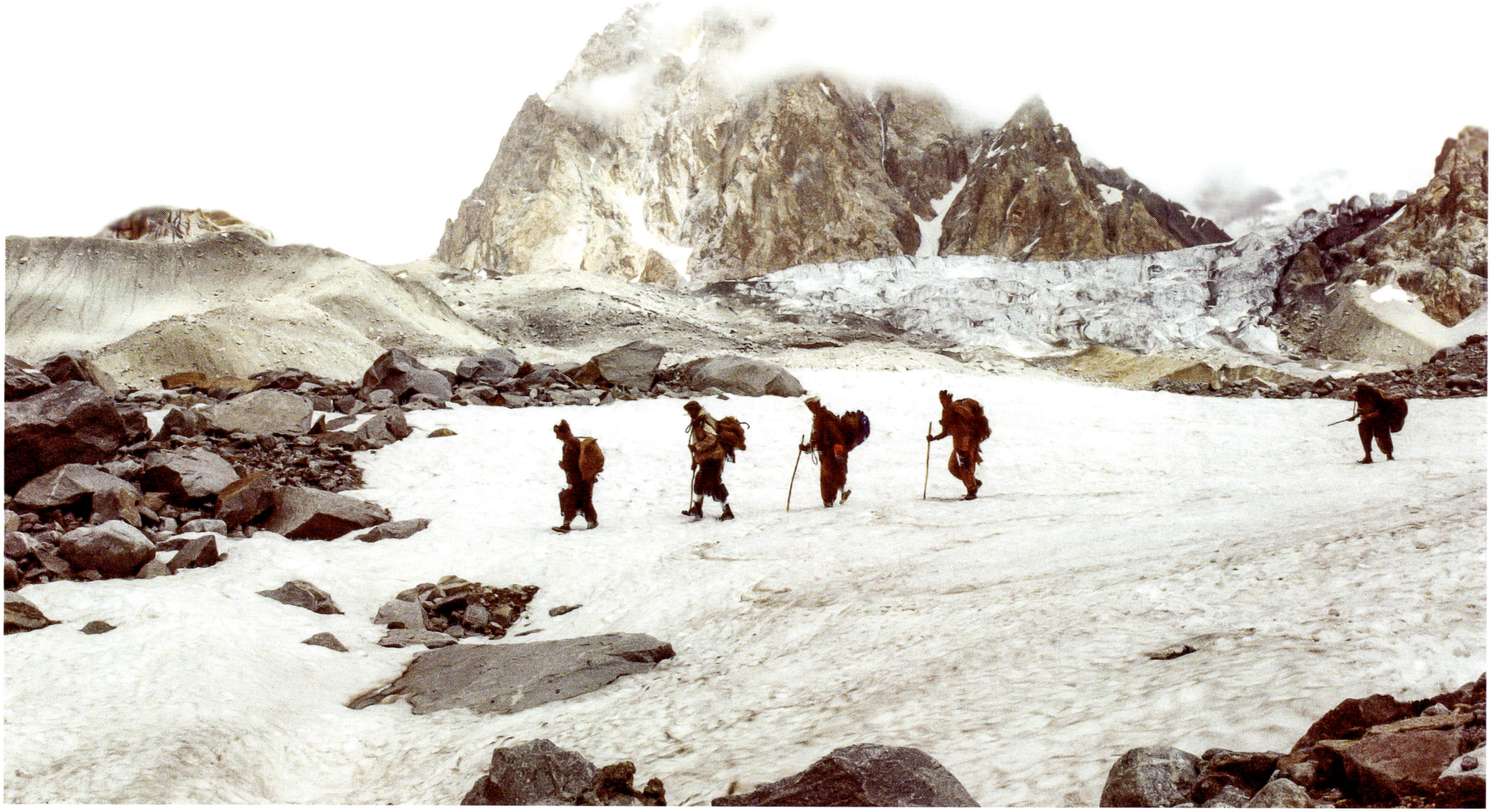

Below Thui An Pass, Pakistan, 1985

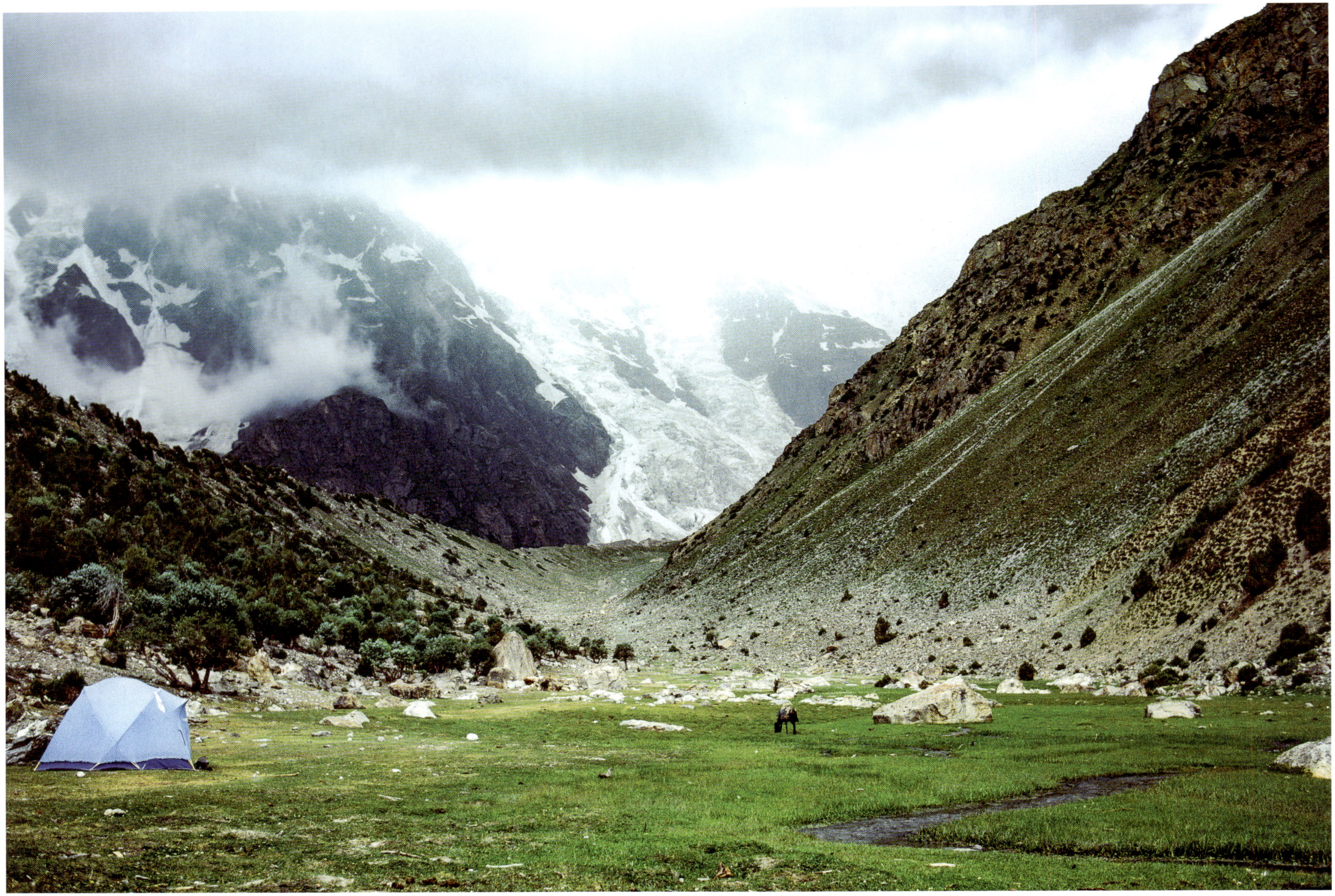

Rupal Peak Base Camp, below Nanga Parbat, 1985
Opposite: Nanga Parbat, ninth-highest mountain in the world, 1985

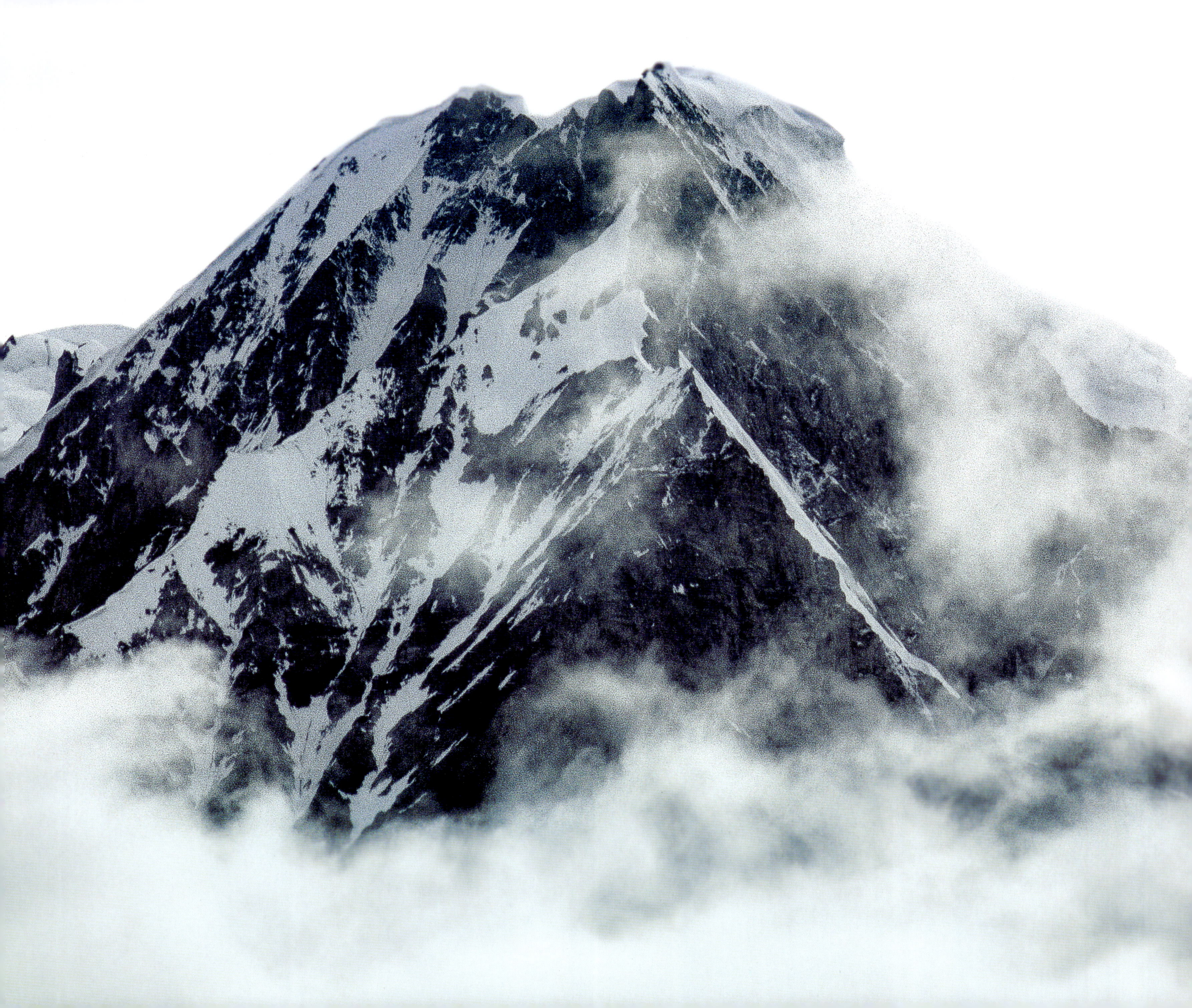

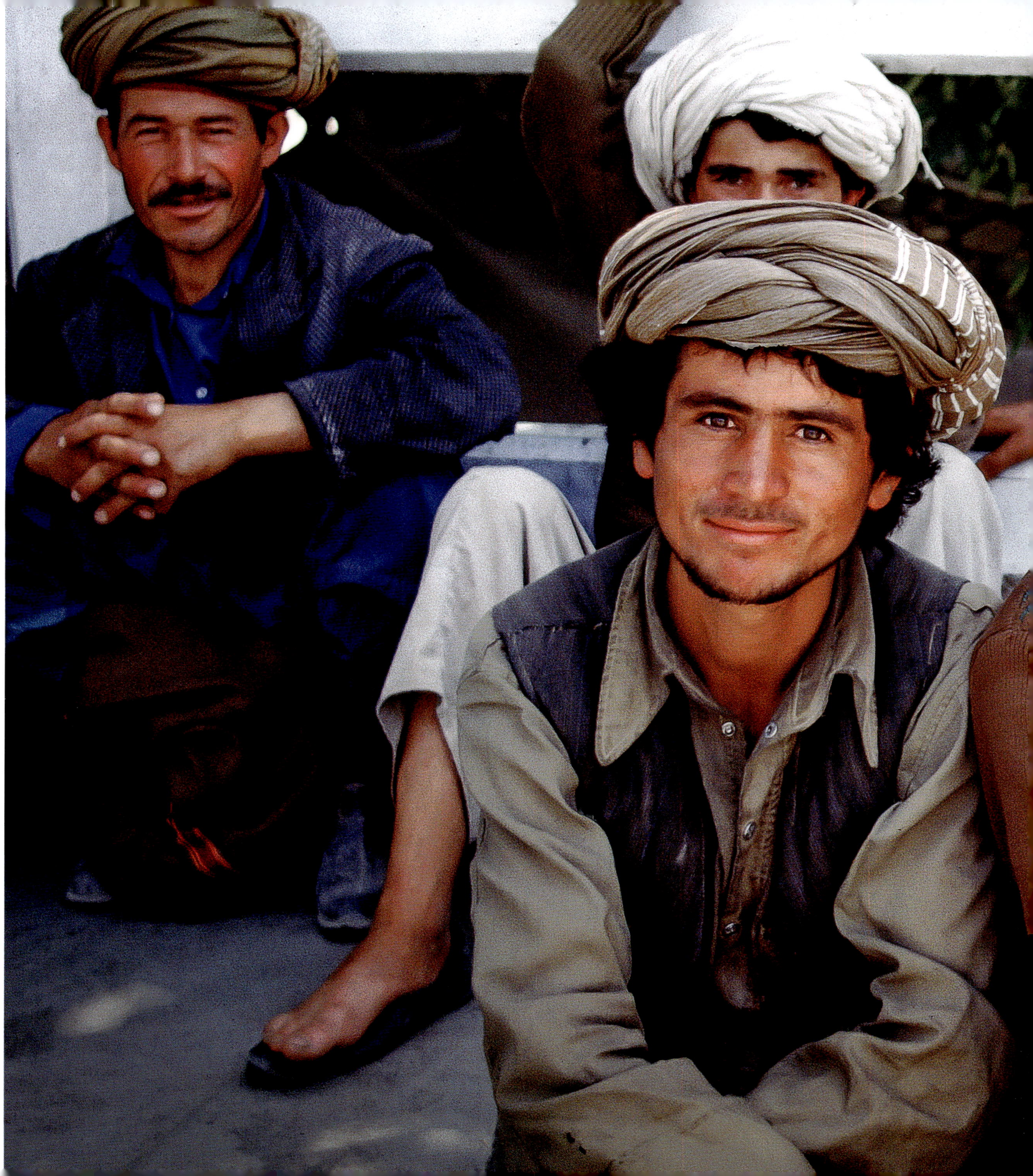

Afghan refugees,
Chitral, Pakistan, 1981

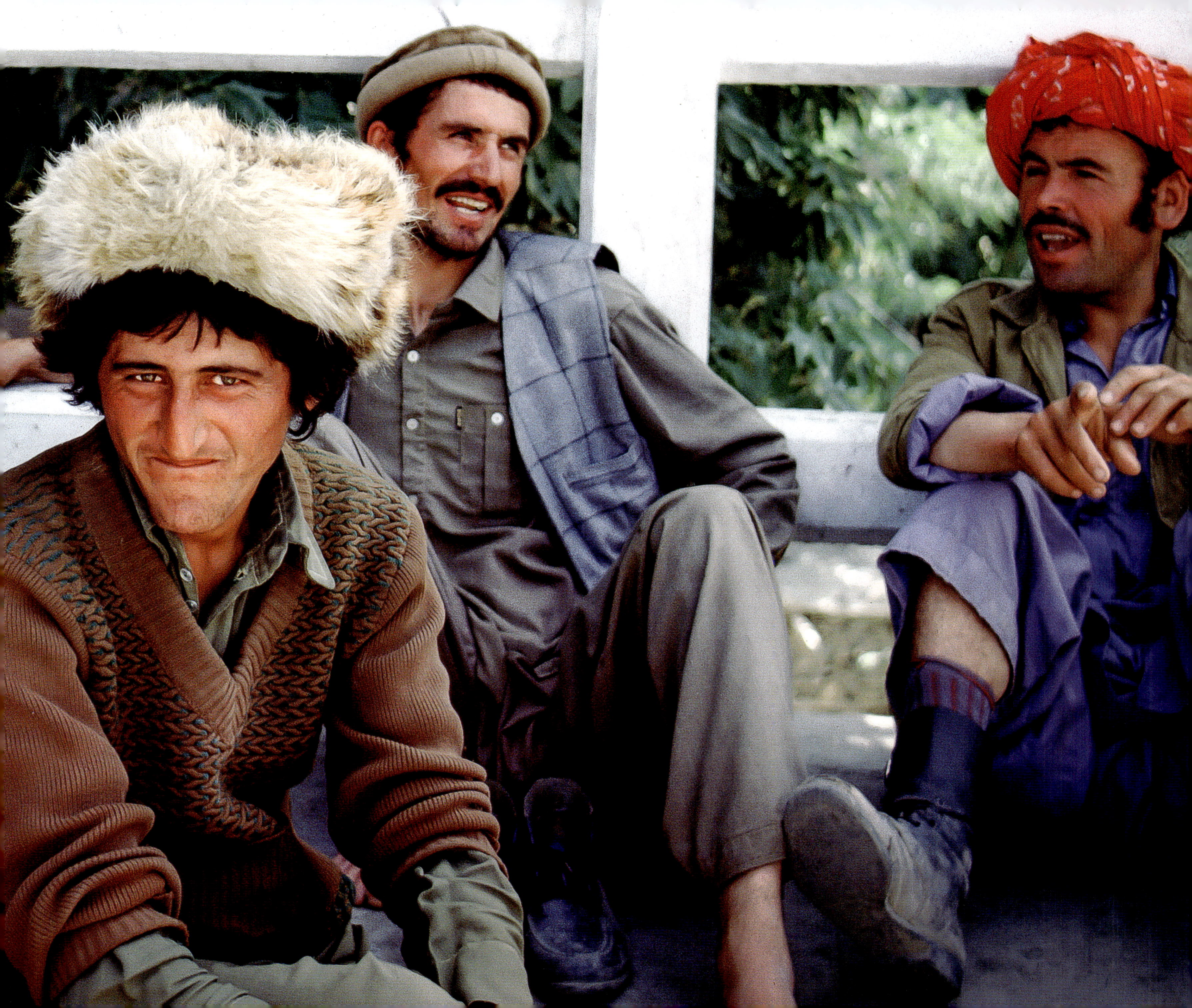

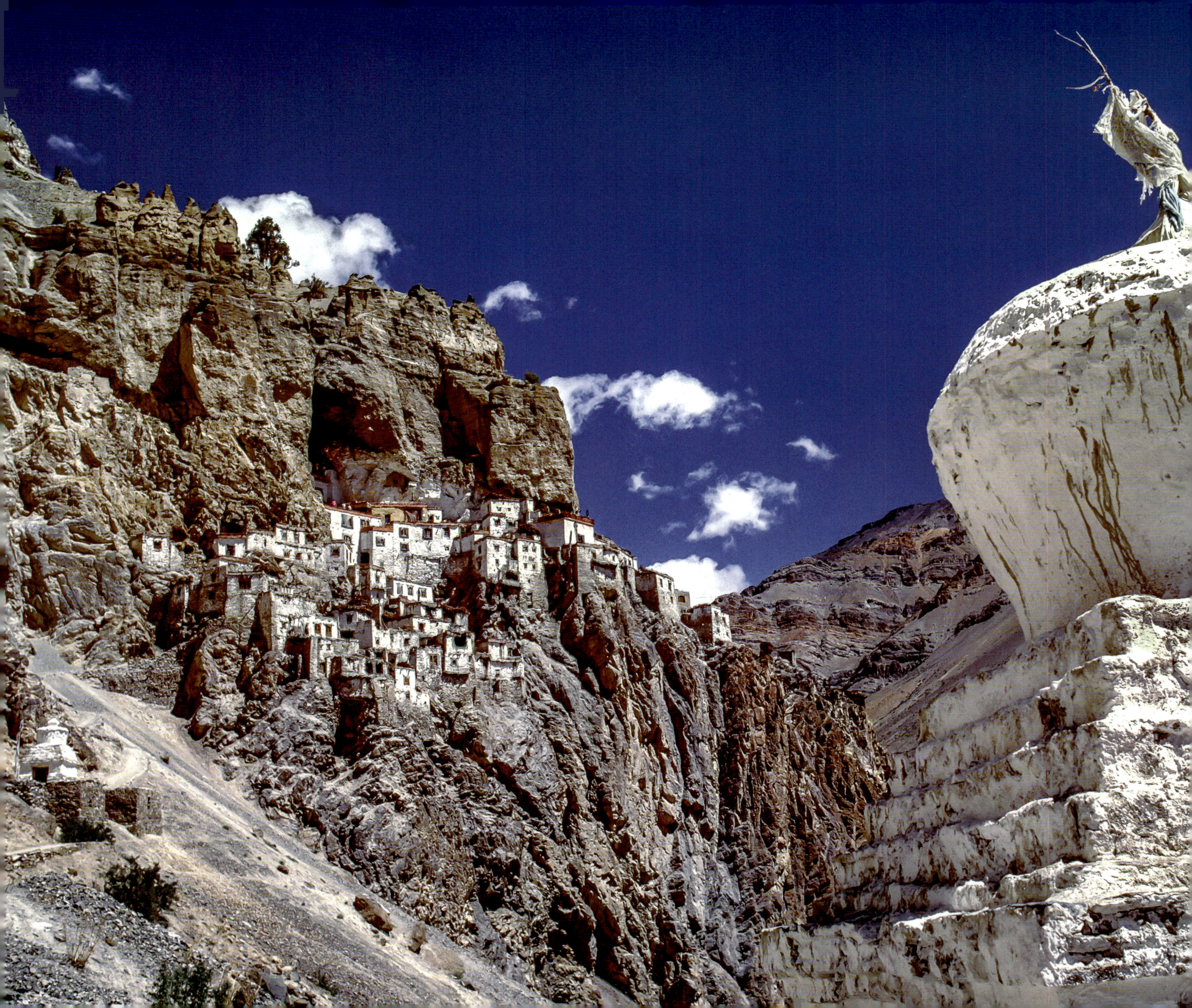

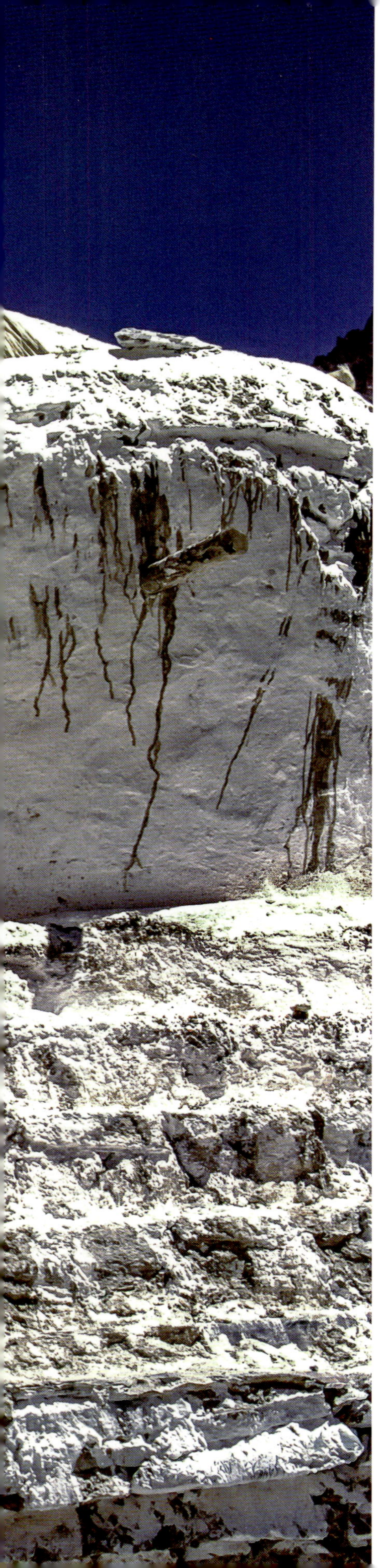

Treks and Travels Through Ladakh, India

AUGUST–SEPTEMBER 1984 AND JUNE 2018

It was a relief stepping off the bus in Darcha, Ladakh. To reach this starting point for our trek in northern India we had just endured a thirteen-hour bus ride from Manali that had been preceded by several days traveling by bus on a round-about route from Delhi, due to mudslides on the roads. Not only were the roads perilous, but our fellow passengers appeared ready to pounce on us when they weren't scuffling with each other. Ram rode on the luggage rack on top of the bus with our bags so they would not disappear at some dark, unscheduled stop. Inside, we tried to stay awake as we pressed on the wallets with our money and passports inside our shirts.

The landscape that greeted us in Darcha was stark and intimidating, like a black-and-white photograph of a moonscape. Ladakh was once part of western Tibet and has a similar high, arid landscape with a vibrant form of Tibetan Buddhist religion and culture undiluted by Chinese intervention, as it is in Tibet. This region has always been isolated, especially during the winter, and we heard stories of trekkers who stayed too late into the season and got trapped there for months until the snow melted. Our trek would traverse the rugged Zanskar District of Ladakh, just north of the main Himalaya Range and in its dry rain shadow. We knew we had at least twenty-one days of hiking ahead of us that would

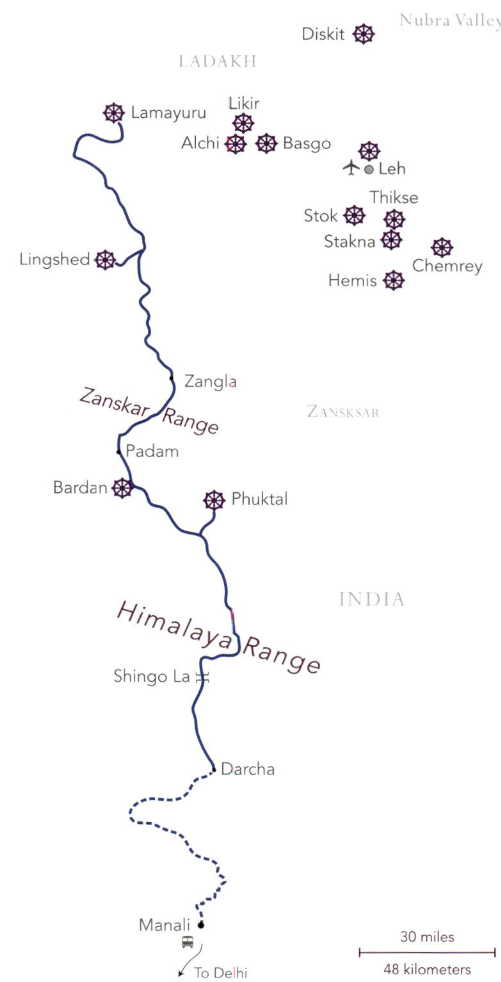

Phuktal Monastery, Lungnak Valley, Zanskar, India, 1985

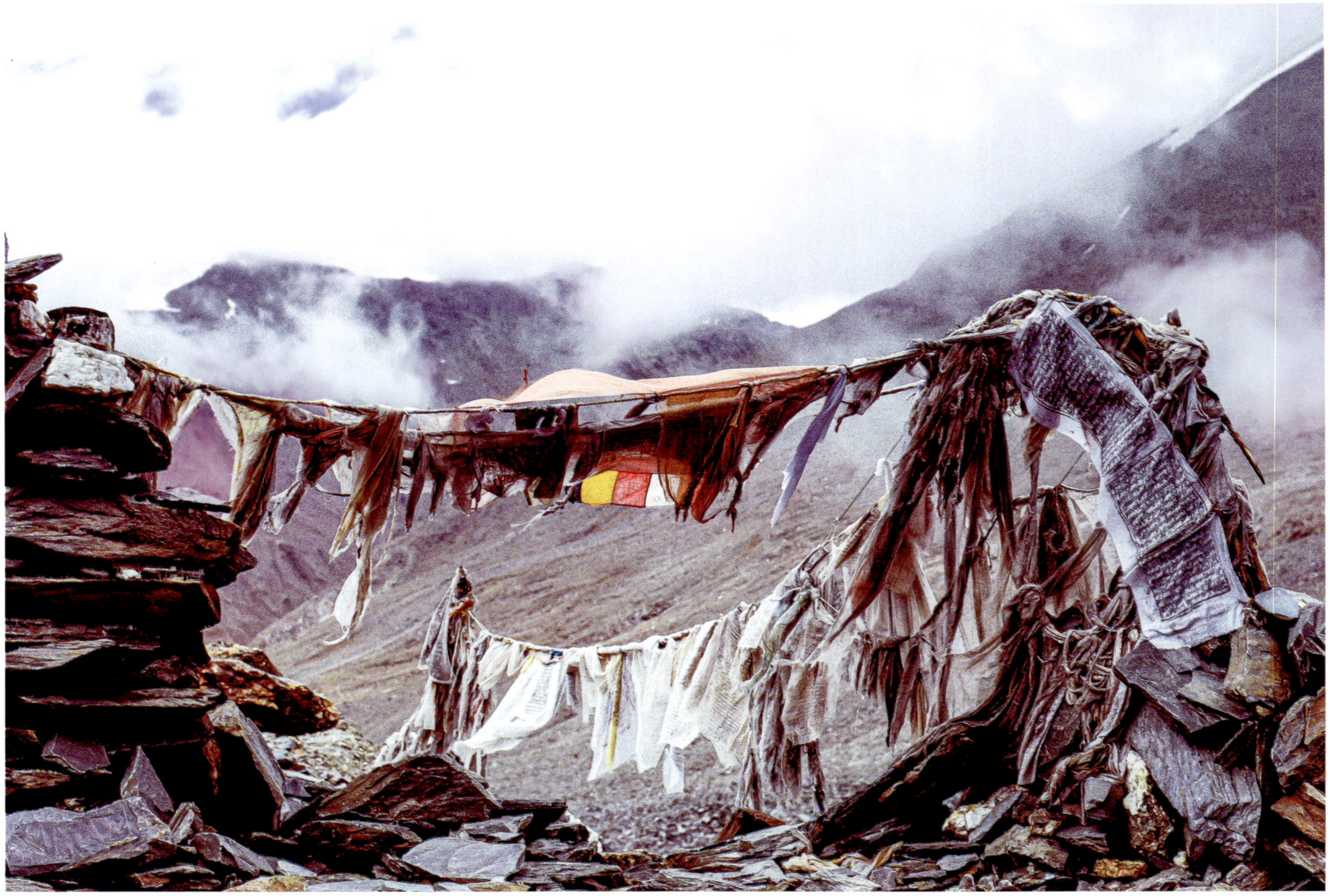

Shingo La, Zanskar, India, 1985

require maneuvering steep, exposed trails and crossing high passes ranging from 12,500 to nearly 17,000 feet in elevation between Darcha and our end point, Lamayuru Monastery.

Our small group included Ram, Bill, our friend, Brian, our Ladakhi horseman, Ramdev, and me plus four small horses to carry our bags, gear, and food. On our first day we grew accustomed to following the narrow, rocky trail that stretched out in front of us far into the distance. We camped at nearly 12,000 feet and the next day went even higher. On day four, with trepidation, we set off for Shingo La, a 16,700-foot pass where Ram had gotten sick the year before, perhaps from altitude sickness. We all arrived at the top of the pass without incident and took the time to place prayer flags on the large chorten marking the top. Unbelievably, a road was built to Shingo La in 2016, but weather conditions and flooding in places below the pass keep it largely unpassable.

Over the next seventeen days we crossed at least nine high passes, walking for long uphill stretches that got steeper and steeper until suddenly we arrived at the top. The only landmark to mark the high point was usually a cairn of stacked stones and prayer flags. Views from the trail varied from magnificent, wide valleys to monotonous, gray hills that looked like mine tailings. In between the passes, at lower elevations, it was hot and dry, and there were few trees. Ram and I both carried umbrellas to protect ourselves from what he called "black sun" because it made his skin so dark.

We encountered few people in villages along the way, but we saw families in their fields working feverishly to harvest crops that would sustain them through the brutal winter. Most activity we encountered was at Tibetan Buddhist gompas, nestled into limestone cliffsides or sitting on top of craggy peaks overlooking valleys. The imposing white-stuccoed structures were almost always strategically situated and would come into view from miles away. Fortress-like, they spread widely over the hillsides they were perched on, with assembly halls, chapels, libraries, and living quarters for the monks and students. Although many monasteries were in poor condition, they still featured a wealth of once brightly painted frescoes depicting religious and cultural scenes and symbols.

After walking up steep flights of steps to a main gate of the gompas we visited, we would peek inside the large assembly halls where monks and students gather to pray. We were always welcomed with an invitation to come inside, join a prayer session, and drink hot buttered tea that had a decided rancid butter aftertaste. At the front of the assembly halls we saw wooden altar tables filled with flickering candles, butter lamps, and piles of flowers, food, and other offerings. Glass cases behind the altars housed images, often in gold, of Buddha in one of his manifestations, as well as important lamas and other disciples. The monasteries we visited in Zanskar are largely associated with the Gelugpa Order, which owes allegiance to the Dalai Lama and is often referred to as the Yellow Hat school of Tibetan Buddhism. Throughout Ladakh a total of four sects is represented: Gelugpa, Nyingmapa, Kagyudpa, and Sakyapa. These repositories of Tibetan Buddhist culture and religion flourished from the tenth to twelfth centuries CE. Each has a unique history based on its sect and the line of spiritual masters with which it is associated.

The most memorable monastery we visited on our trek—based on its remote location, huge size, and large number of monks—was Phuktal Monastery. Built into a natural cave in the cliffs above the Niri Chu River, it can only be reached by foot, even today. The monastery, which is believed to have been founded during the early twelfth century CE, featured a beautiful Buddha statue illuminated by an ancient skylight.

On the twenty-first day of our trek we arrived at our destination, Lamayuru Monastery, one of the largest and oldest gompas in Ladakh. The monastery sits atop a barren, stone outcropping, and its three-story, white-stucco, main gompa building is visible for miles. It was the middle of the day when we arrived, and the monks were in their living quarters so the gompa was quiet. A small boy wanted to charge us ten Indian Rupees to go inside, so we just stuck our heads in and then walked down the hill to the highway, which follows the course of the old Silk Route, and sat down to wait for a bus to take us to Leh, the capital of Ladakh.

Thirty-three years later, when we returned to Lamayuru Monastery for the annual Yuru Kabgyat ceremonies, it was brimming with people. Rows of families with young children, old ladies in long woolen robes, young men in leather

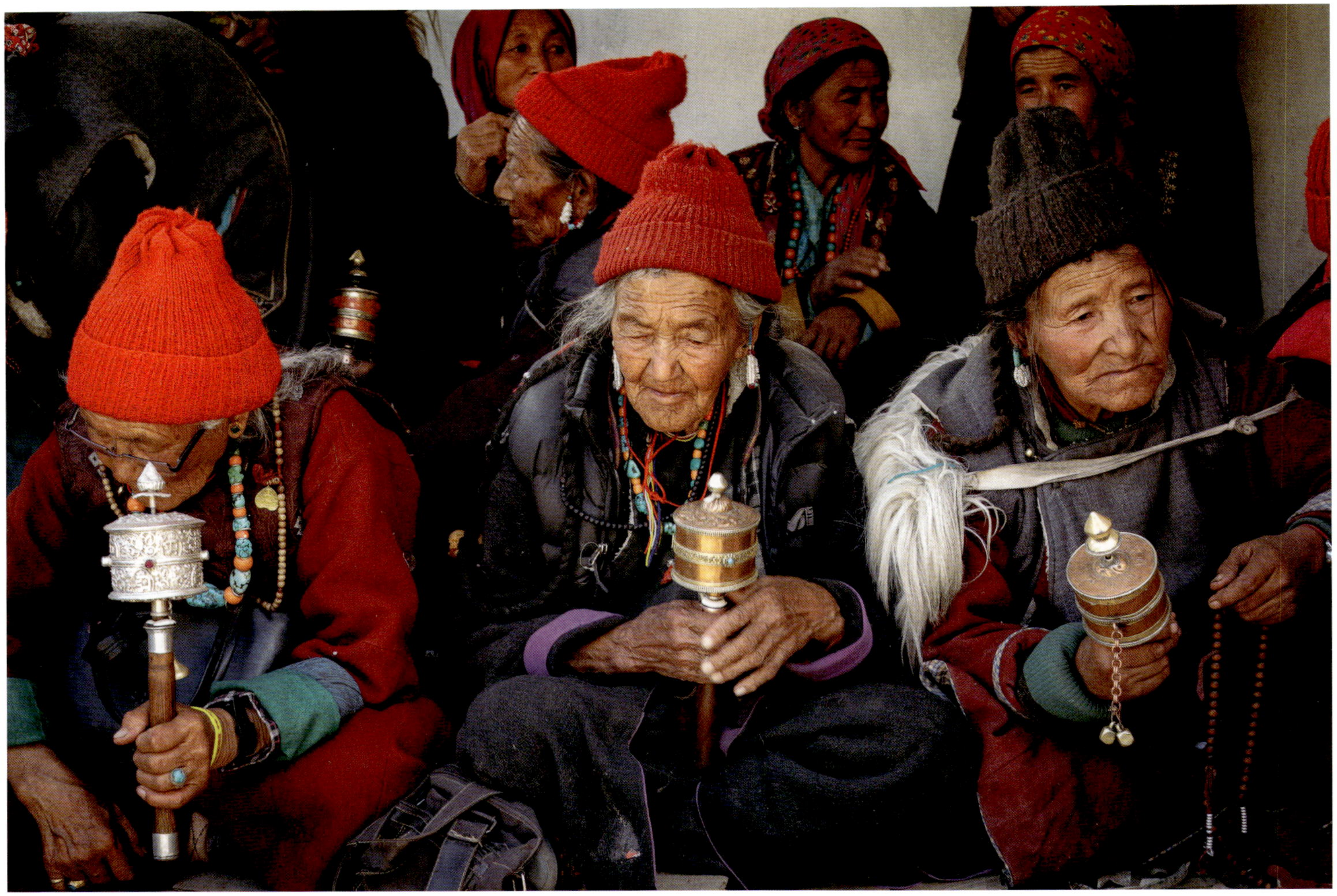

Ladakhi women at the Yuru Kabgyat ceremony, Lamayuru Monastery, Ladakh, India, 2018

jackets and sunglasses, as well as a few tourists like us lined the monastery's big courtyard. Bill moved around constantly taking photographs, but I sat in the dust with the crowd and watched from the sidelines. The action was non-stop over two days. Monks in large, crescent-shaped red hats came out in a line, accompanied by cymbals and the deep, animal-like sounds from the long Tibetan horns known as *dungchen*. Dancers in masks, depicting deities, demons, and animals and dressed in layers of brightly colored silk brocade, spun around the courtyard in unison.

Several weeks later we attended the ceremony at Hemis Monastery, which celebrates the life of Lord Padmasambhava, known also as Guru Rinpoche, one of the founders of Tibetan Buddhism. Over two days we watched ritual dance performances starting with the Black Hat Dance. At the end of the second day a huge *thanka* (scroll painting) was unfurled. As at other sacred celebrations we have attended over the years, both Buddhist and Hindu, we understood the underlying religious philosophy and that the purpose of the ceremony is to bestow health and pacify disease and negative forces, but the specifics remained a mystery.

Over the course of several weeks we visited more than twenty Tibetan Buddhist monasteries in Ladakh, all beautiful testaments to the tenacity of this centuries-old religion. In the idyllic small village of Alchi, set alongside the Indus River in a narrow valley, we visited a monastery complex believed to date from the eleventh century CE. The murals in the main temple, called the Sumtsek, and in the Dukhang assembly hall are rare, surviving artworks of a style that originated in the region of Kashmir, just several hundred miles west of there. The graceful, elongated figures painted on these richly colored murals are reminiscent of Kashmiri bronze sculptures from the tenth to eleventh century CE, and it is conjectured that the painters used them as a model for the murals.

From the back of the Sumtsek we glimpsed a row of white chortens high above the village, so we walked over. An intriguing square structure among the chortens had a sign that told us it was the Shangrong Temple, an assembly hall dating to around 1400 CE, with some of the few remaining paintings of an early regional school of art. The doors were locked, but we found a friendly woman in a nearby house who opened them for us and let us look inside as she lit butter lamps and burned incense. Like an image from an ancient time, she wore a long, dark maroon dress, a colorful fringed shawl, and the traditional Ladakhi top hat with upturned rim known as a *perak*.

We returned to Thikse Monastery, which we visited in 1985, and found it to be now a thriving place. Located about halfway between Hemis Monastery and the town of Leh, it is a complex of square, white-stucco structures spread all over a hillside. At 6:00 one morning we walked up the hill to the roof of the main building to witness the daily ritual of monks blowing on conch shells. It was reassuring to see that the gompas in Ladakh were in much better condition than they had been during the mid-1980s and are still active religious centers.

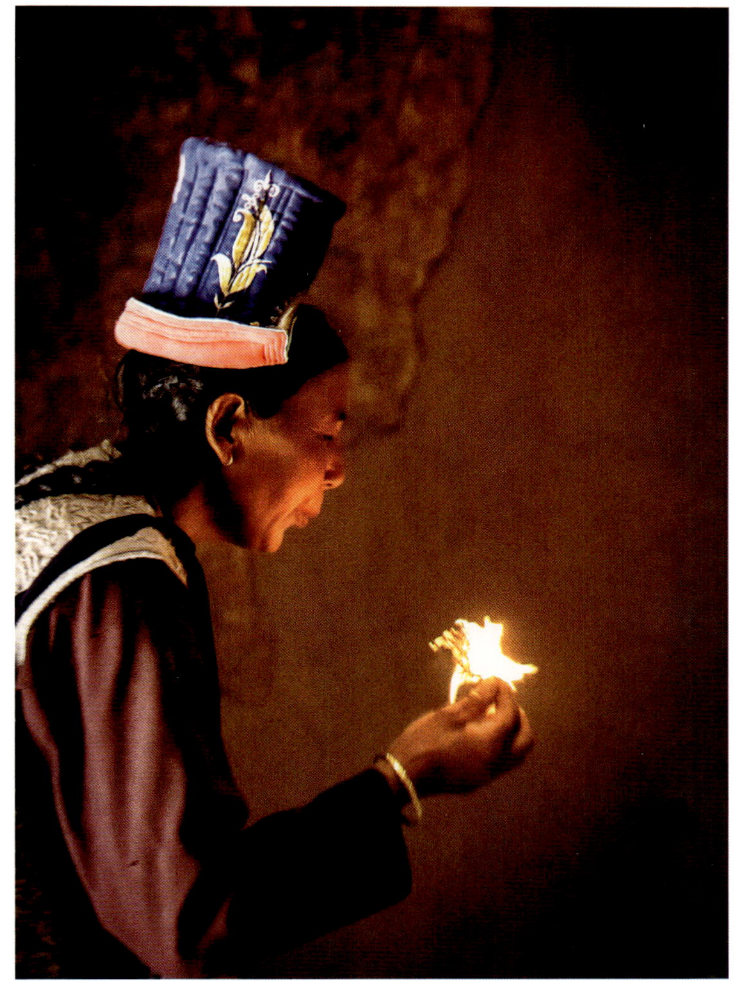

The caretaker of the Shangrong Temple above Alchi village, Ladakh, India, 2018

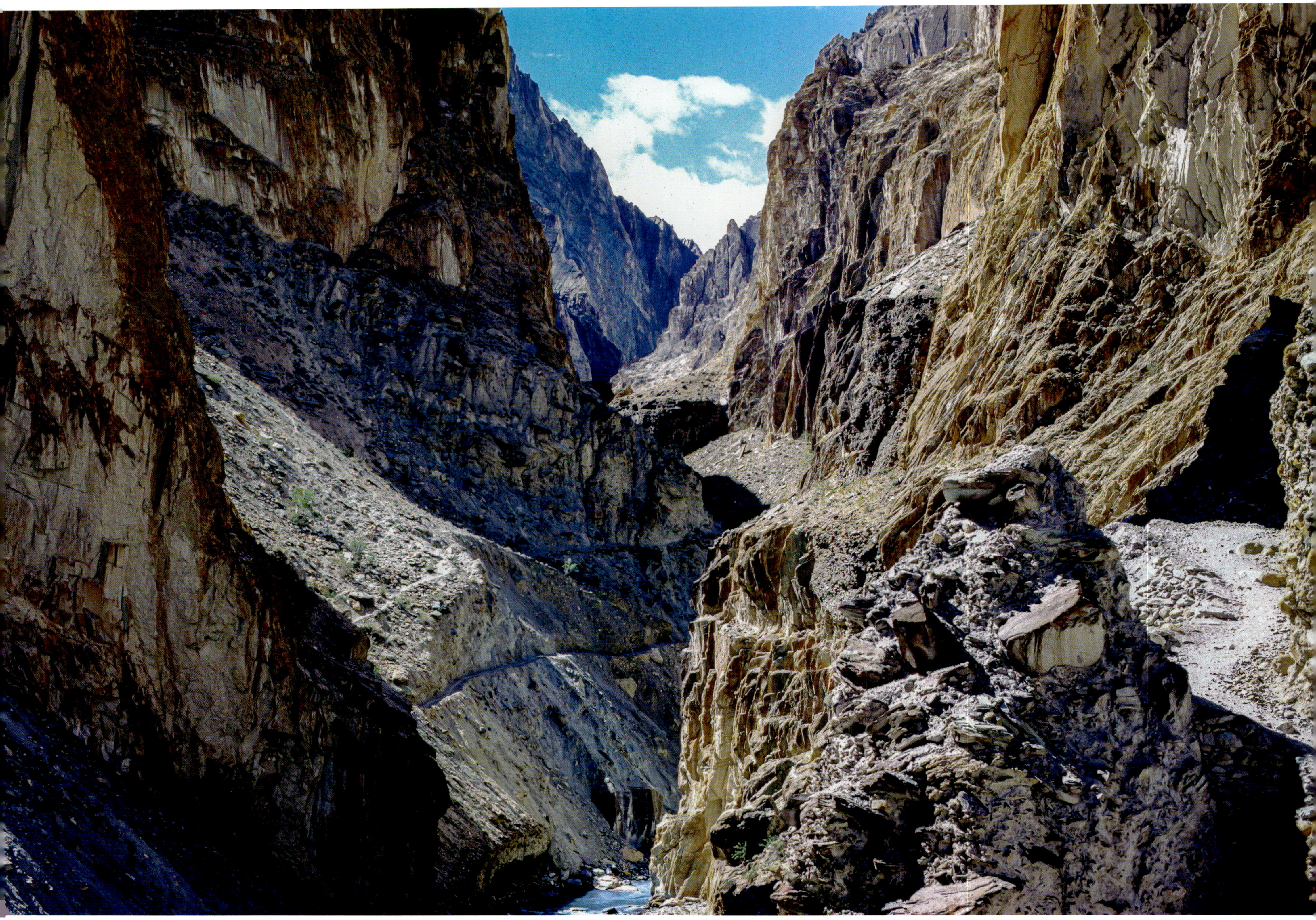

Zanskar, India, 1985

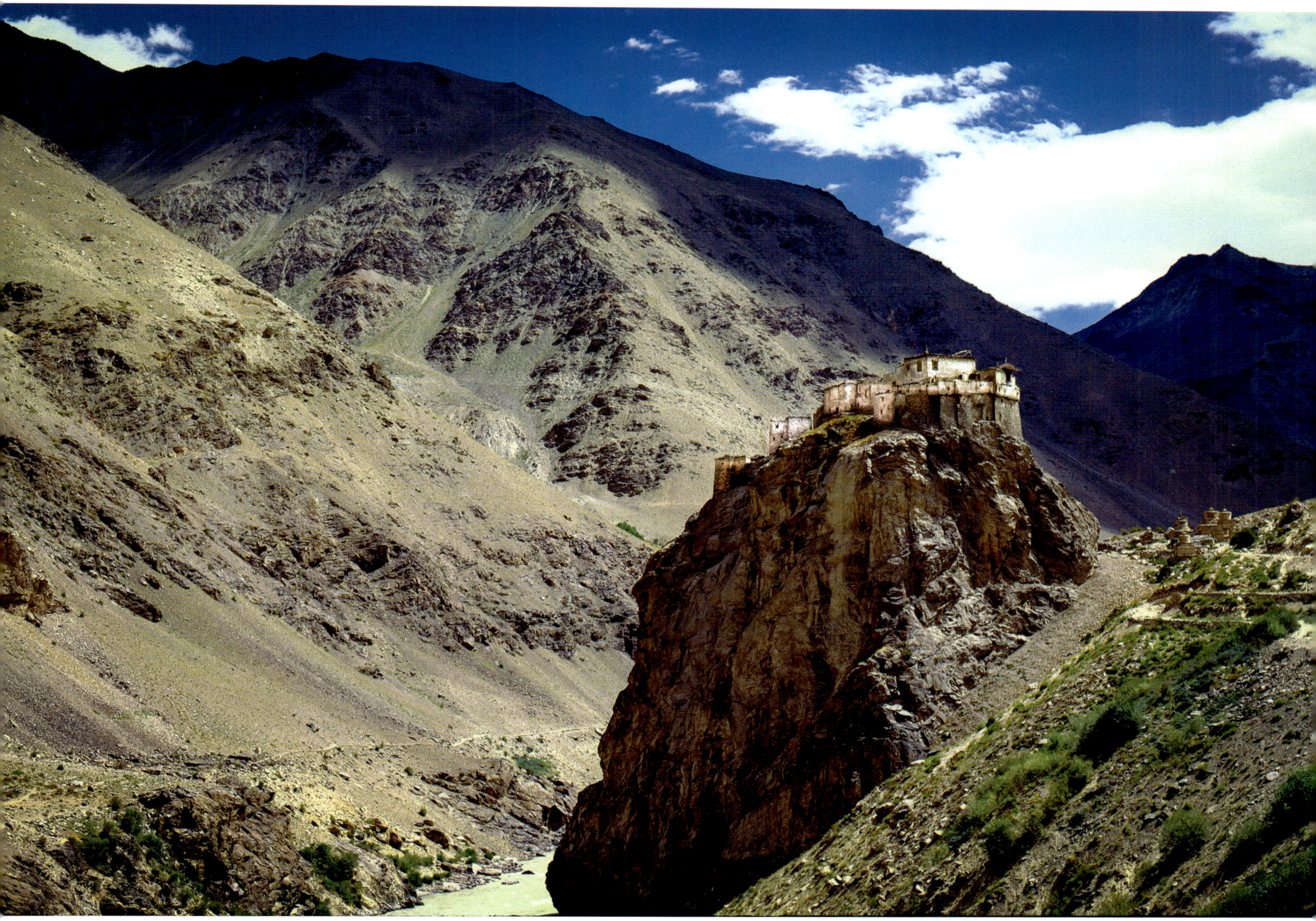

Bardan Monastery, Zanskar, India, 1985

Zangla Fort, Zangla village, Zanskar, India, 1985

Far right: Zanskar landscape, India, 1985

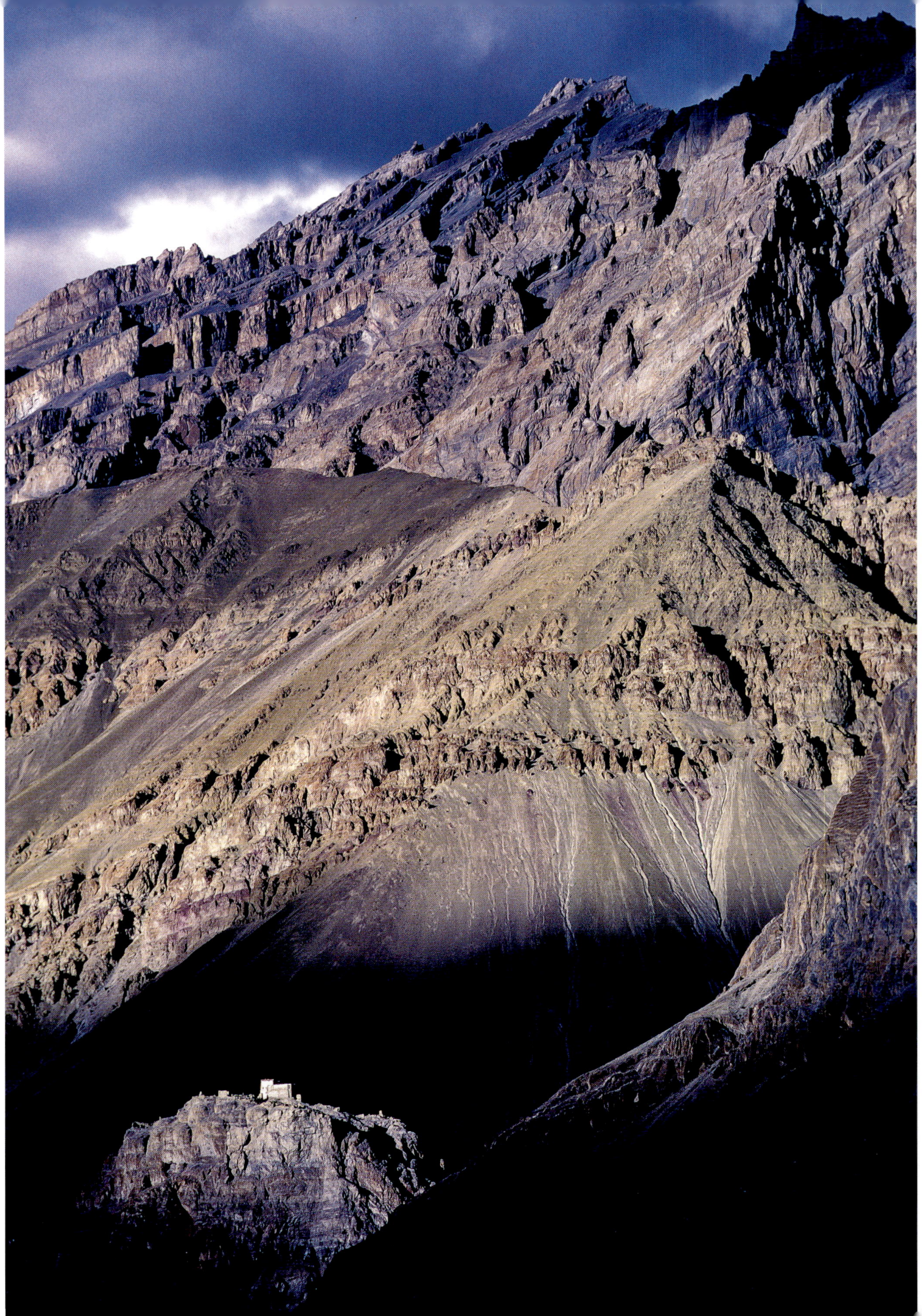

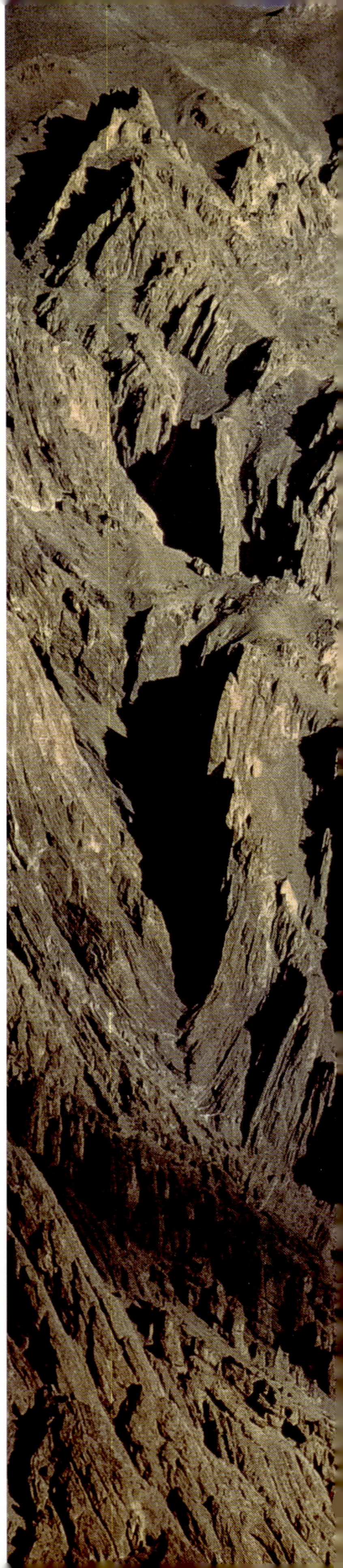

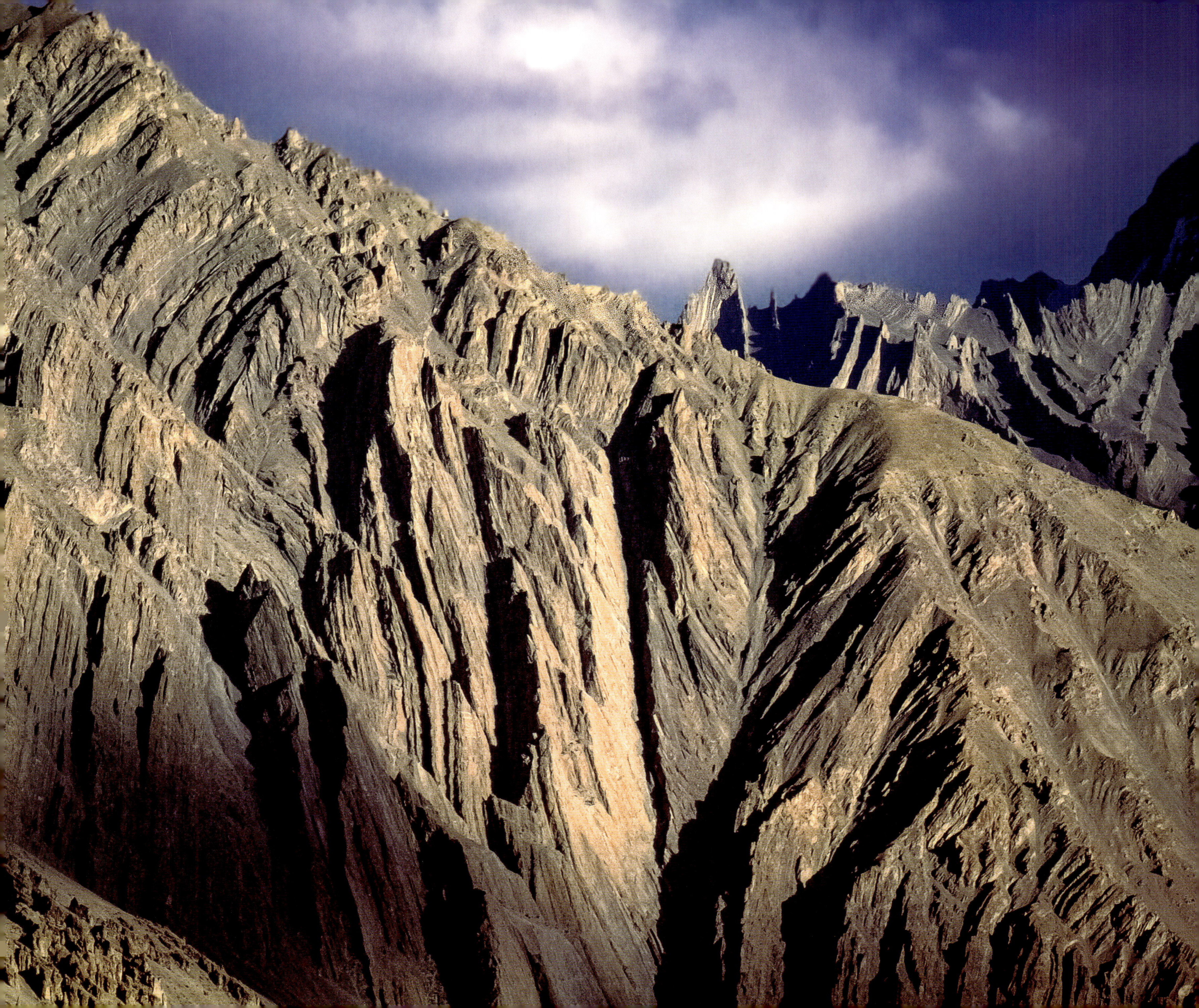

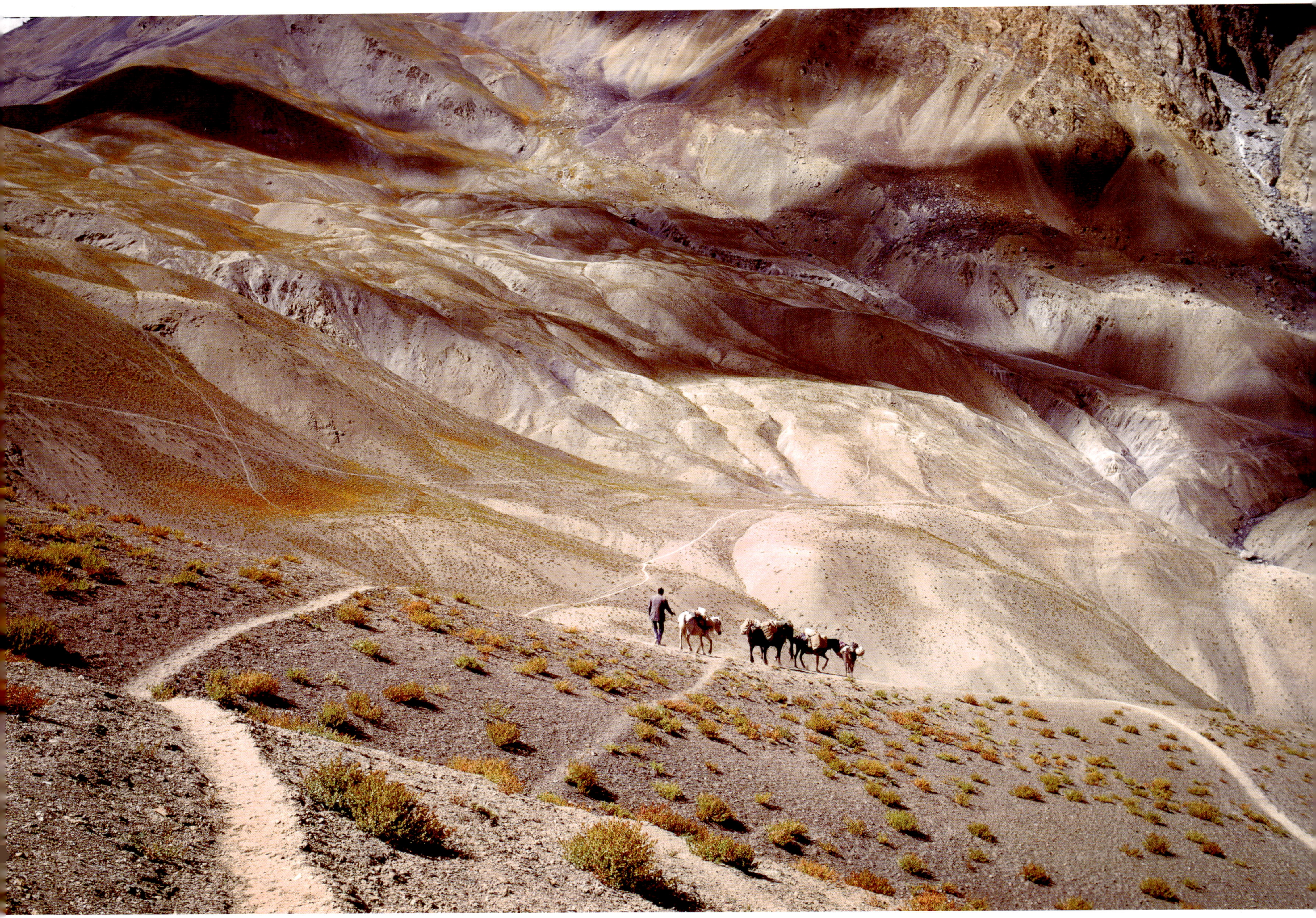

Ramdev, our horseman, below Shingo La, Zanskar, India, 1985

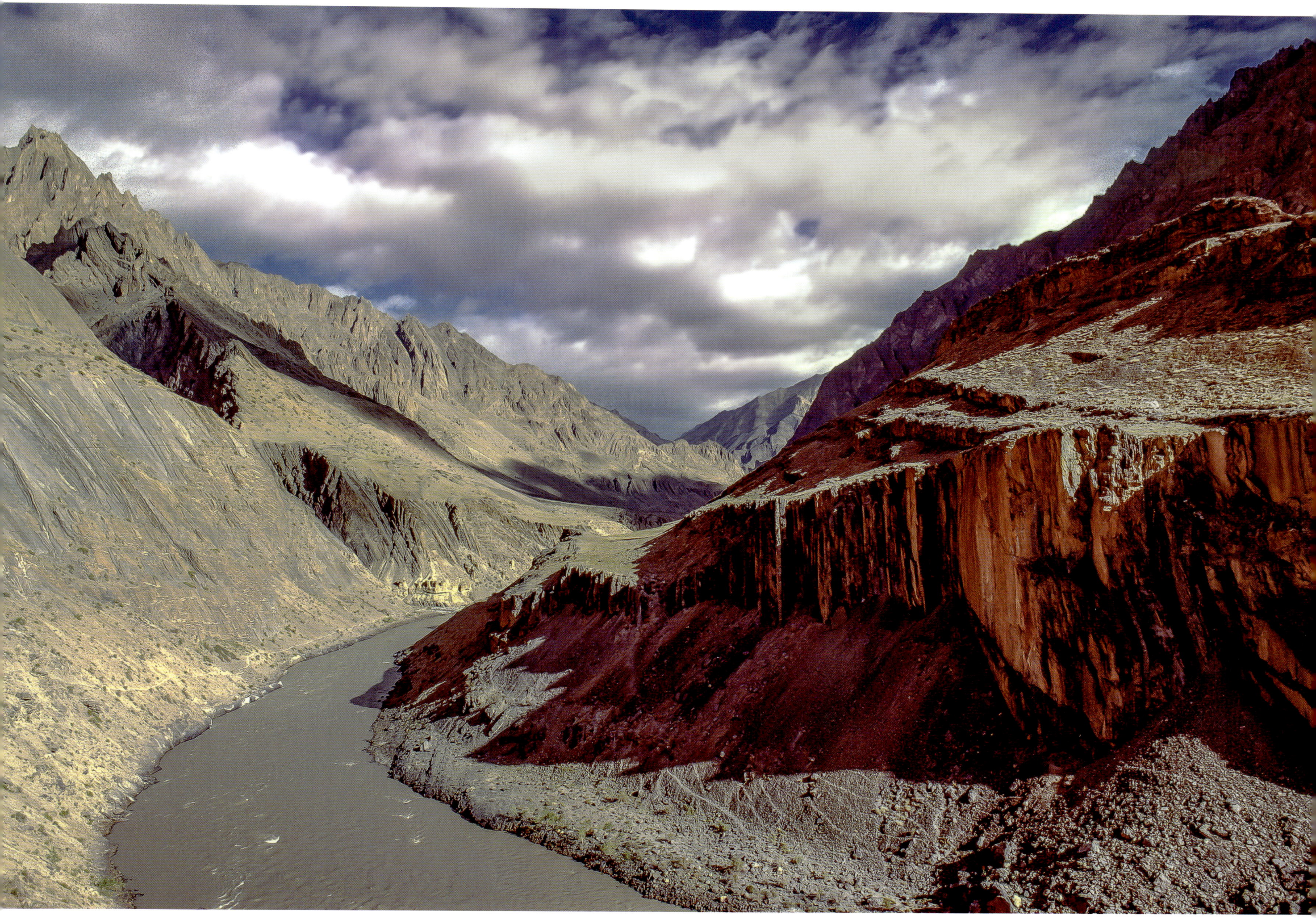

Zanskar River, Zanskar, India, 1985

Following spread: Stakna Monastery, above the Indus River, Ladakh, India, 2018

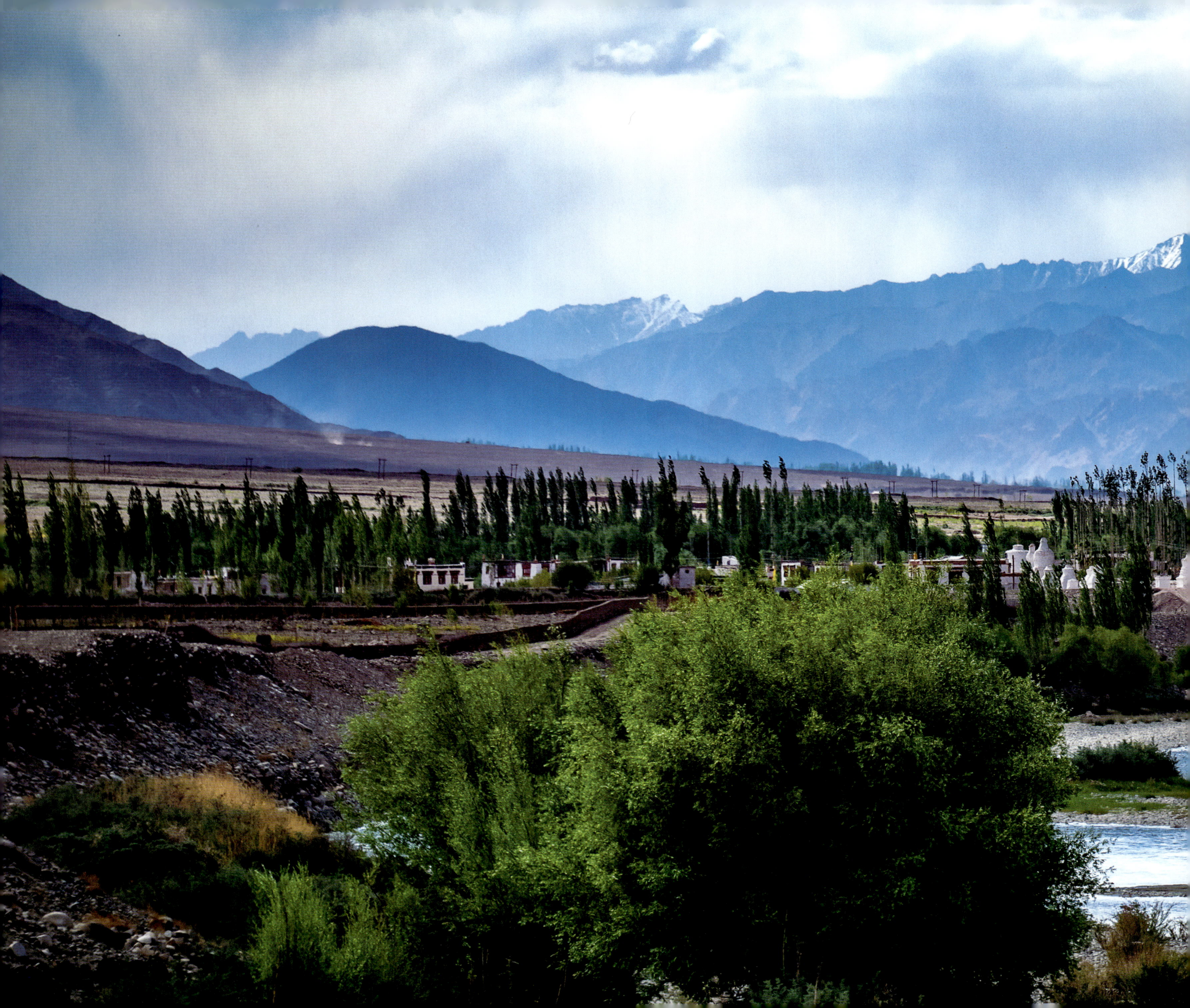

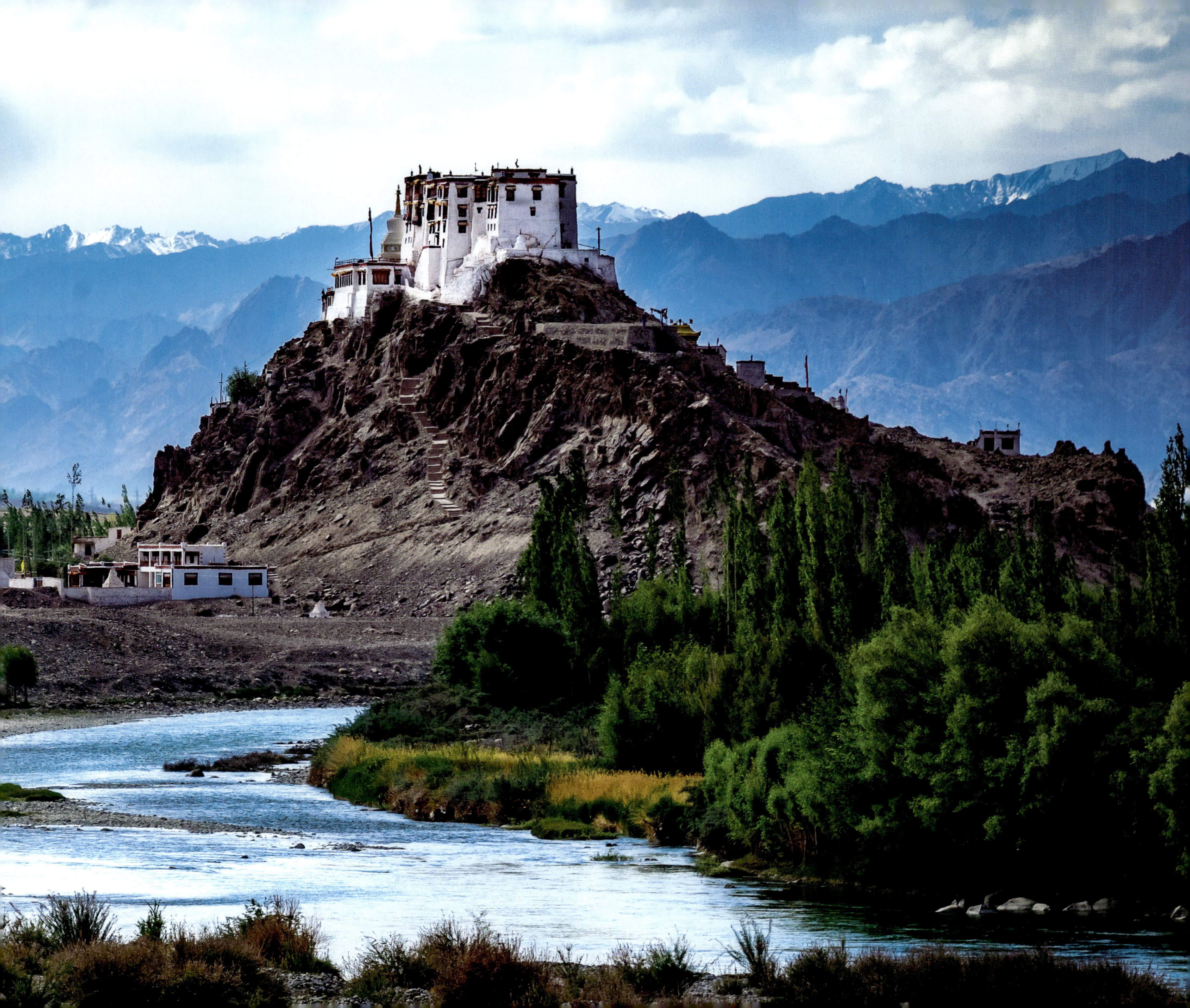

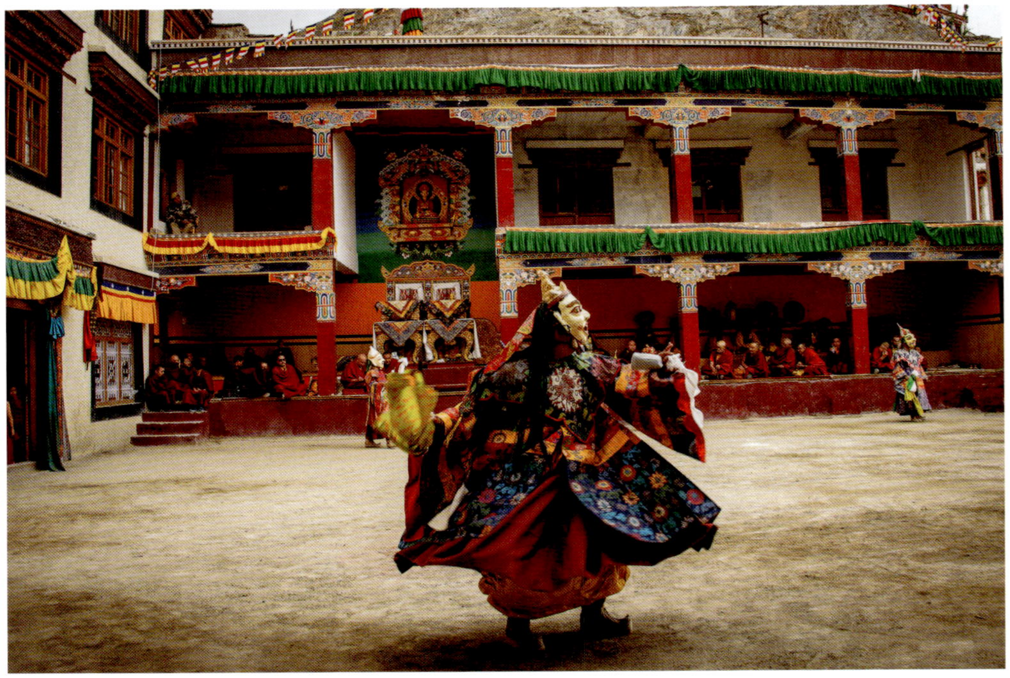

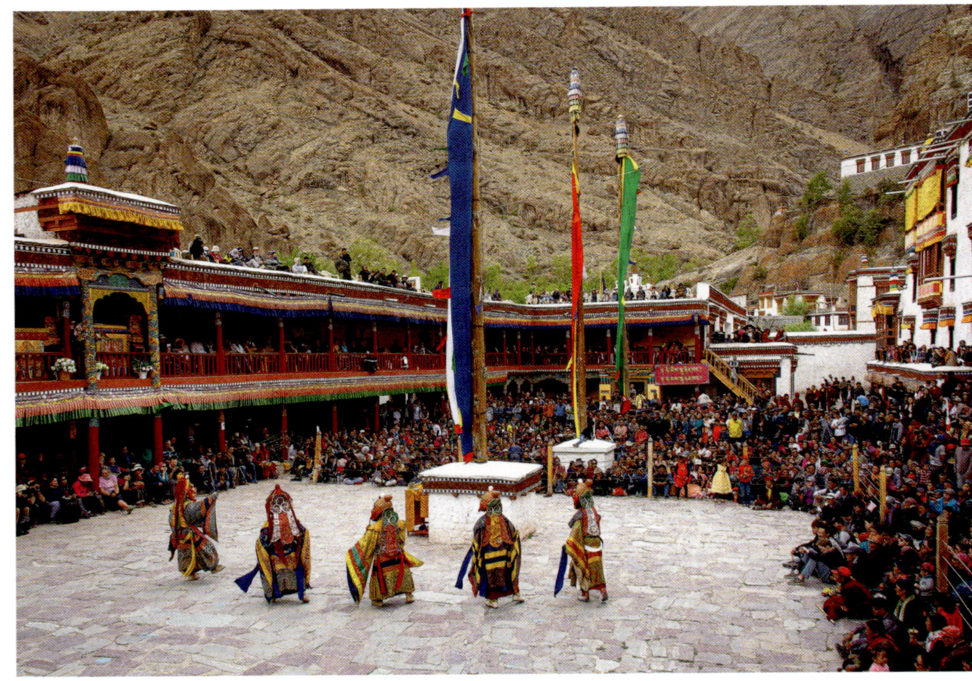

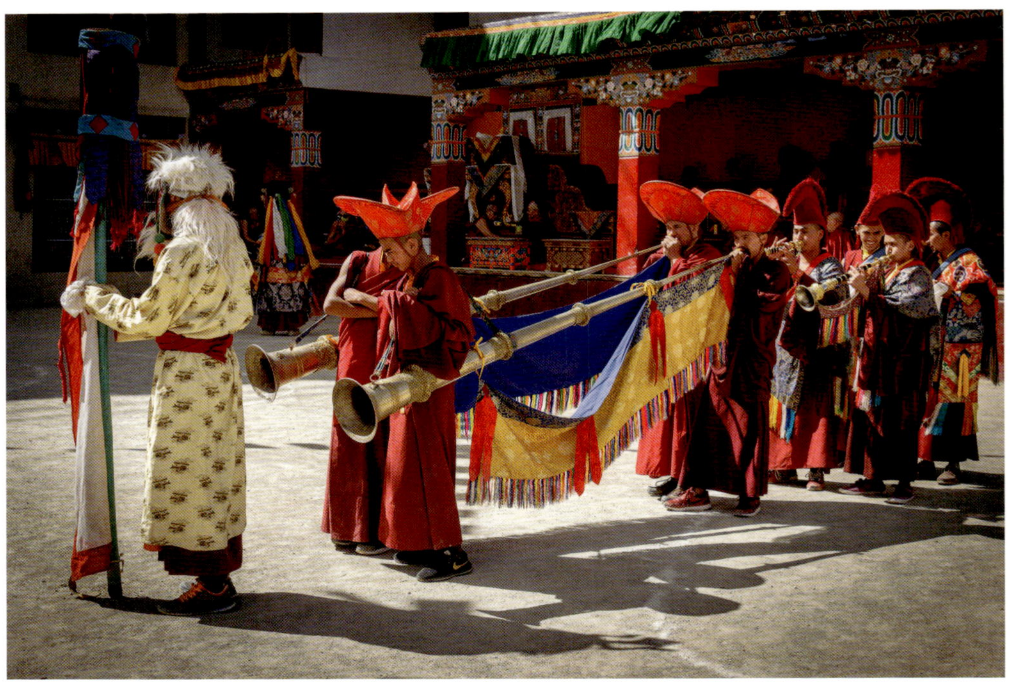

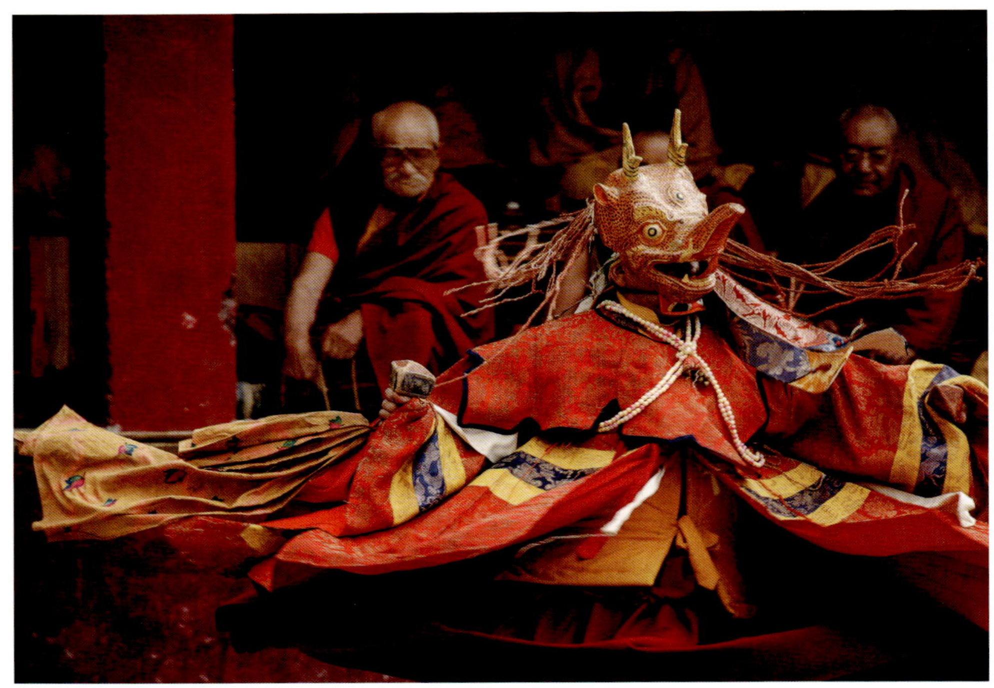

Top left, bottom left, and bottom right: Yuru Kabgyat ceremony, Lamayuru Monastery, Ladakh, India, 2018

Top right: Hemis Monastery ceremony dedicated to Lord Padmasambhava, Ladakh, India, 2018

Opposite: Yuru Kabgyat ceremony, Lamayuru Monastery, Ladakh, 2018

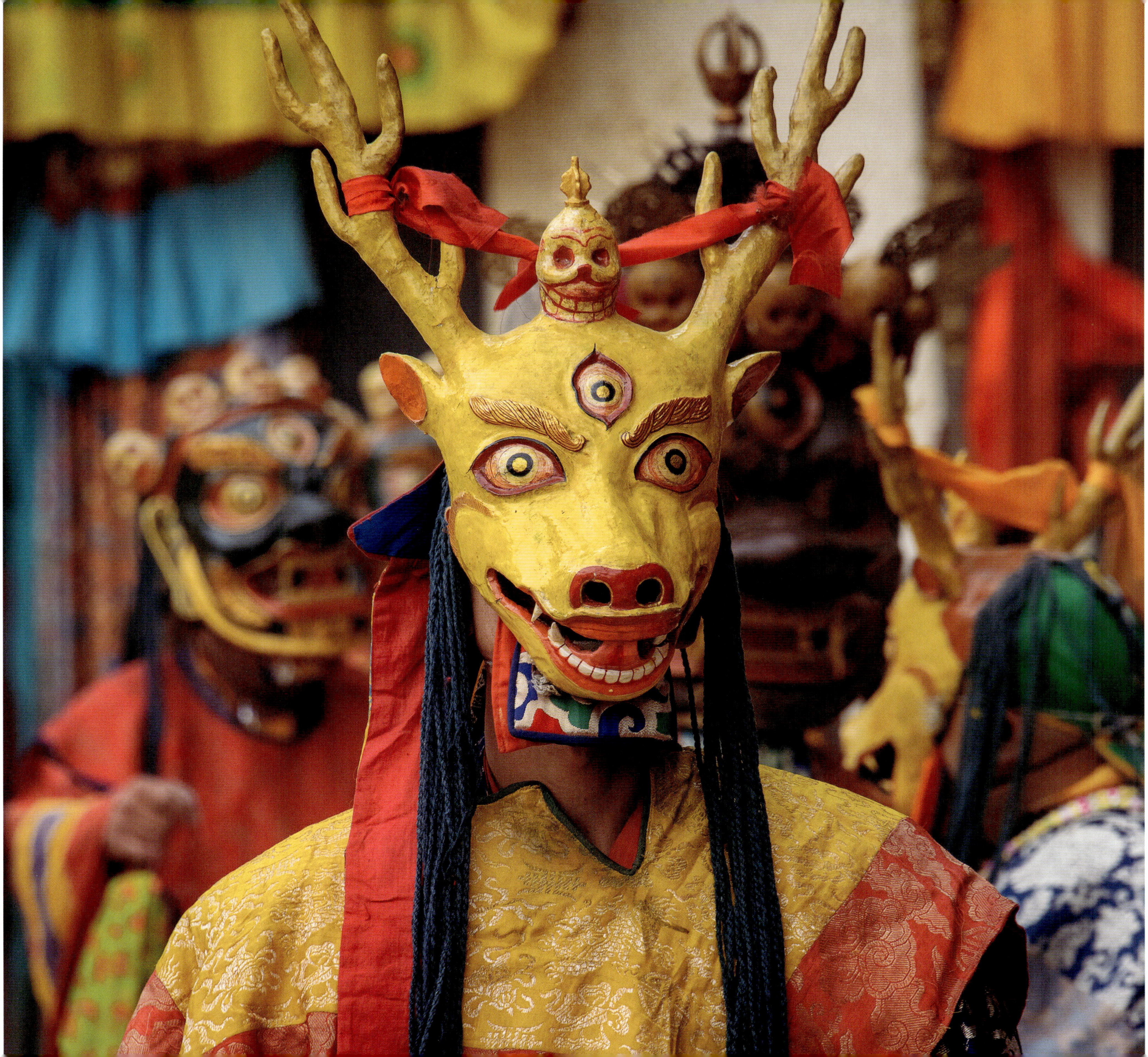

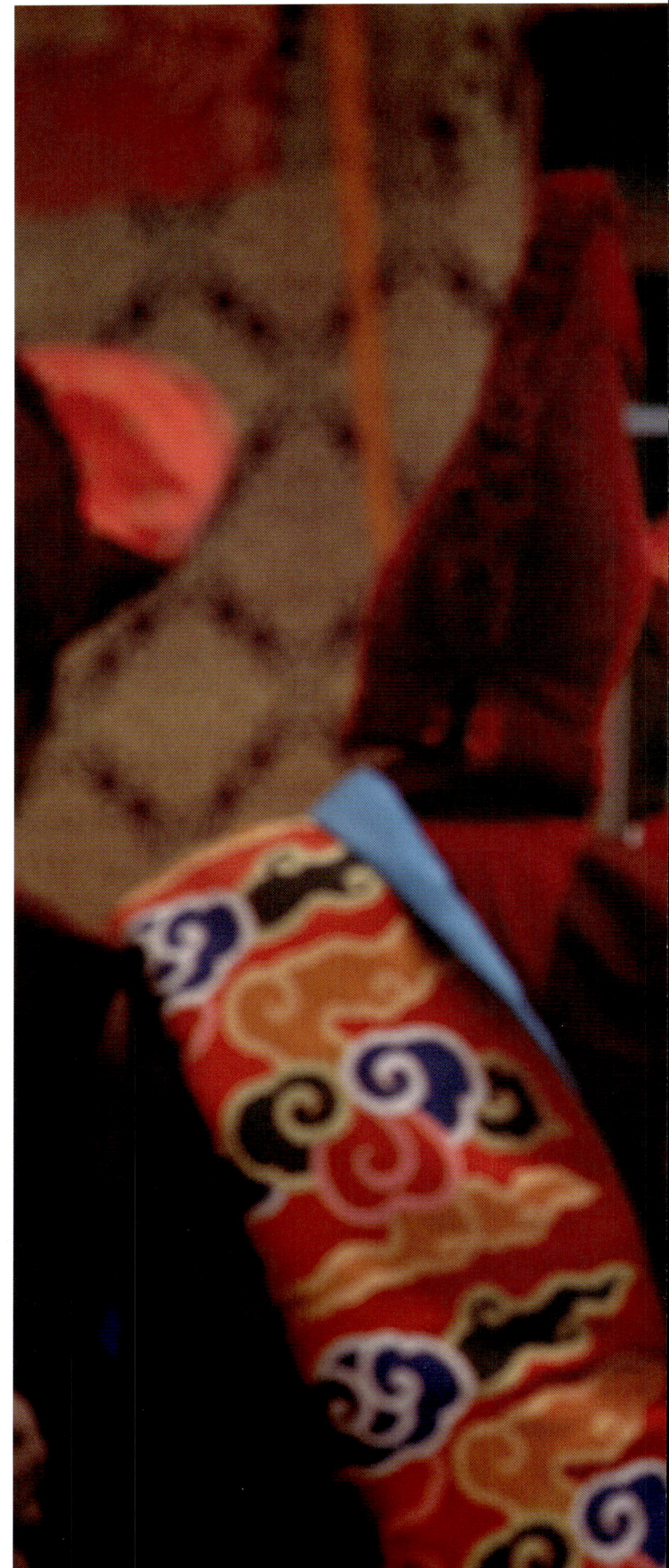

Lamayuru Monastery, Ladakh, 2018

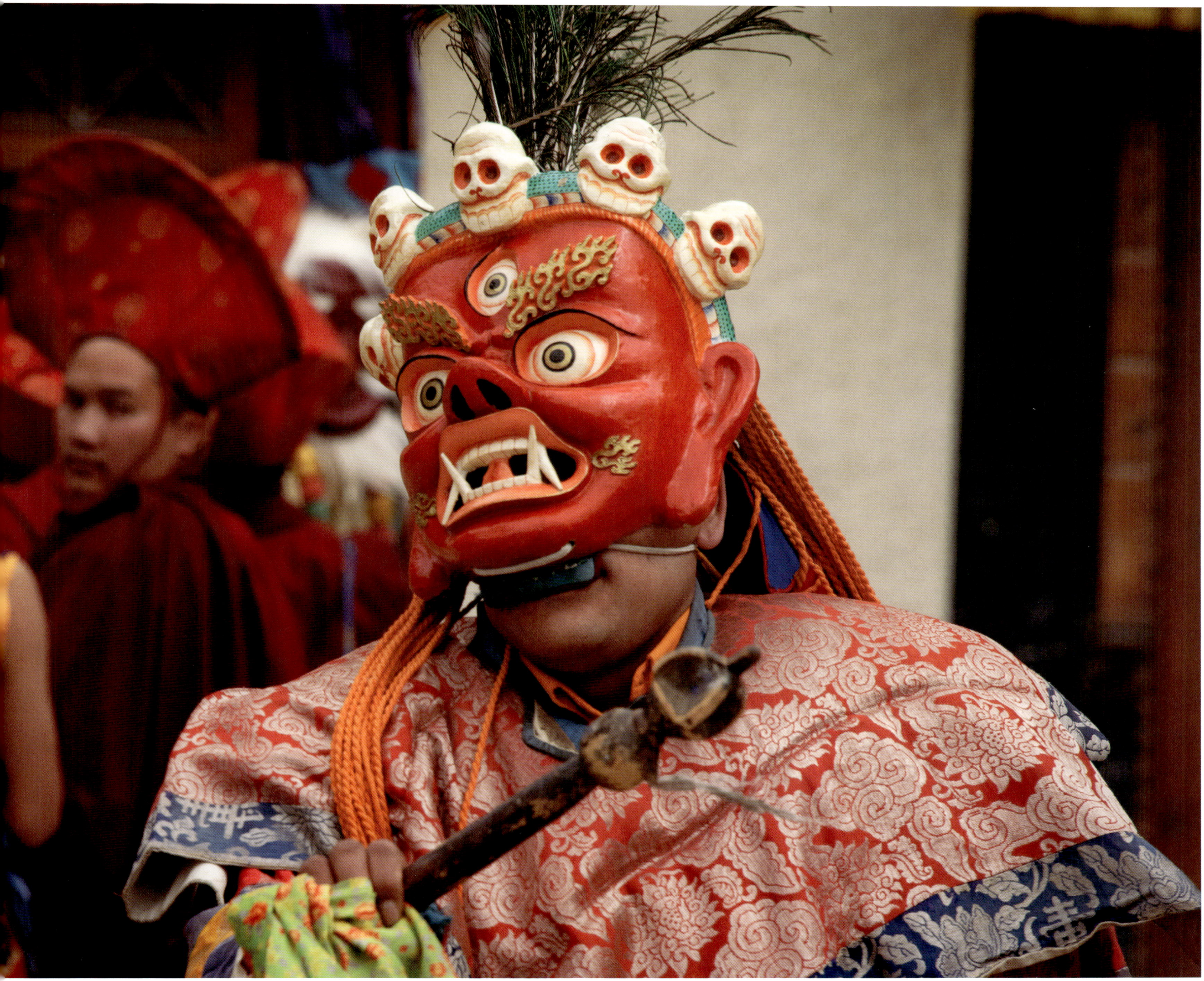

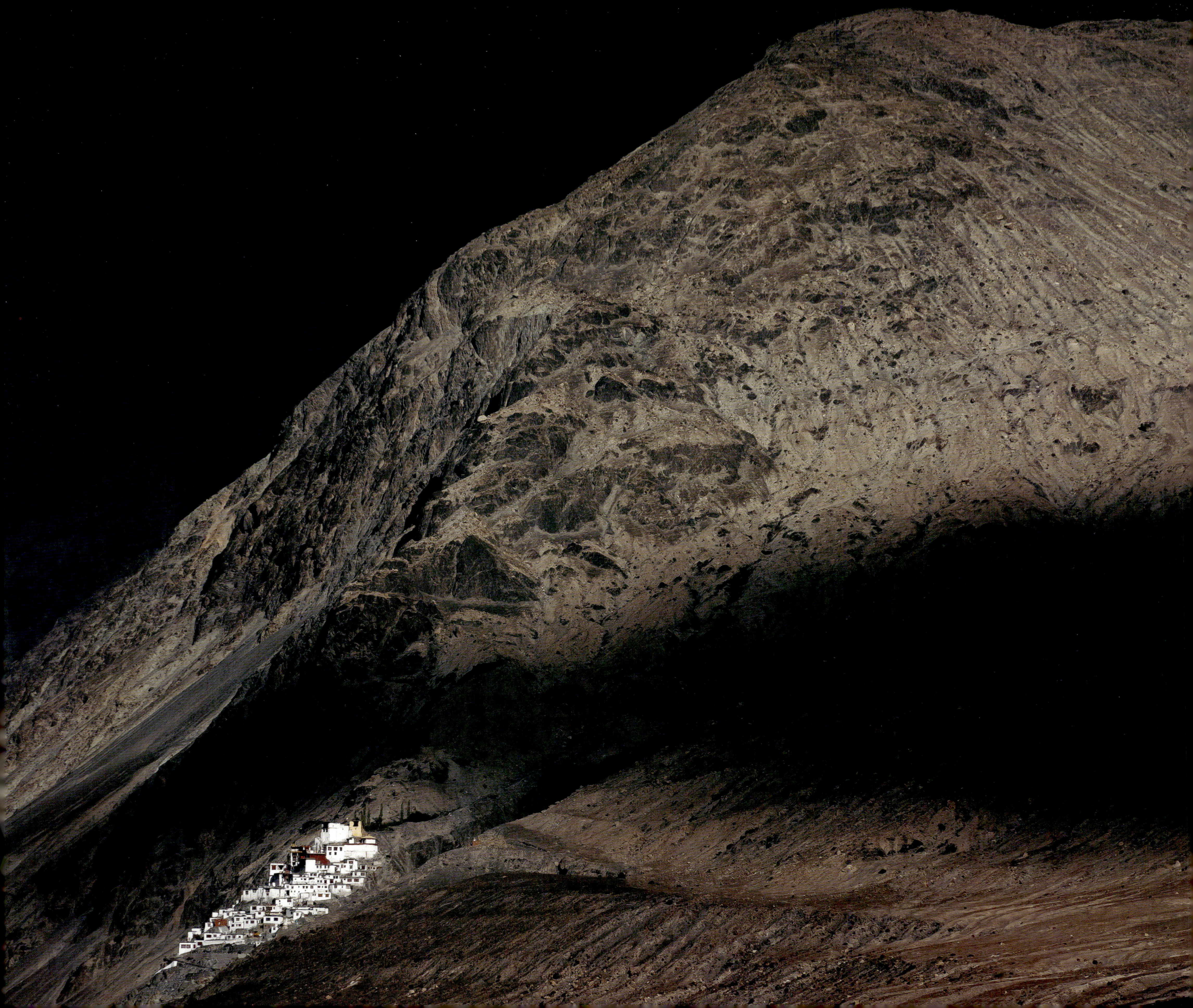

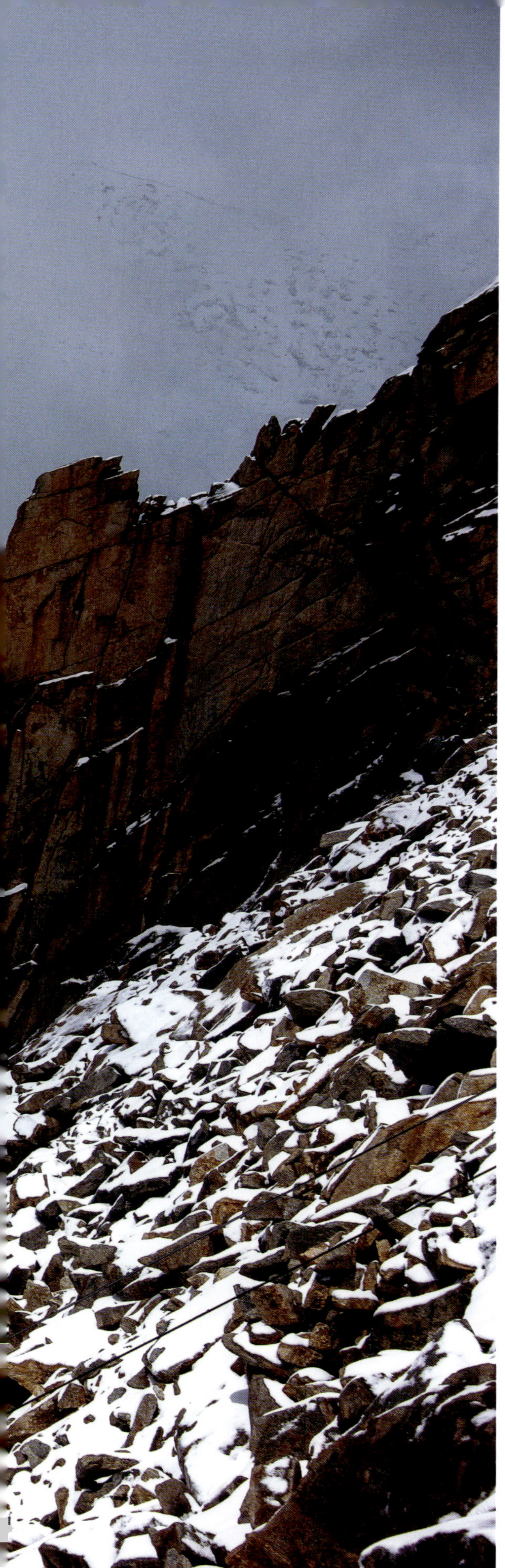

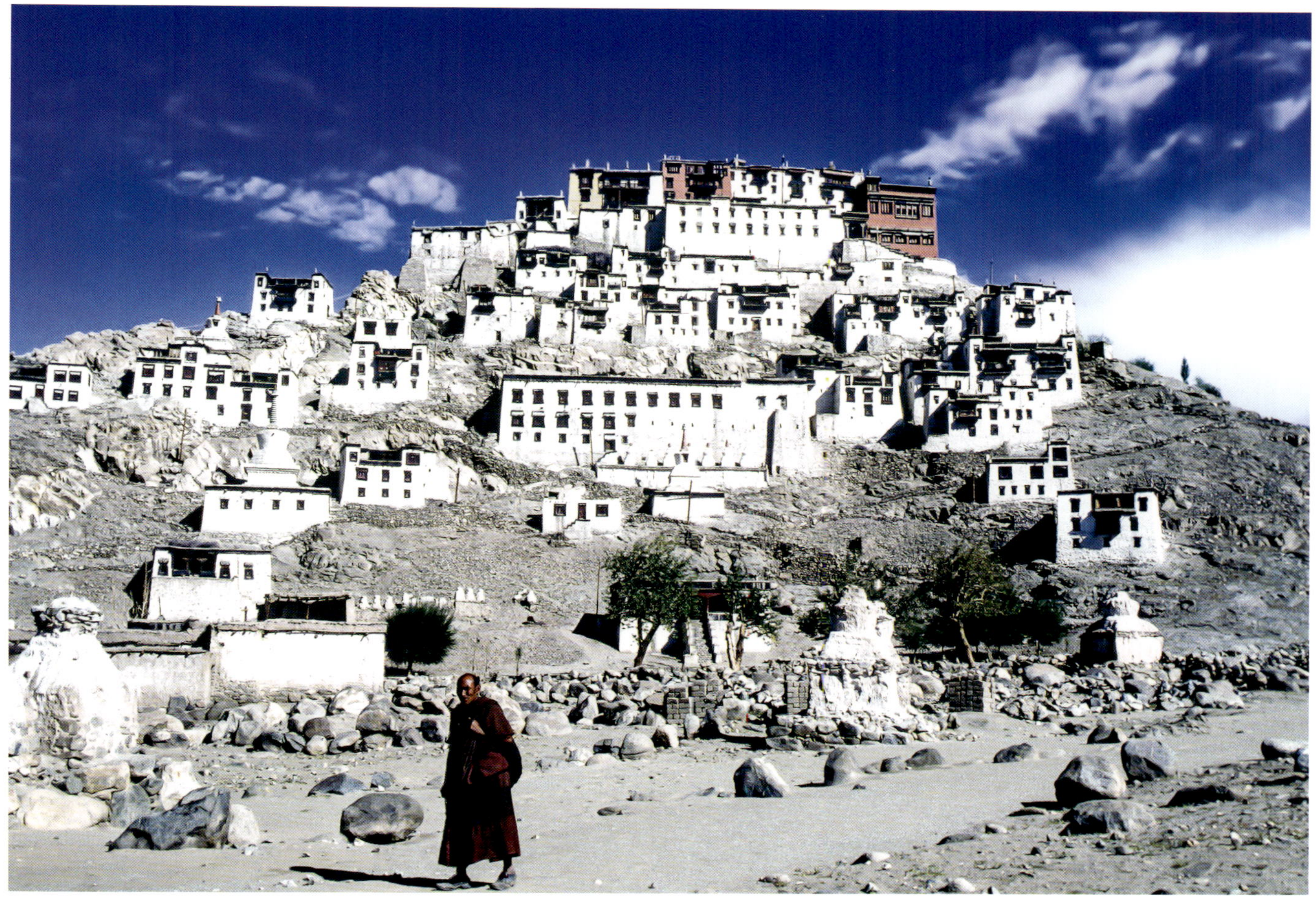

Thikse Monastery, Ladakh, India, 1985

Kardung Pass, Ladakh, India, 2018

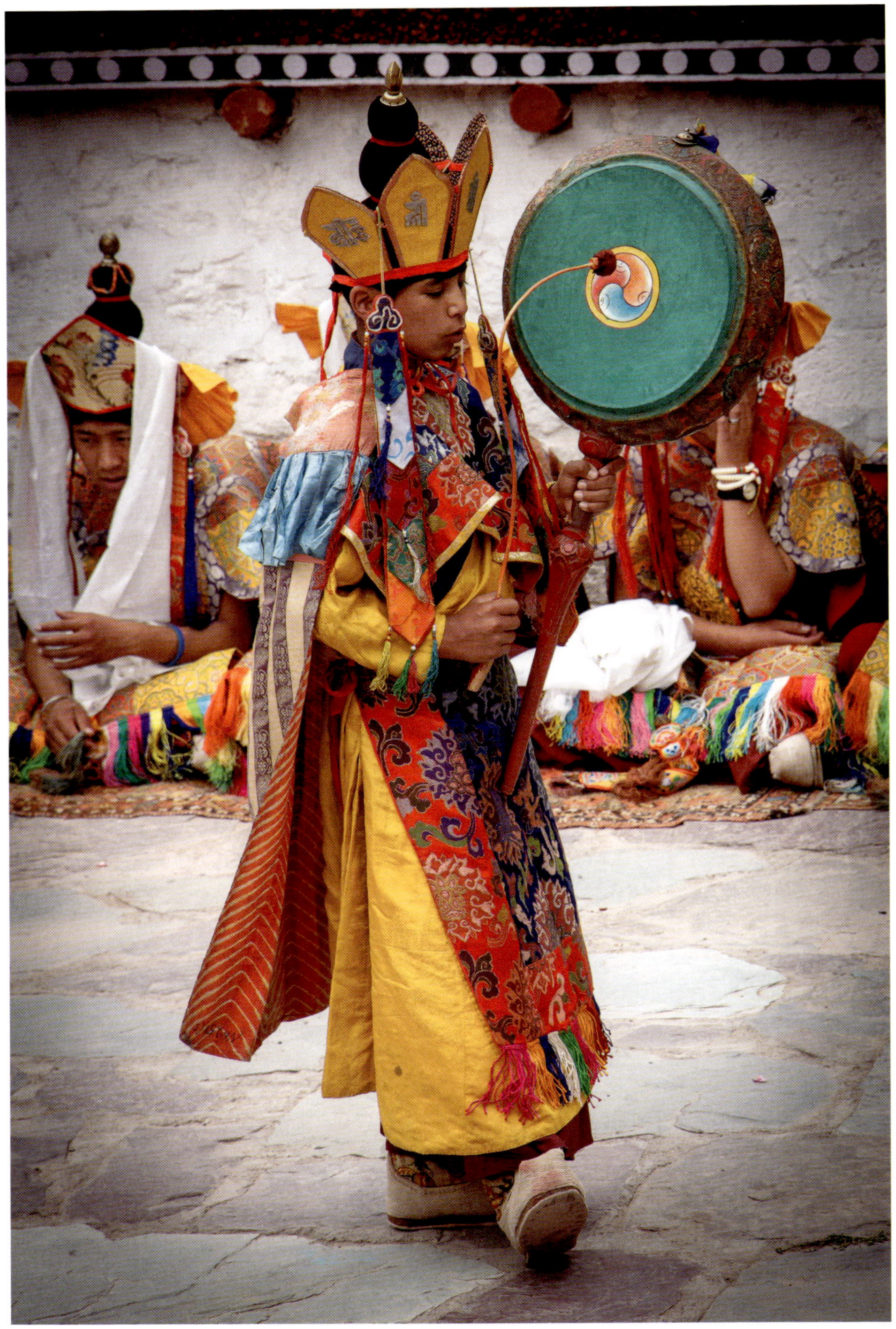

Hemis Monastery ceremony dedicated to Lord Padmasambhava, Ladakh, India, 2018

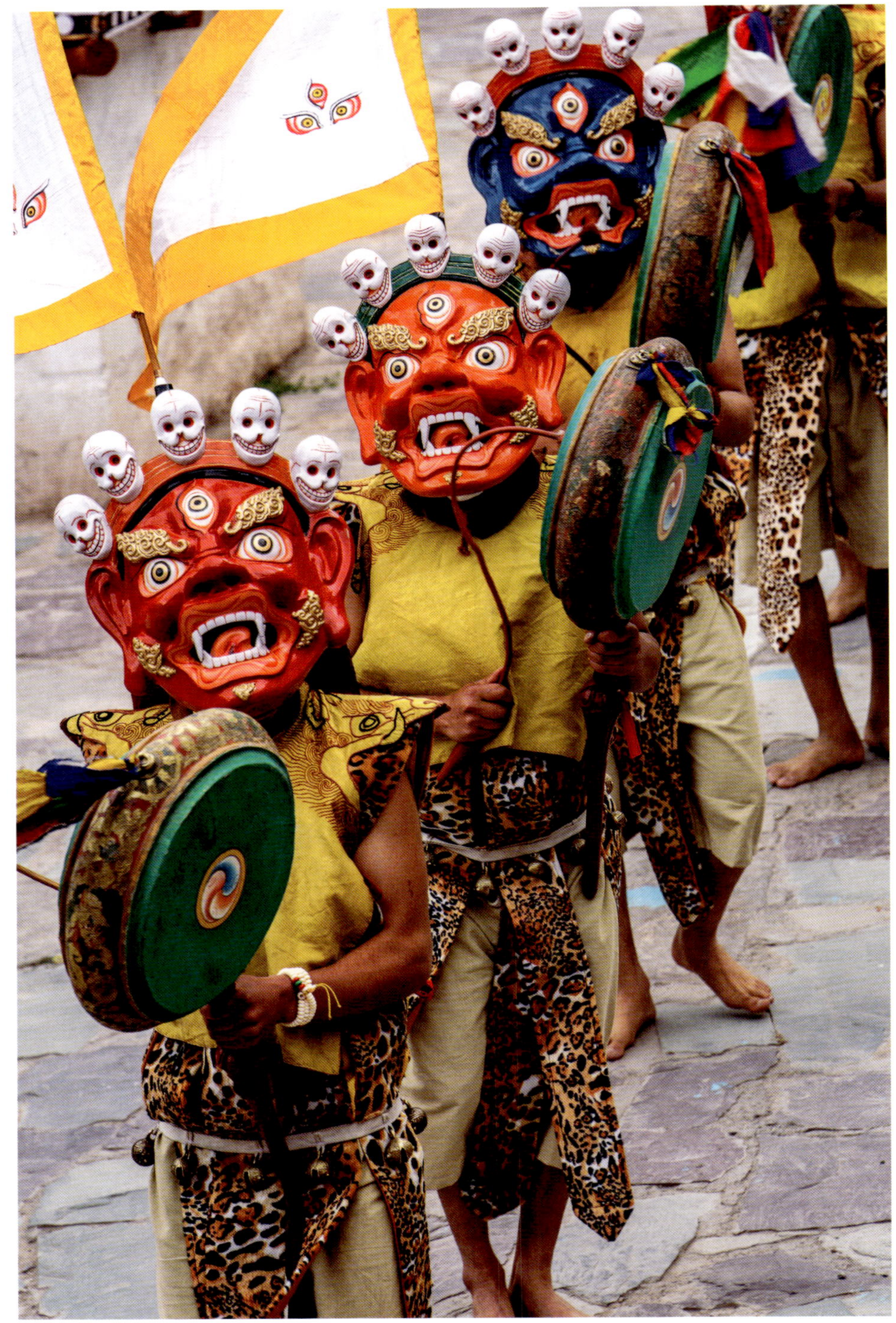

Hemis Monastery ceremony dedicated to Lord Padmasambhava, Ladakh, India, 2018

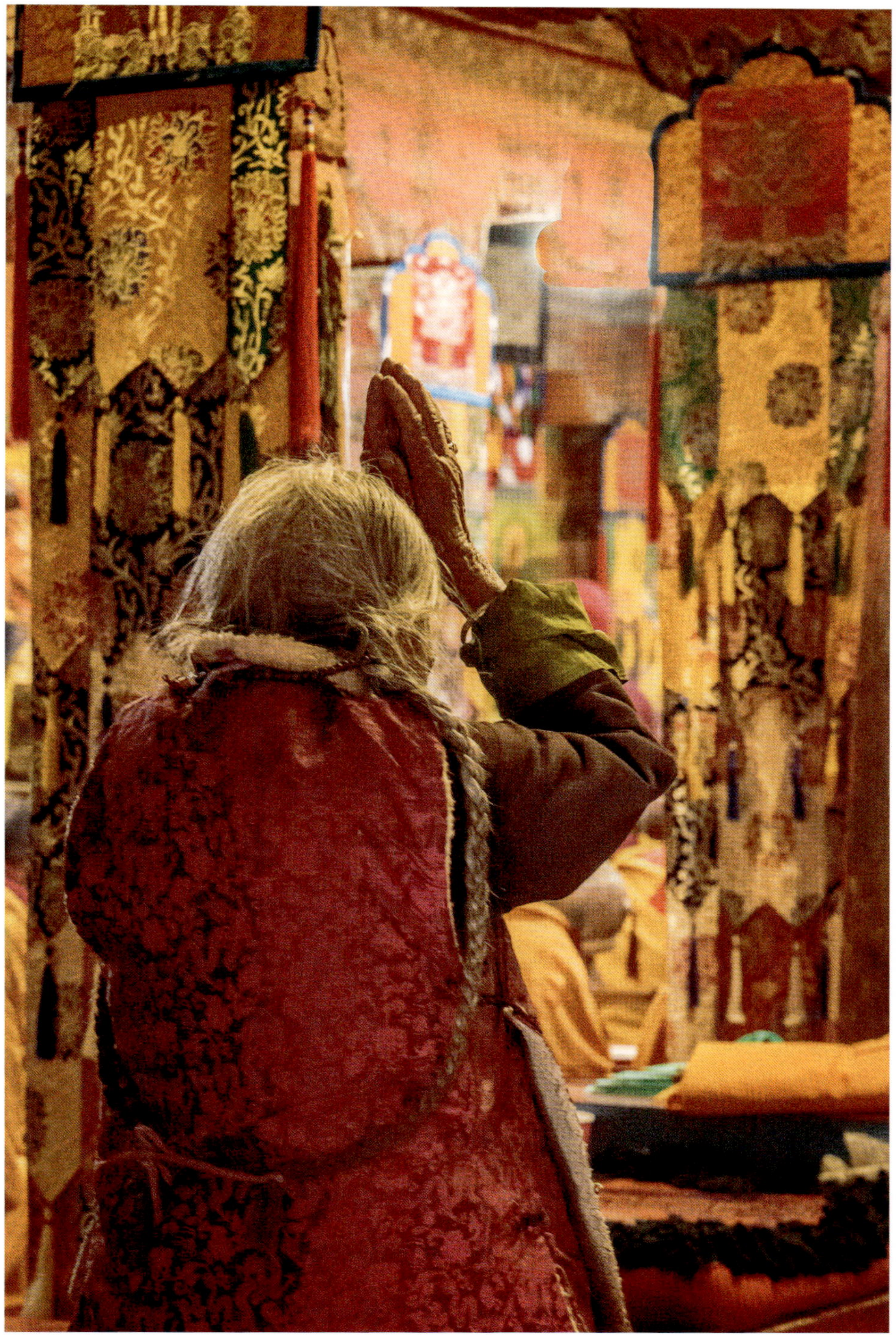

Ladakhi woman praying in Hemis Monastery, 2018

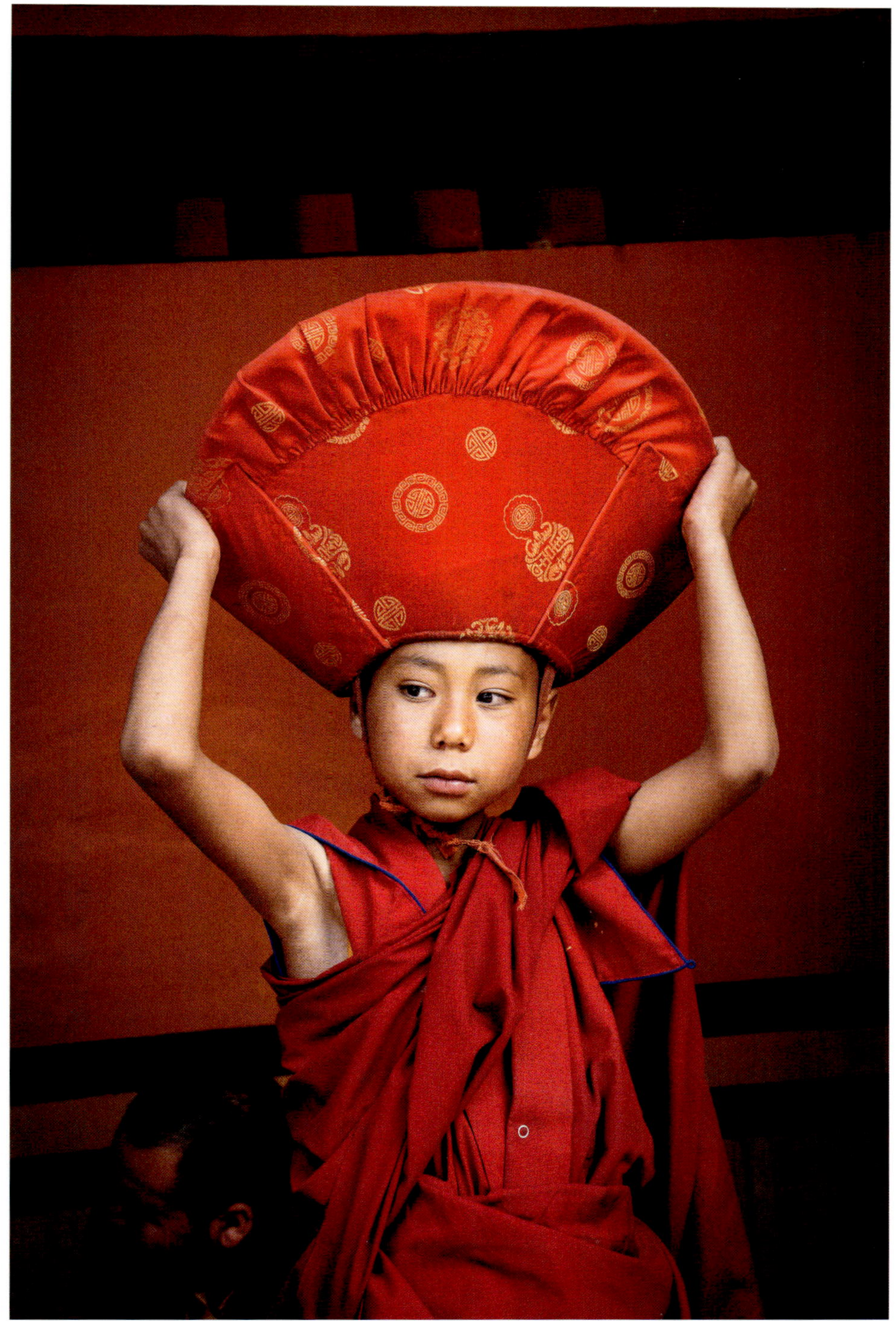

Young monk, Yuru Kabgyat ceremony, Lamayuru Monastery, Ladakh, India, 2018

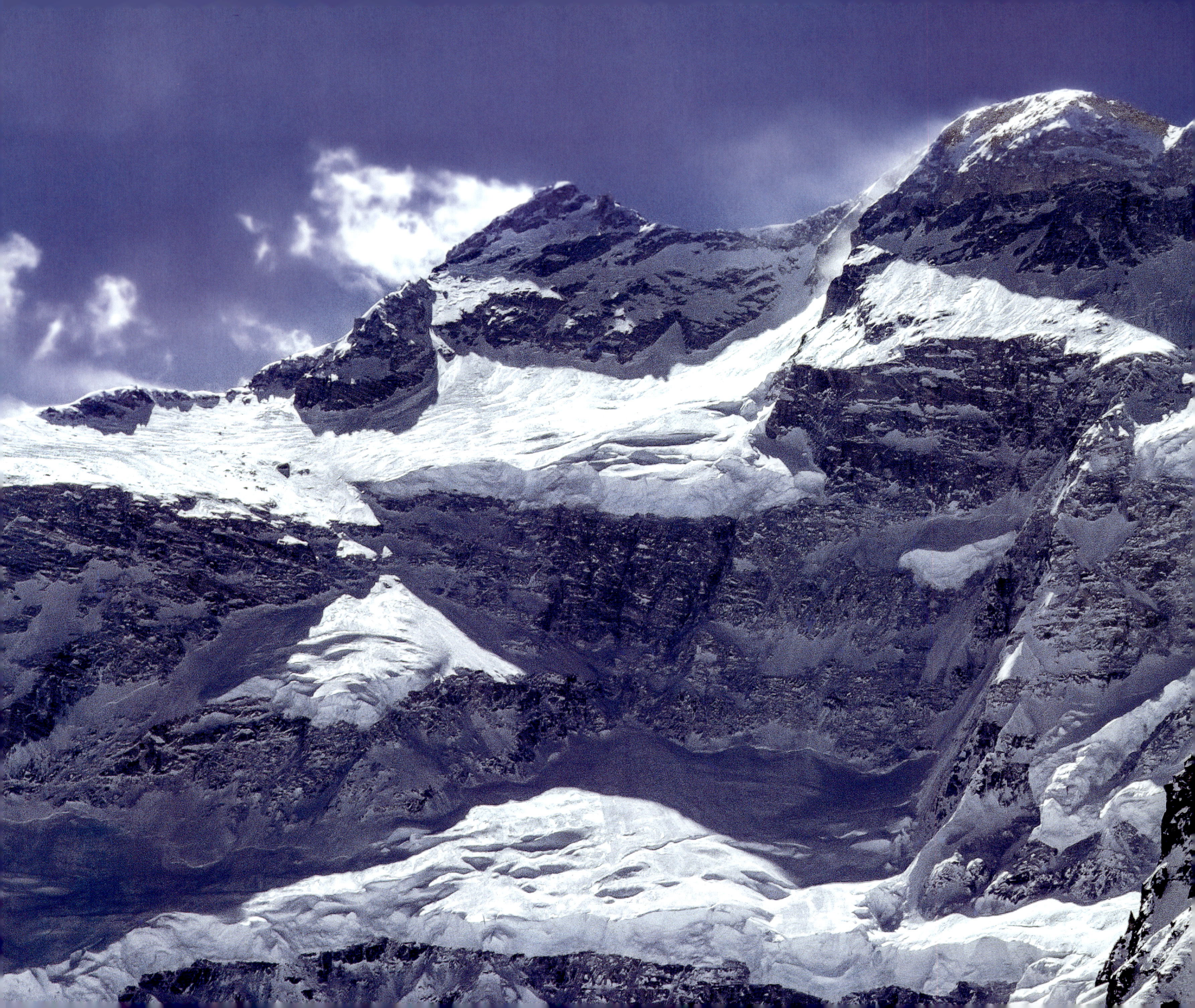

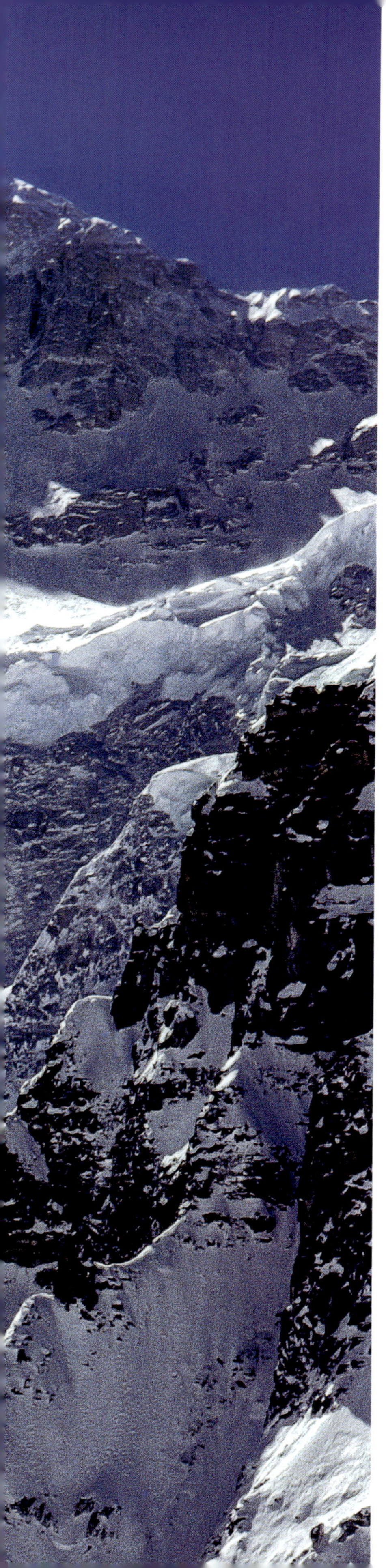

From Kangchenjunga to Everest, Nepal

OCTOBER–DECEMBER 1985

It can be bitterly cold in Nepal's high-mountain areas in the winter months, but, with the monsoon rains over, the views are crystal clear. We took advantage of these conditions starting in mid-October, spending fifty-four days hiking to the North and South Base Camps of Mount Kangchenjunga and from there all the way across the eastern half of Nepal to the Khumbu Region south of Mount Everest.

Mount Kangchenjunga, also known as "Kangchendzonga," is the third-highest peak in the world (elevation 28,169 feet). It sits on the border of eastern Nepal and India and is claimed by both countries. The mountain is a huge massif with sixteen peaks higher than 23,000 feet that can be seen clearly from miles away.

Our trek to the foot of Kangchenjunga began in the town of Ilam, not far from the border with the Indian state of Sikkim. Our crew for the two of us included Sila, our cook/guide, and six porters. In Kathmandu we outfitted everyone with shoes, jackets, sunglasses, and blankets, but they started complaining about the weather on the first day on the trail and often got lost or showed up to camp late in the evening. We needed them though, since there were absolutely no tourist facilities in the region surrounding Kangchenjunga in those days. Occasionally, we were able to obtain eggs and vegetables along the way, and once we bought a chicken that fed all nine of us, because every part, including its feet and comb, was used, but for the most part we relied on the canned food we carried from Kathmandu.

From Ilam we walked for nearly two weeks toward Ramser Lake through a green countryside of rice terraces, forests filled with rhododendron trees, and stands of bamboo. Kangchenjunga came into view on the third day, but we knew that the best views of the mountain lay days ahead from vantage points near the glaciers at its base. It was the multi-day

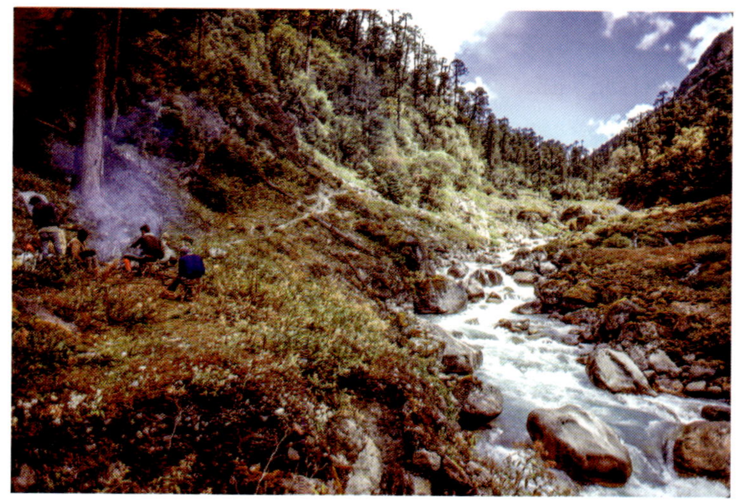

Camp in the Yalung Valley enroute to Mount Kangchenjunga, Nepal, 1985

Kangchenjunga North Face, from Pangpema, Nepal, 1985

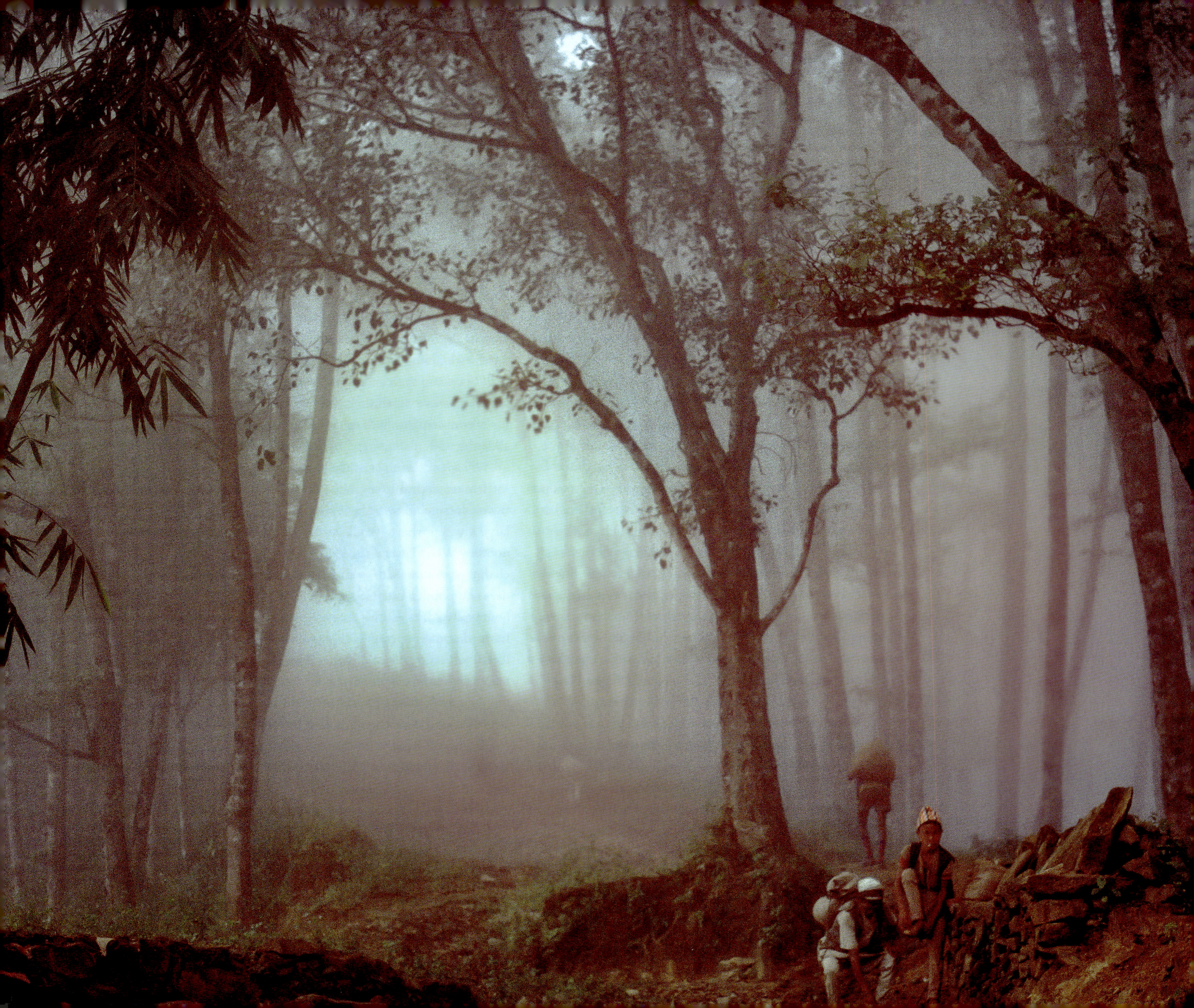

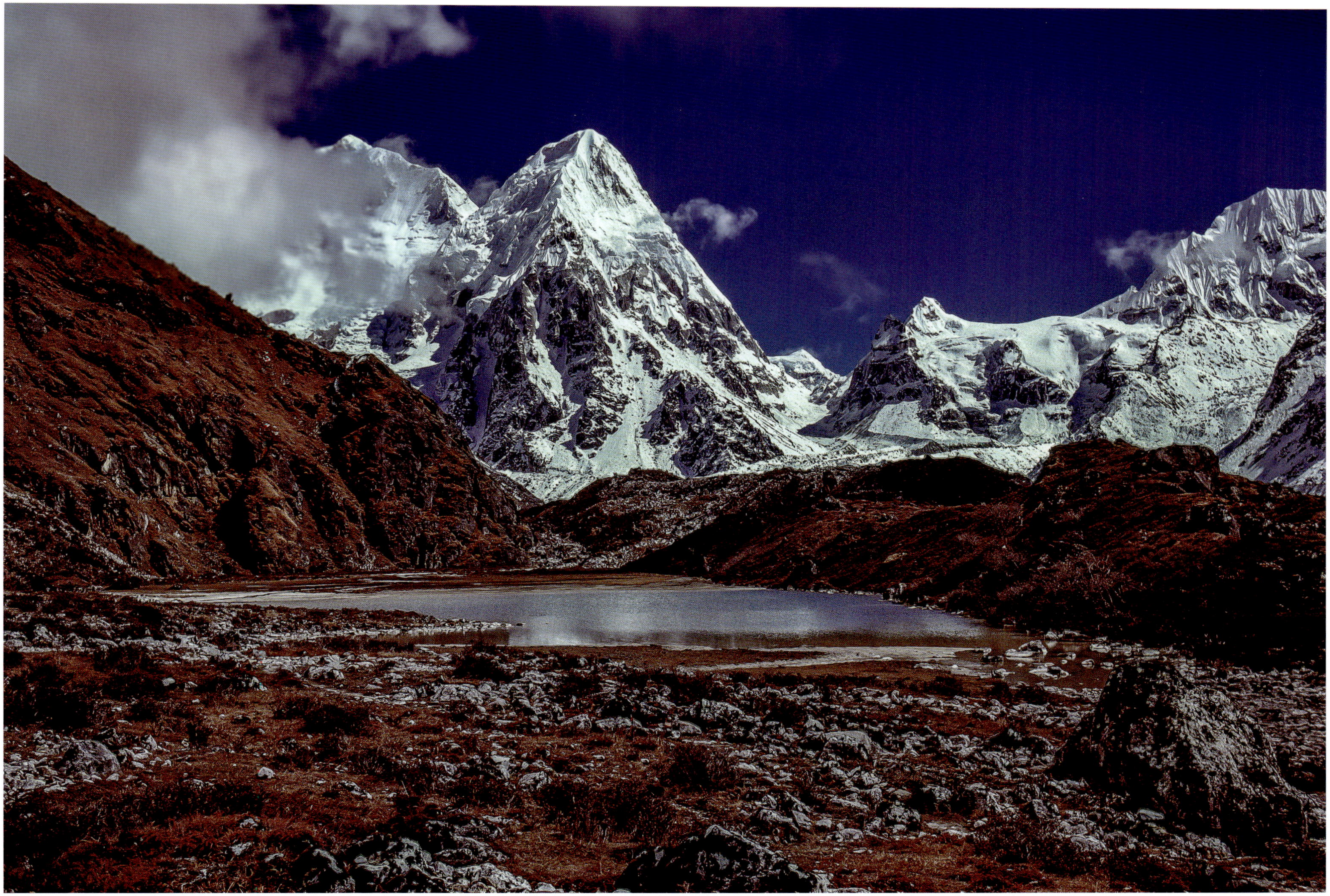

Yalung Glacier moraine, camp below Kangchenjunga, Nepal, 1985

Opposite: Start of trek to Kangchenjunga, Nepal, 1985

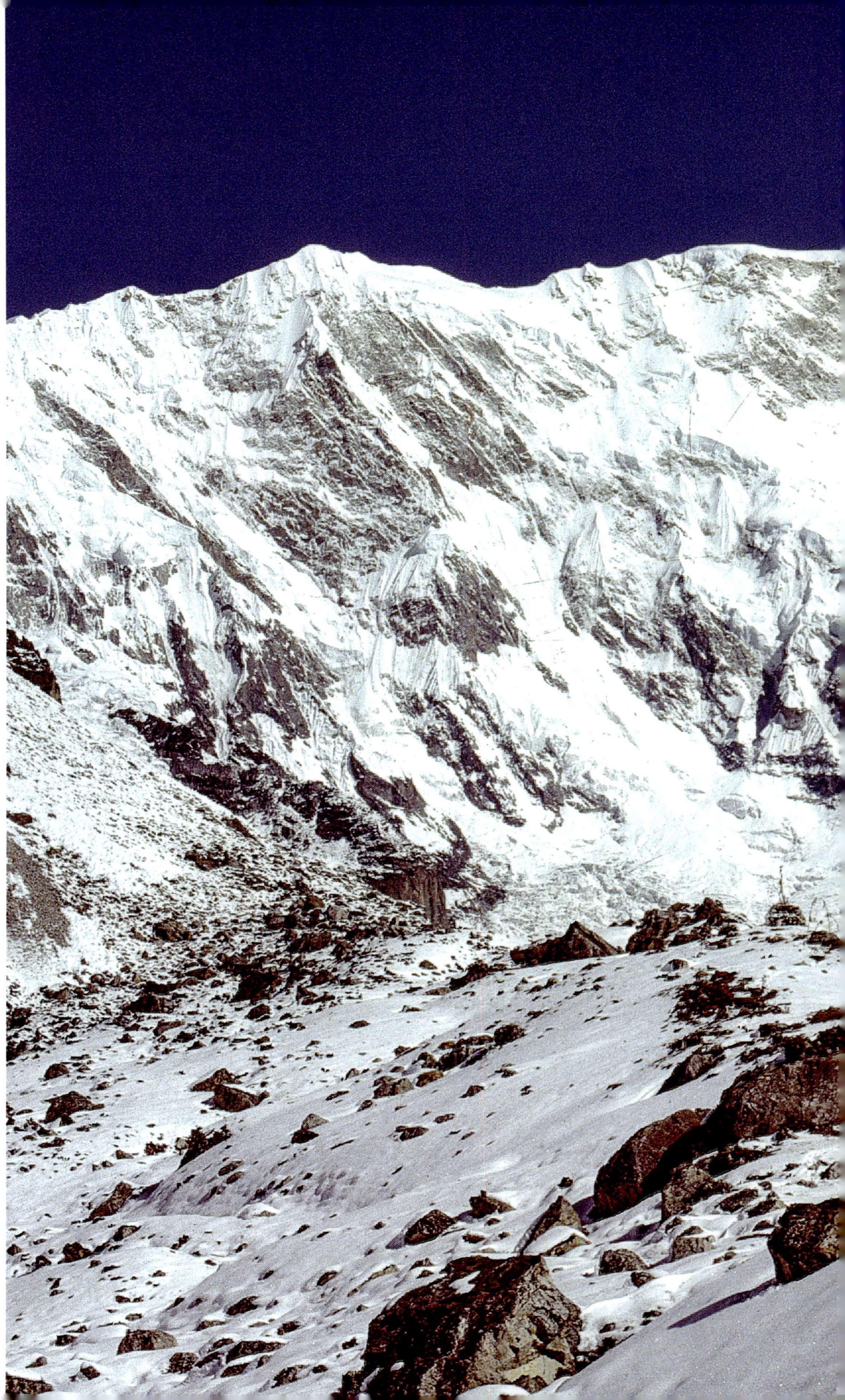

Kangchenjunga, Nepal, world's third highest mountain, 1985

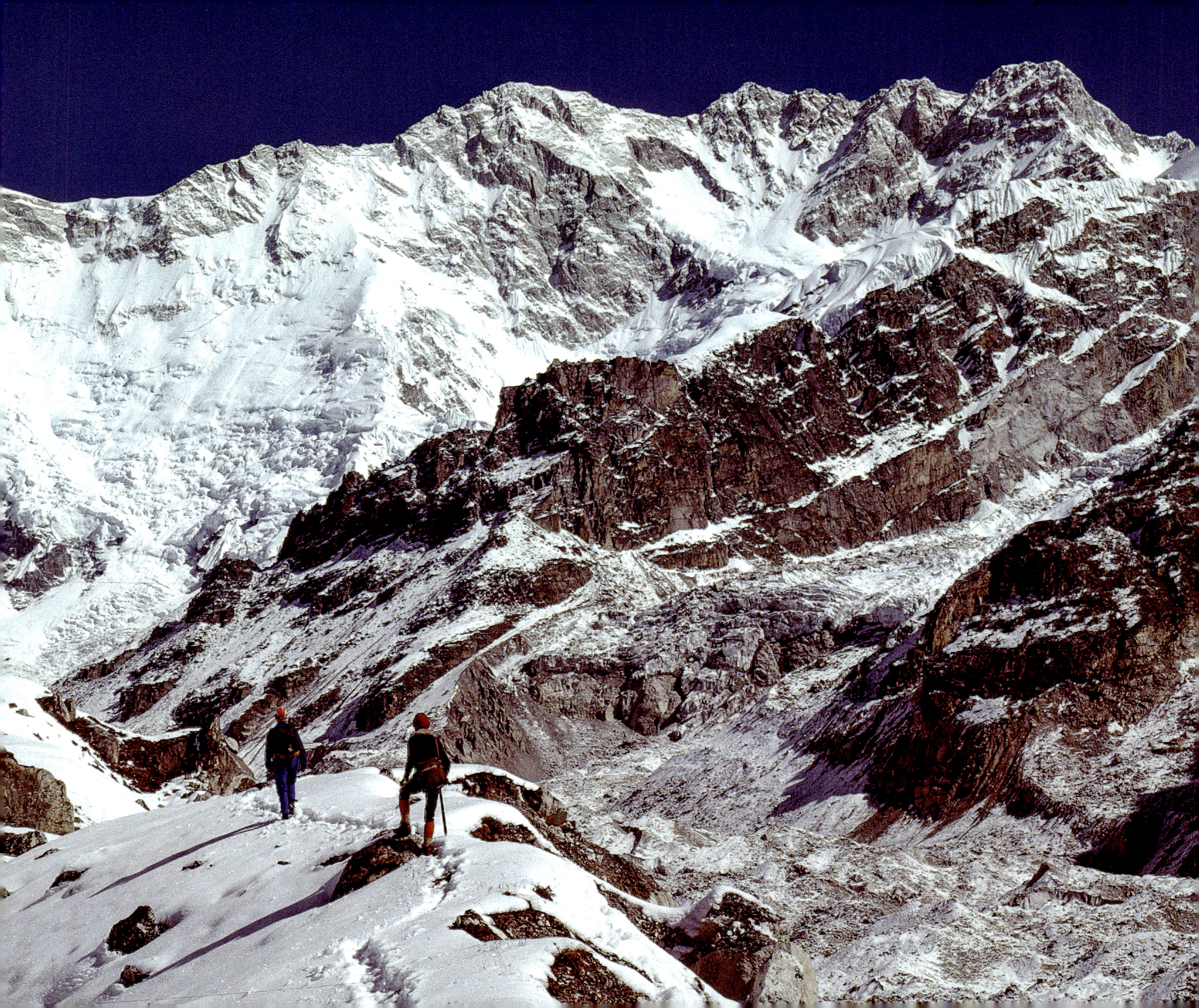

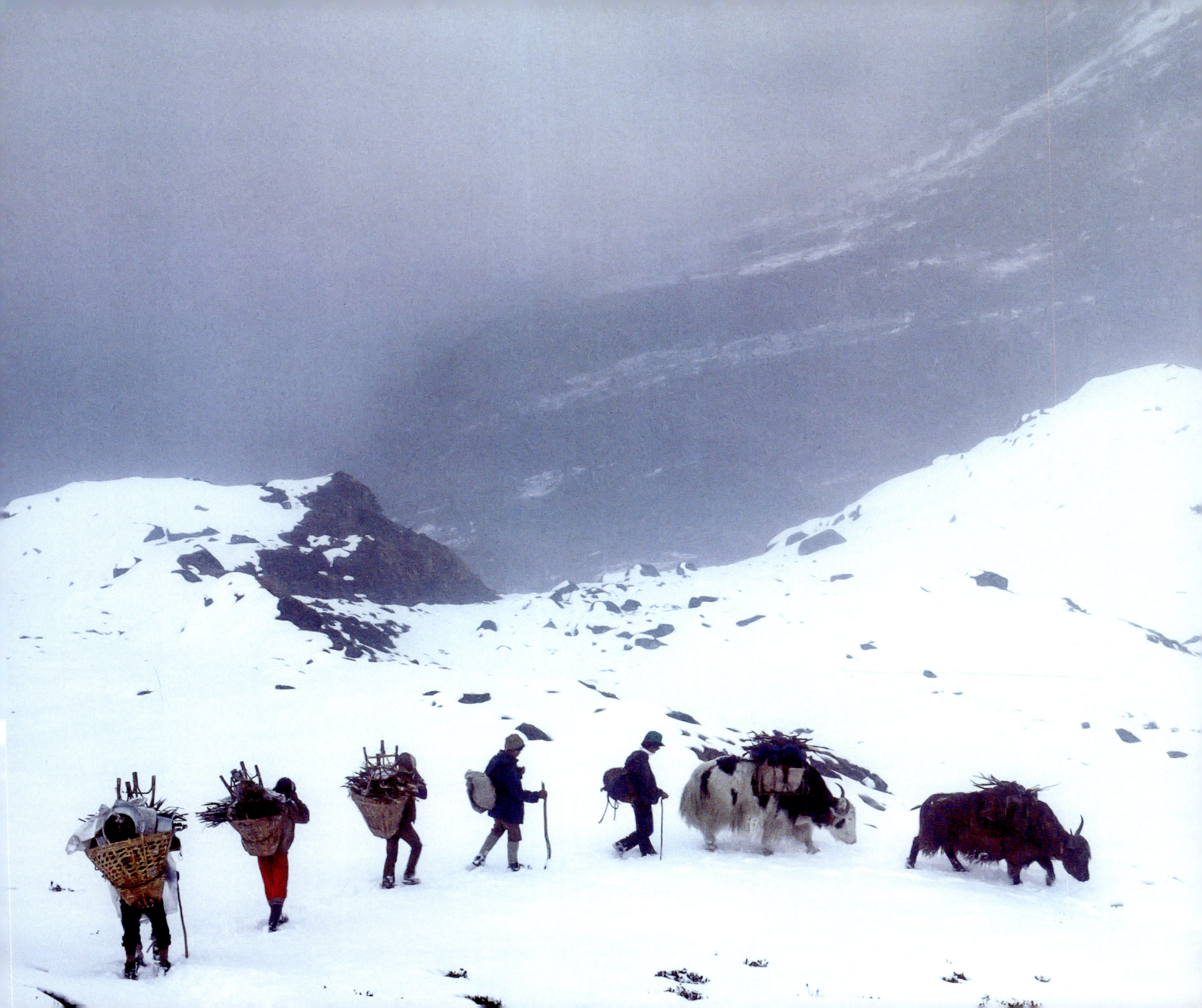

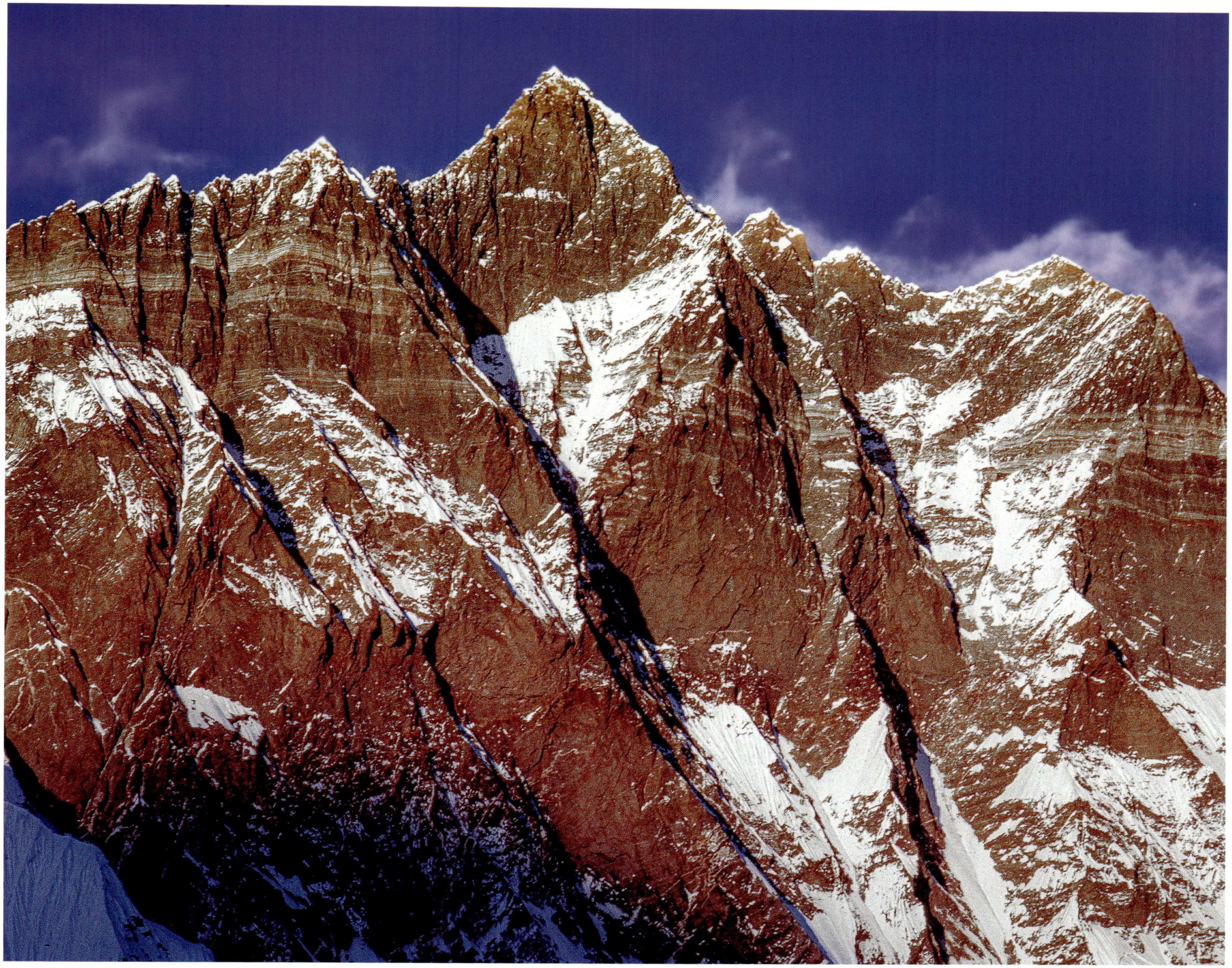

Lhotse, fourth highest mountain in the world, Khumbu, Nepal, 1985

Opposite: Our yaks and porters crossing Mirgin La, Nepal, 1985

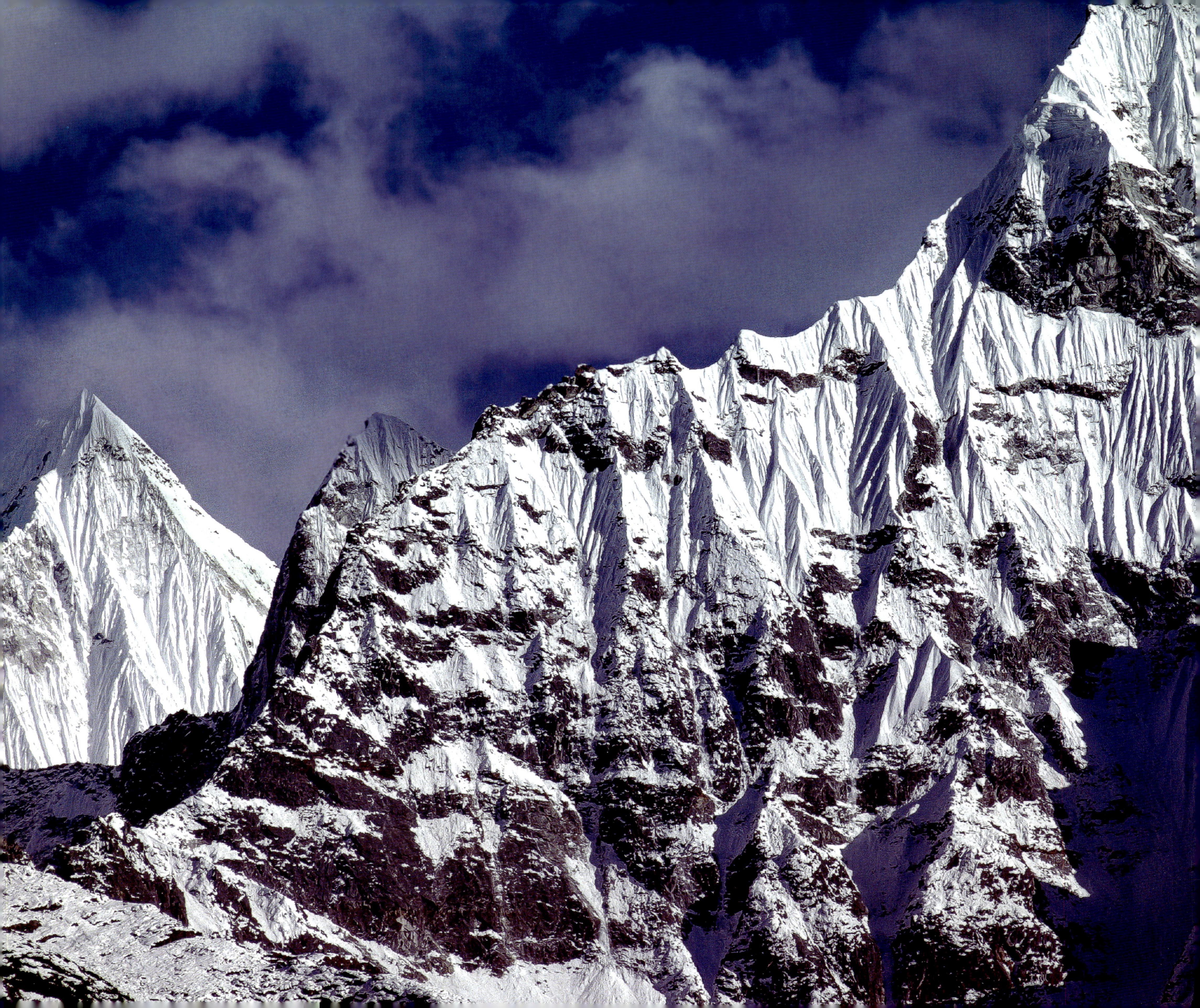

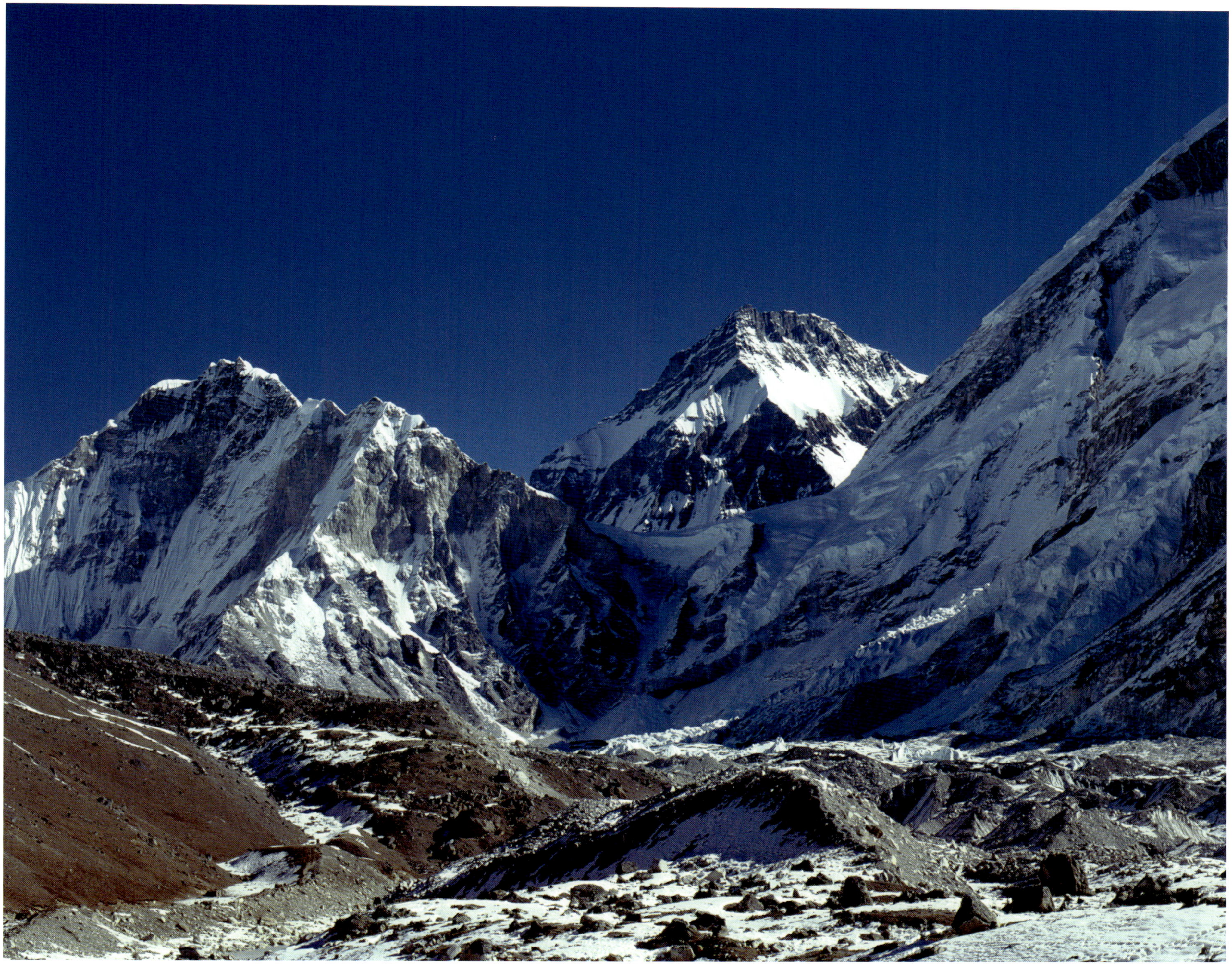

Above Everest Base Camp, Khumbu, Nepal, 1985

Opposite: Fluted peaks from Chhukung Ri, Khumbu, Nepal, 1985

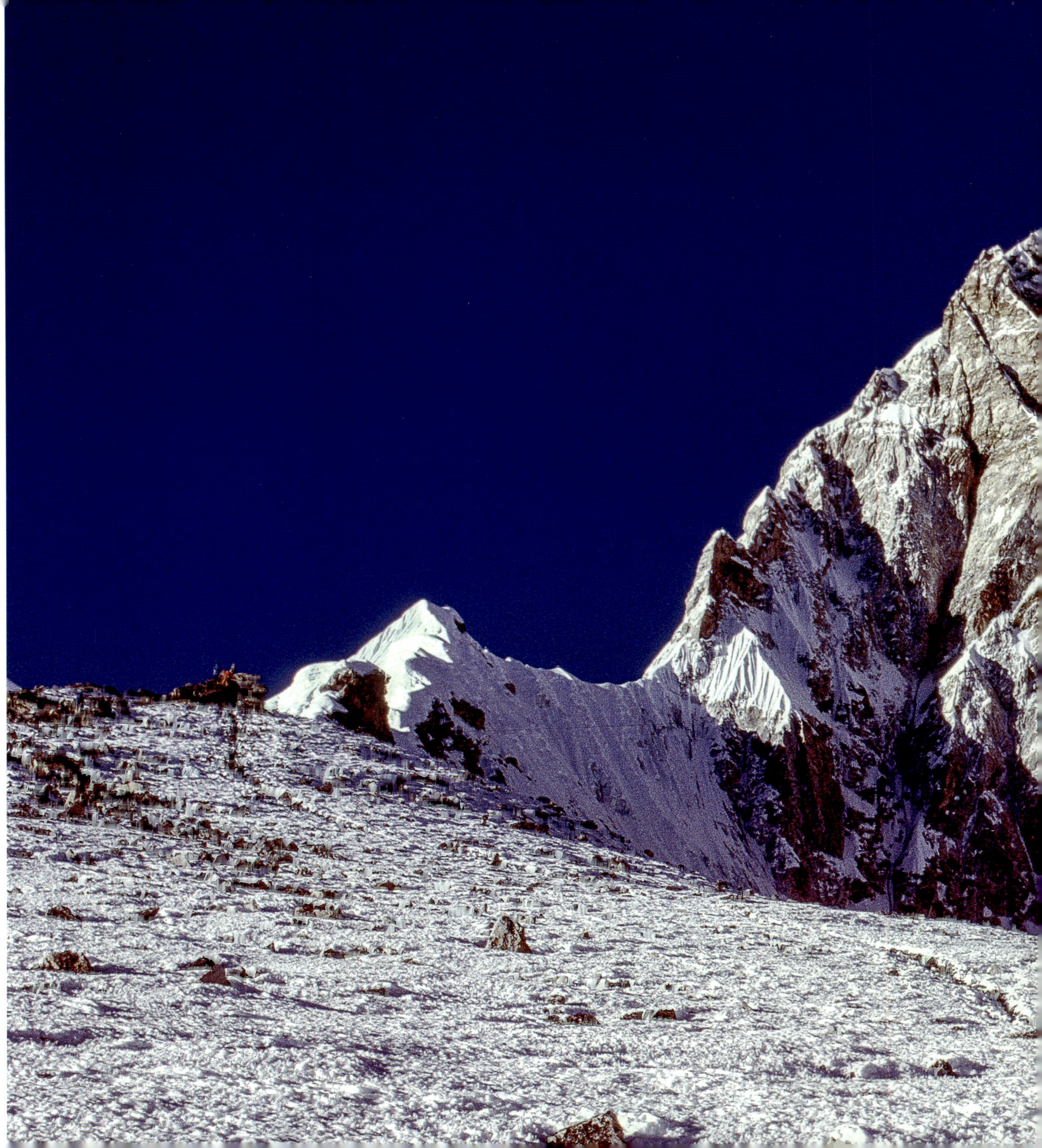

Anne below Pumori enroute to Kala Patthar, 1985

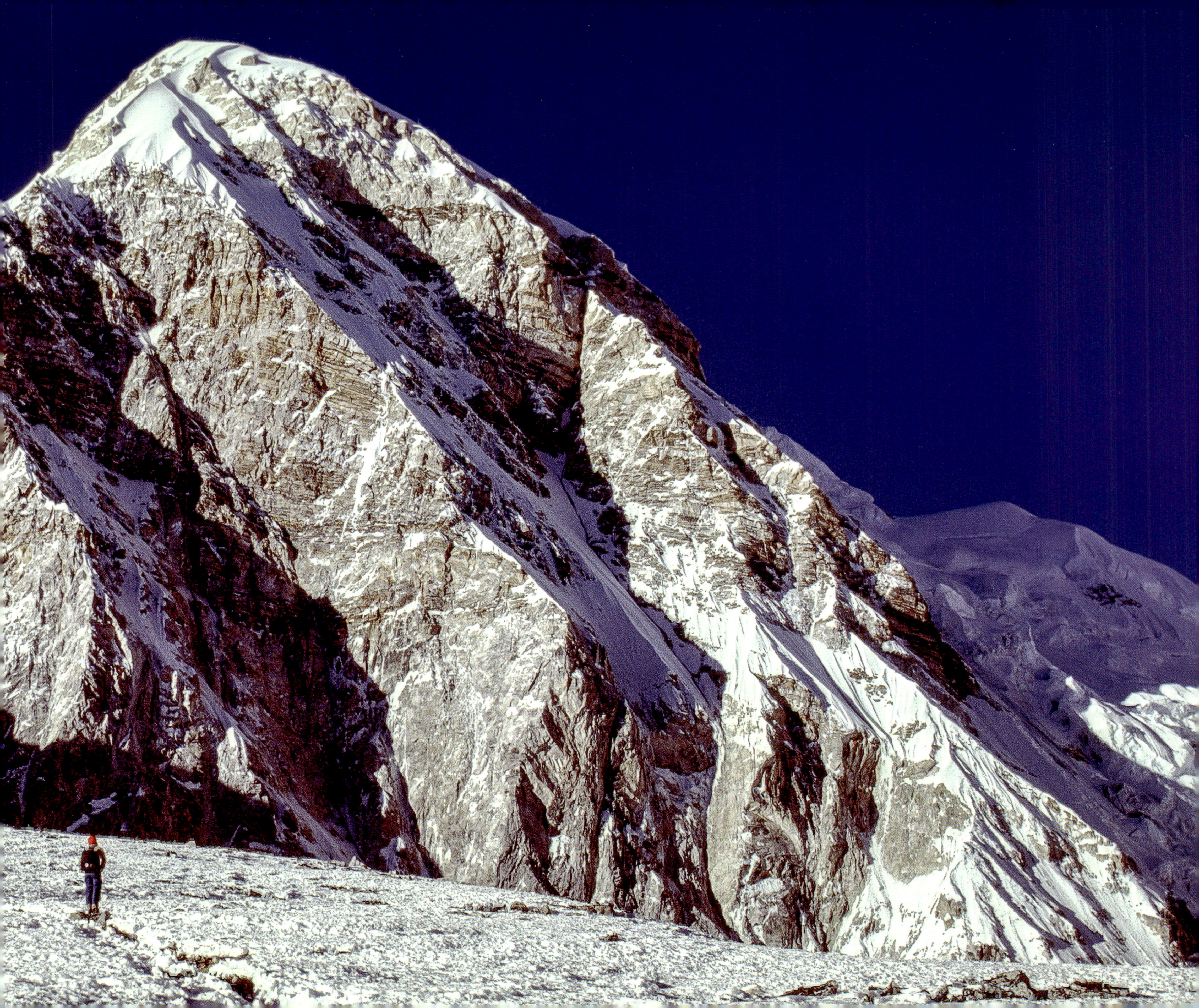

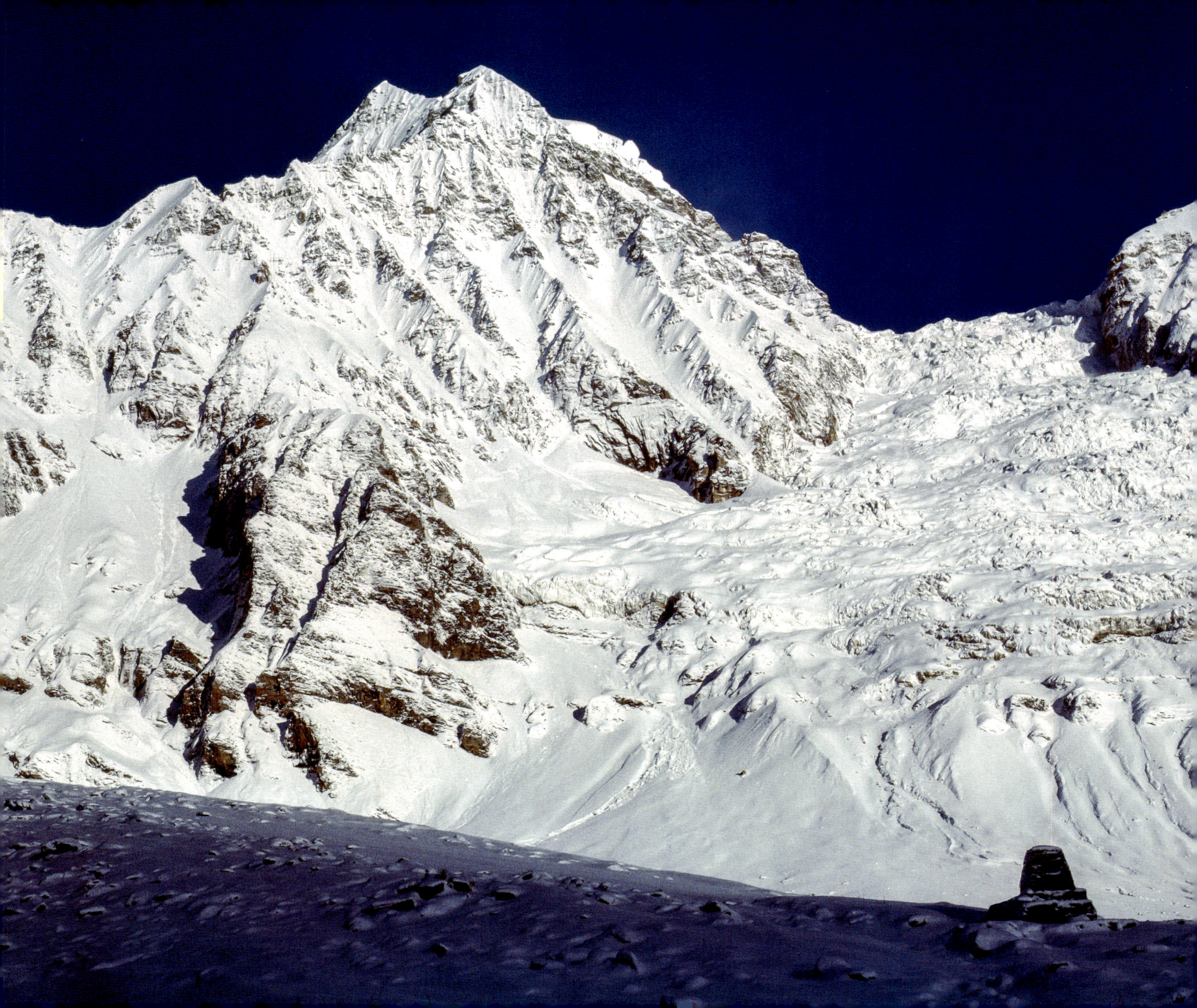

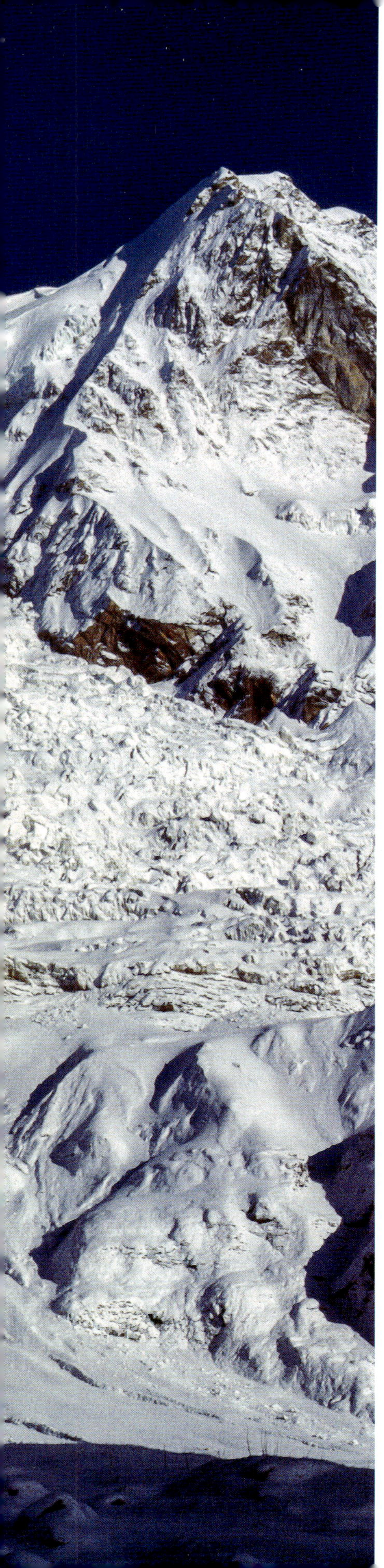

Below Dhaulagiri Icefall, Nepal

APRIL 1986

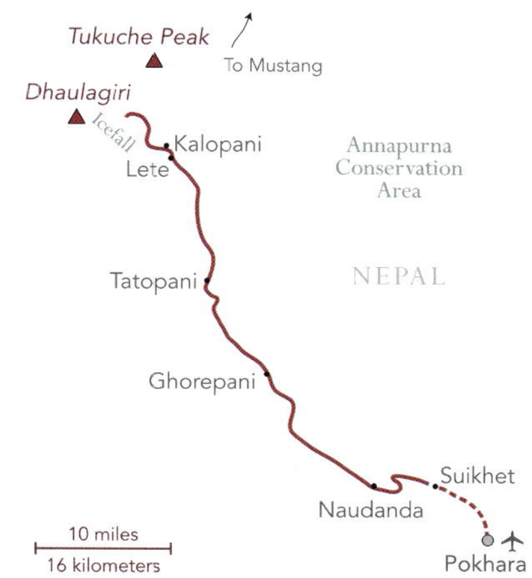

Mount Dhaulagiri, the world's seventh-highest peak (elevation 26,795 feet) is known for high winds and unpredictable weather that whip across its upper reaches. After seven unsuccessful attempts to reach its summit, eight members of a Swiss-Austrian-Nepali team made the first ascent on May 13, 1960. That expedition made the unique use of a small airplane dubbed the "Yeti" to ferry supplies and climbers from the town of Pokhara to a high pass on the mountain, which cut several days off the approach. Alas, after almost twenty daring high-altitude flights, the plane crashed on take-off. Fortunately, no one was injured.

For our short trek to the base of Dhaulagiri in April, we first traveled from Kathmandu to Pokhara by bus. We started walking in Suikhet and from there spent five days hiking the popular Annapurna trail and enjoying the comforts of trekker teahouses. Leaving the tourist circuit in Kalopani, we spent a solitary day hiking up to the camp below the Dhaulagiri icefall, a massive glacier at the foot of the mountain. We reached there in late afternoon and put up our tent. Immediately, the customary Dhaulagiri fog and snow rolled in, and we spent hours inside listening to the sound of avalanches on the mountain and reaching out the door every few hours to shake off the heavy, wet snow so the tent would not collapse on us. In the morning, the clouds broke and Dhaulagiri appeared for a few memorable minutes. The sight lived up to its name, which comes from Sanskrit and means "dazzling white mountain." Afterward, we headed down, struggling to find the trail now hidden under a foot of new snow. We twice ended up at the edge of cliffs and had to turn back but eventually found our way to the main trail. We stopped at a stone memorial to the five American and two Sherpa climbers who had died in an avalanche on the southeastern glacier of Dhaulagiri seventeen years earlier to the day. Two years later in Jakarta, Indonesia, where we now were living and working, we met the mother of one of the climbers at a friend's house. She was touched to hear that we had left a small bouquet of alpine grasses there in their memory.

Dhaulagiri and Dhaulagiri Icefall above the memorial to American and Nepali climbers, 1986

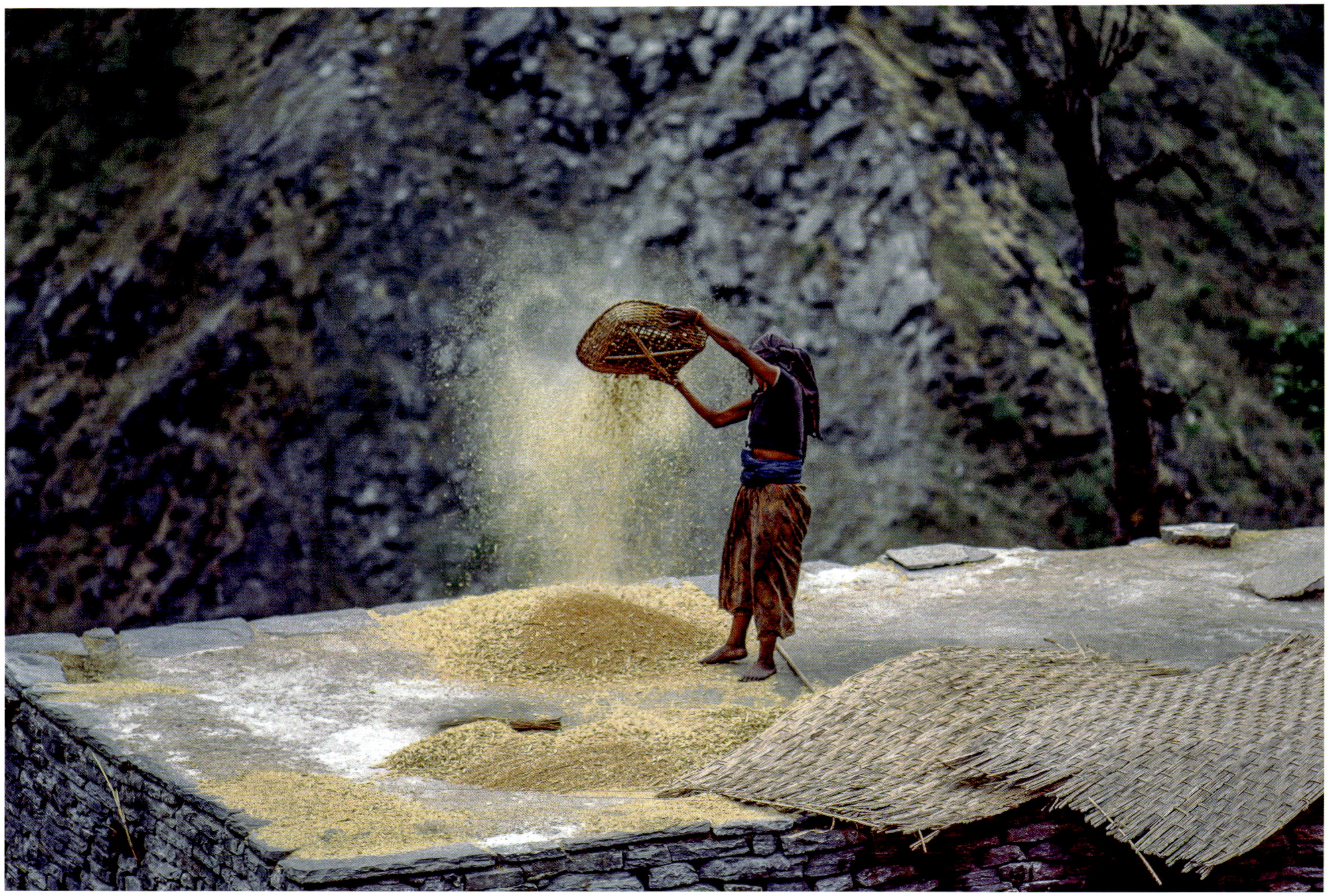

Sheathing rice above Kalopani, Nepal, 1986

Rice terraces enroute to
Dhaulagiri Base Camp, Nepal, 1986

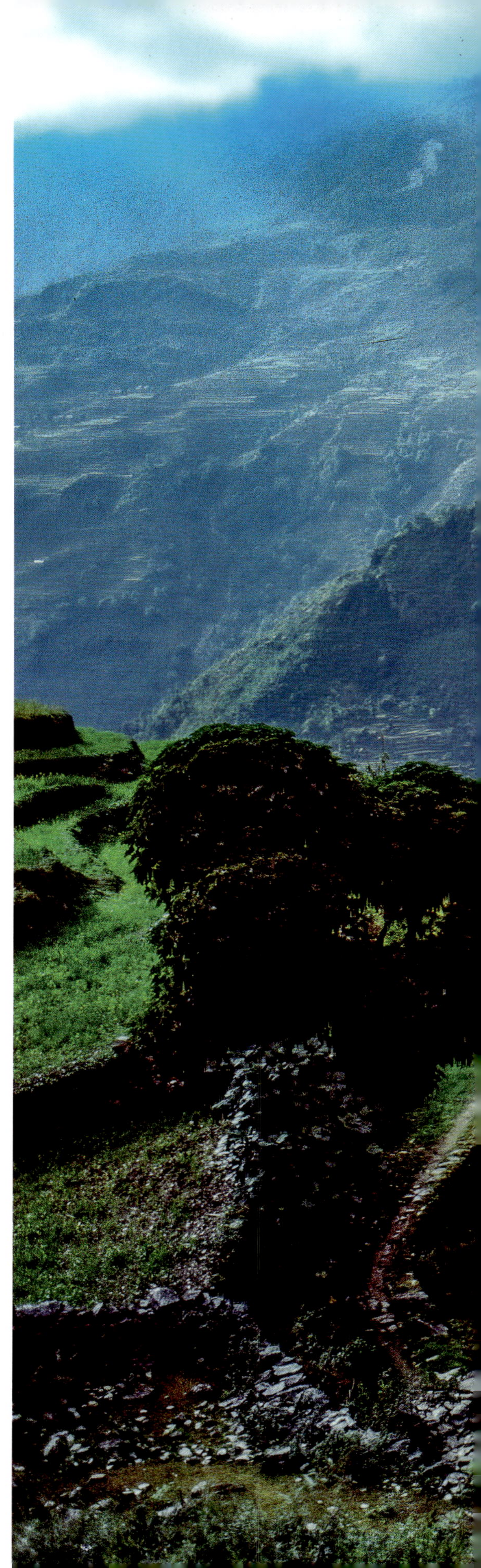

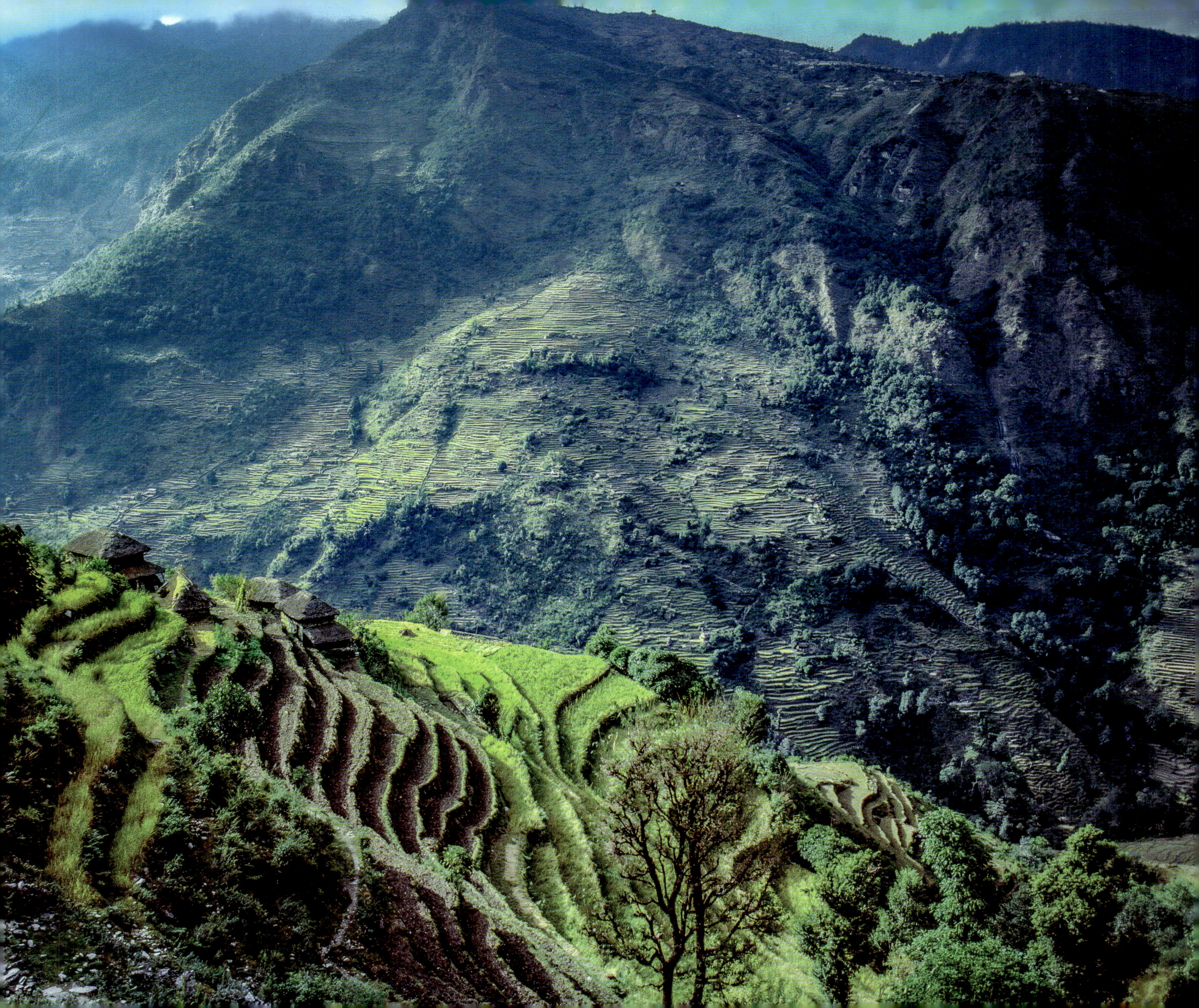

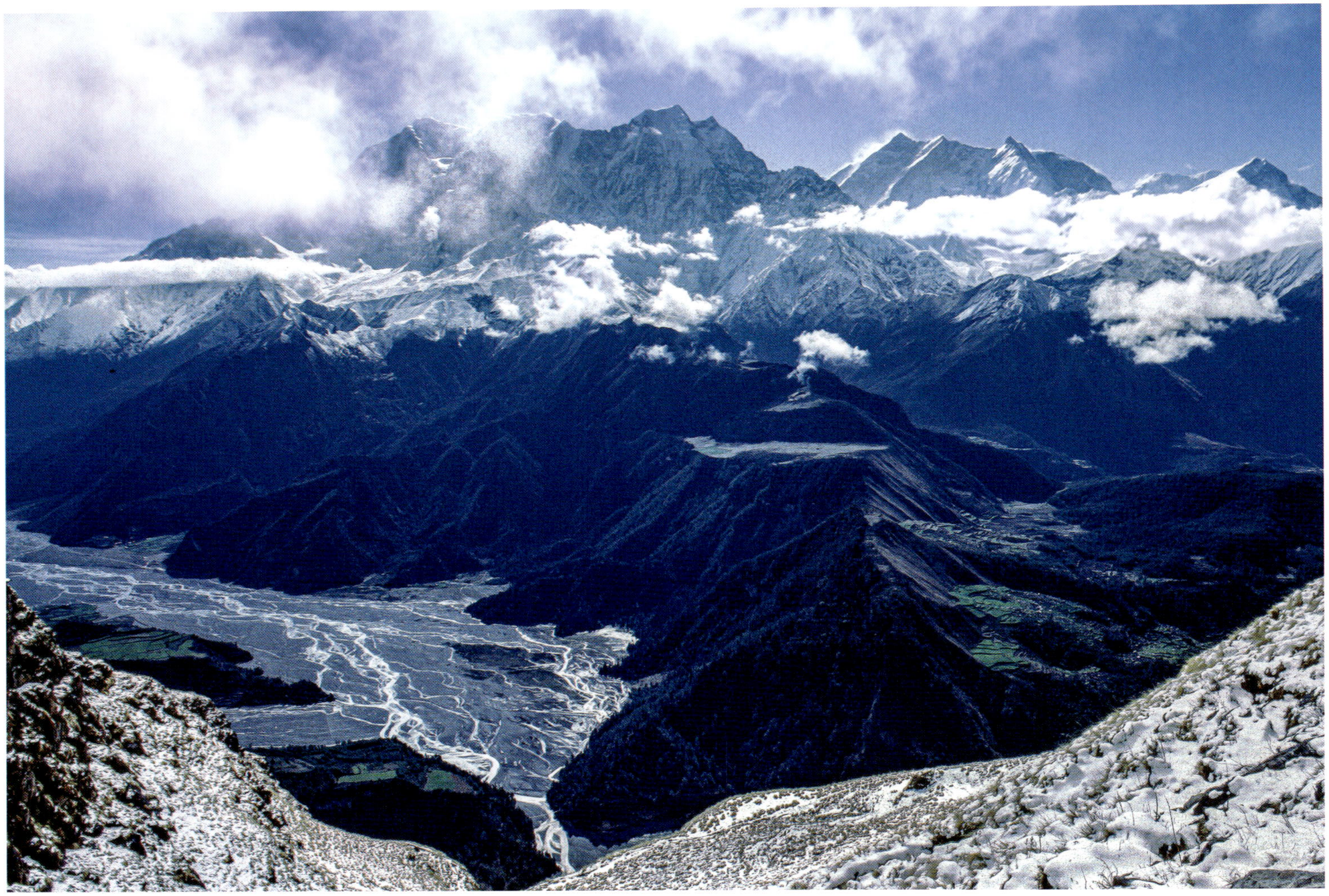

Above the Kali Gandaki River basin with views of the Annapurna Range enroute to Dhaulagiri, Nepal, 1986

Southeast ridge of Dhaulagiri, Nepal, 1986

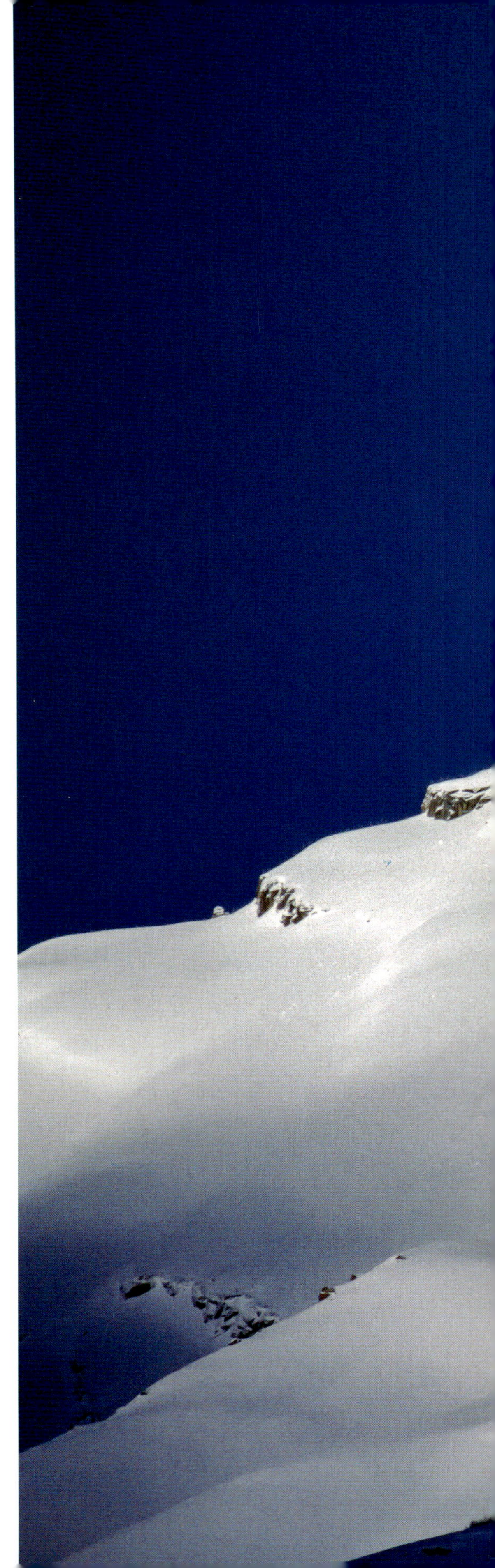

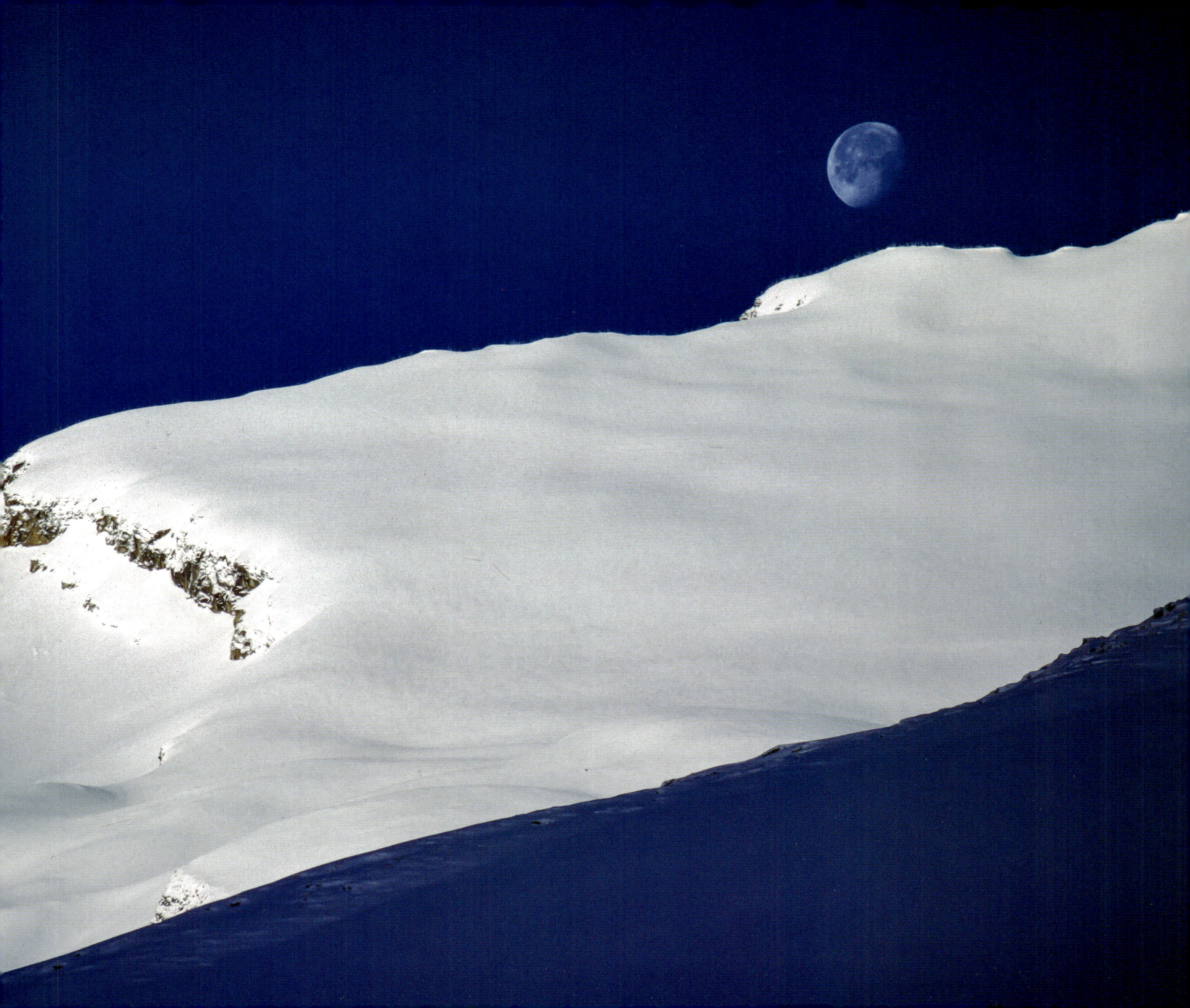

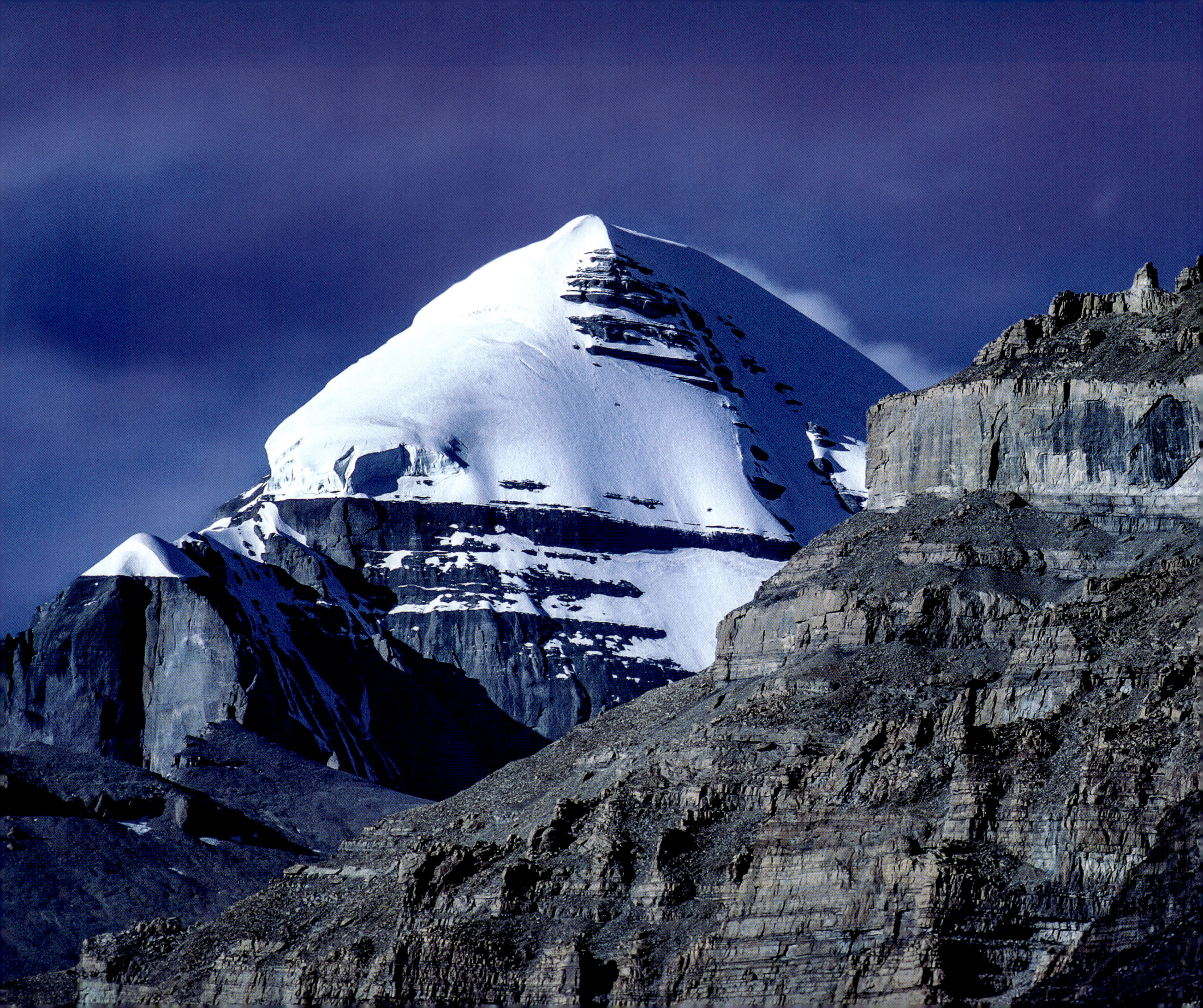

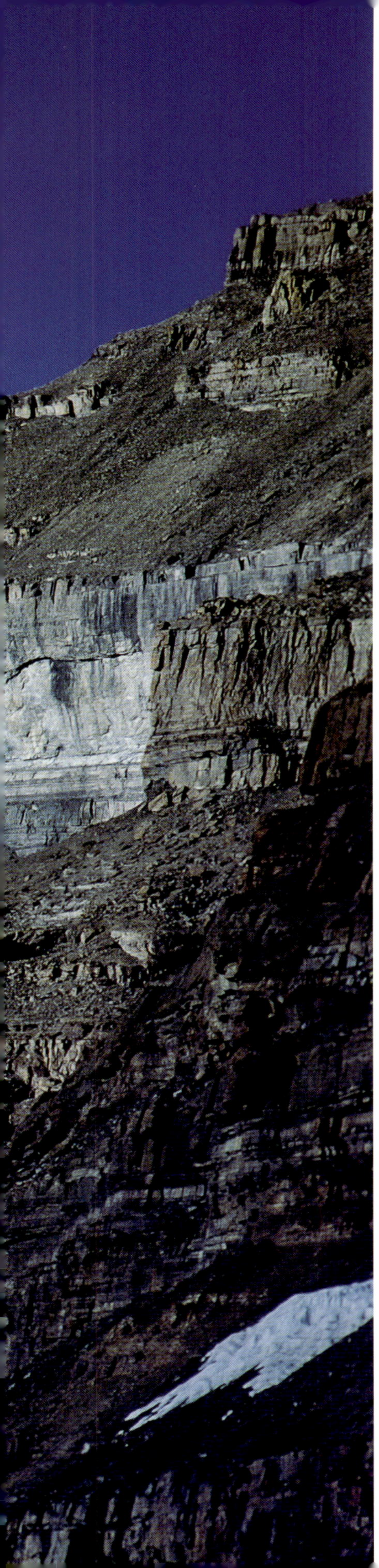

Across Western Tibet to Mount Kailas, Tibet

JUNE 1986

Mount Kailas, the apogee of our twenty-month trans-Himalayan journey, is a holy mountain in western Tibet revered by Hindus, Buddhists, Jains, and followers of the Bon religion. Mount Kailas is considered by many to be the mythical Mount Meru, the cosmic center of the universe. Tibetans know it as Kang Rinpoche and consider it to be the axis of the world. Hindus consider it the holy abode of Lord Shiva, Jains call the mountain Astapada, and followers of Bon, Tibet's pre-Buddhist religion, call it Tise. A walk around the mountain, known as a *kora* or *parikrama,* is considered an act of deep respect, with each step a prayer. Some pilgrims do more than one kora; thirteen is considered especially auspicious. And some even do prostrations for an entire circuit, an act of reverence and respect that can take more than a month.

Mount Kailas (elevation 21,778 feet) is less than 100 miles north of the border in far western Nepal. During the mid-1980s, it was almost impossible, except for the most daring, to cross the border there because it was officially closed and government officials took a very dim view of foreigners attempting this shortcut. So we were forced to choose a longer and more complicated approach that required a flight from Beijing to Lhasa via Chengdu as the first leg and then a 700-mile trip overland by truck across the Tibetan Plateau. In those early days of travel in China, after severe restrictions on foreign tourism were loosened during the late 1970s, the planes and hotels were filled with a jumble of budget travelers, international business people, and newly affluent Chinese citizens. On the flight from Chengdu to Lhasa, the Tibetan woman seated next to me unwrapped the packet of butter on her meal tray and carefully spread it on her face as a moisturizer. Others boarded with burlap bags of plants and herbs that they placed squarely in the middle of the aisle.

In Lhasa we spent two weeks acclimatizing to the high elevations that we would encounter on our trip across the Tibetan Plateau to Mount Kailas. Lhasa, the capital of Tibet, sits at almost 12,000 feet; elevations throughout the Tibetan Plateau average nearly 15,000 feet Each morning we walked to the market for breakfast provisions of yak yogurt and canned peaches, and most days we joined the throngs of Tibetans walking clockwise around the sacred Jokhang Temple in the center of the city and holding spinning prayer wheels in their hands. We visited the Potala Palace, previously the home of the Dalai Lamas and now a museum, and the Norbulingka, the Dalai Lama's summer palace. Most of our fellow visitors were Tibetans who surreptitiously placed photographs of the beloved 14th Dalai Lama, their spiritual leader, and small amulets into niches in the rooms we walked through. We spent afternoons drinking green tea at our

Mount Kailas, western Tibet, 1986

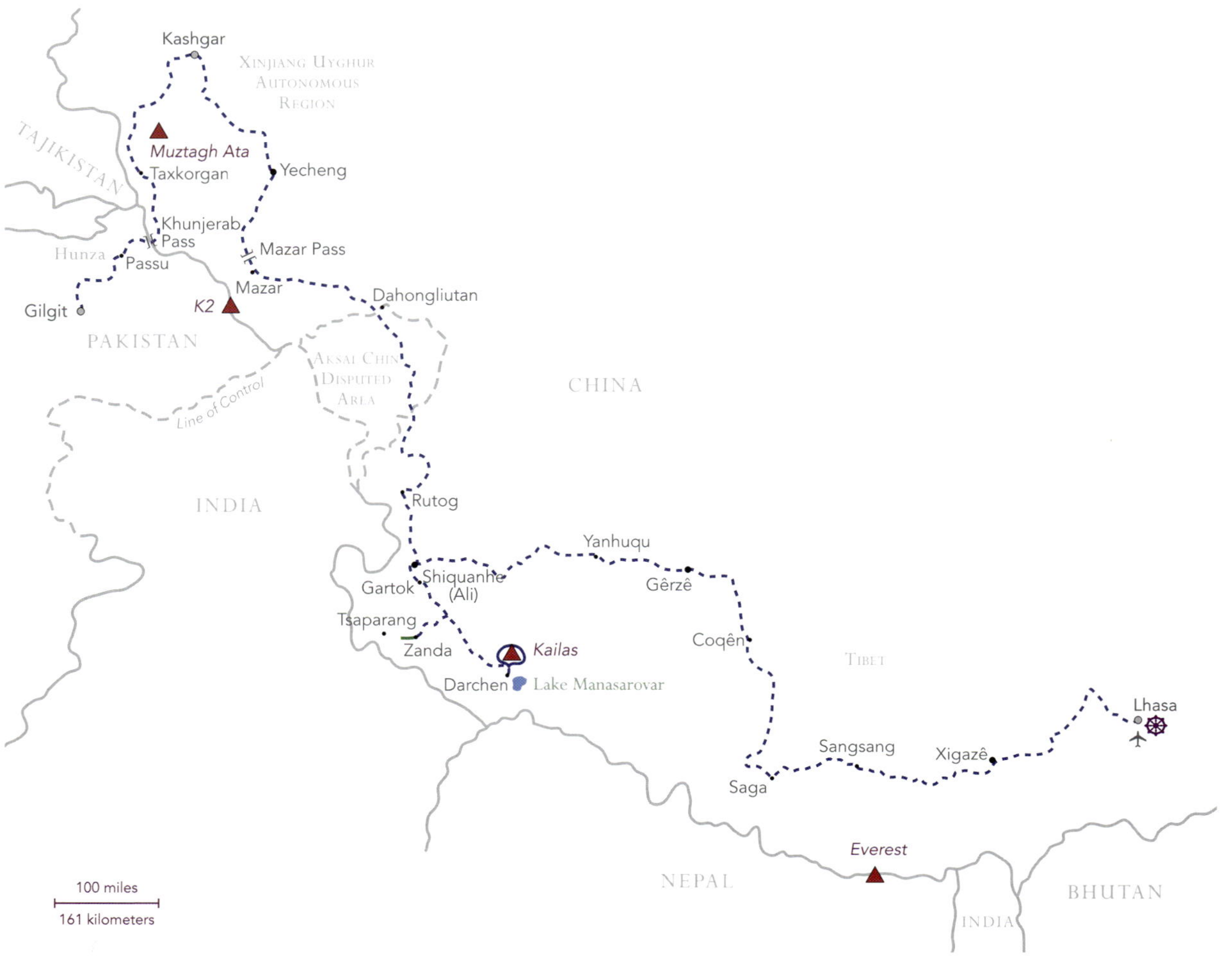

favorite Muslim teahouse and ate our evening meals at one of the newly minted restaurants where we picked ingredients from bowls on a table in the kitchen and watched as everything was stir-fried in a wok over a fire.

Tibet had just opened to tourists a few years earlier, and the tourist facilities were basic. The only modern hotel in Lhasa was a Holiday Inn outside the city center, to which we rode bicycles to eat yak burgers in an upscale setting. Most of the tourists then were backpackers like us who stayed in simple, rustic accommodations. Today, Lhasa has a wide variety of international standard hotels and restaurants, and the annual number of foreign and domestic tourists to Tibet has grown to more than 5,000,000 (official figures of a much higher number are considered unreliable). The completion of the Qinghai-Tibet Railway in 2006 greatly facilitated access to this once-closed region from all over China and has contributed to the increasing numbers of visitors. At the same time, travel within Tibet has become much more regulated by the Chinese government. It is now necessary to join a group tour and obtain travel permits far ahead of time. The days of individual travelers freely hitchhiking through the countryside are over.

In 1986, there was almost no information on what lay beyond the edges of Lhasa. For information on Mount Kailas we relied on advice from Hugh Swift, who had been there the year before, when it first opened to tourists, and on *Kailas Manasarovar*, a book by Swami Pranavanda, originally published in 1949 and republished in 1983. The swami was a Hindu holy man who began making pilgrimages to western Tibet during the late 1920s. Not only did he travel extensively throughout the region, but he completed twenty-three circumambulations of Mount Kailas and twenty-five of nearby Lake Manasarovar. Armed with the notes from Hugh and a photocopy of a map from the book by Swami Pranavanda, we set out to find a ride across western Tibet to Mount Kailas.

We had no luck finding a ride in Lhasa, so we took a bus to the town of Shigatse (now officially spelled Xigazê) where we figured there would be more trucks filled with cargo heading west that might have room for us. After a day of

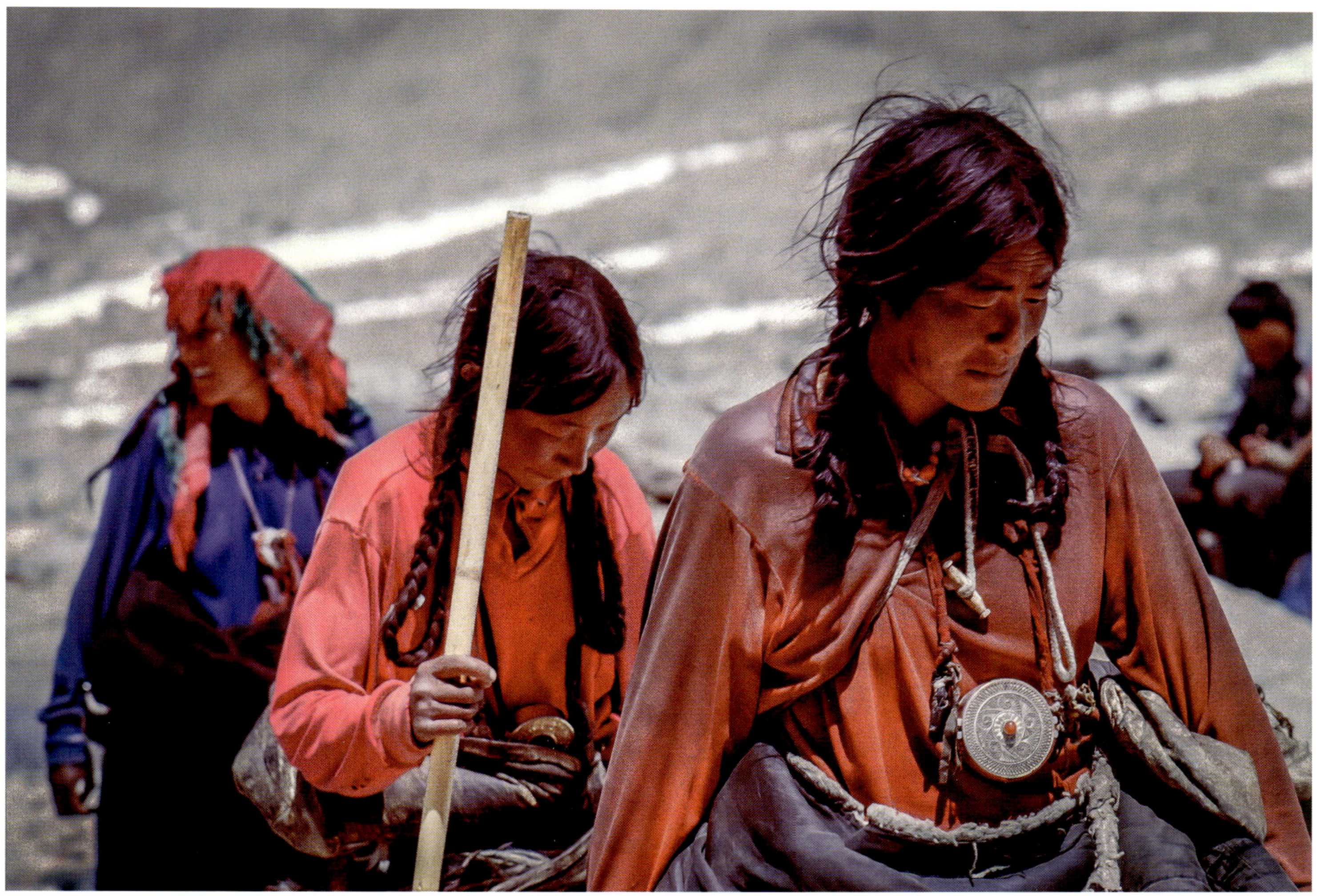

Tibetan pilgrims circumambulating sacred Mount Kailas, western Tibet, 1986

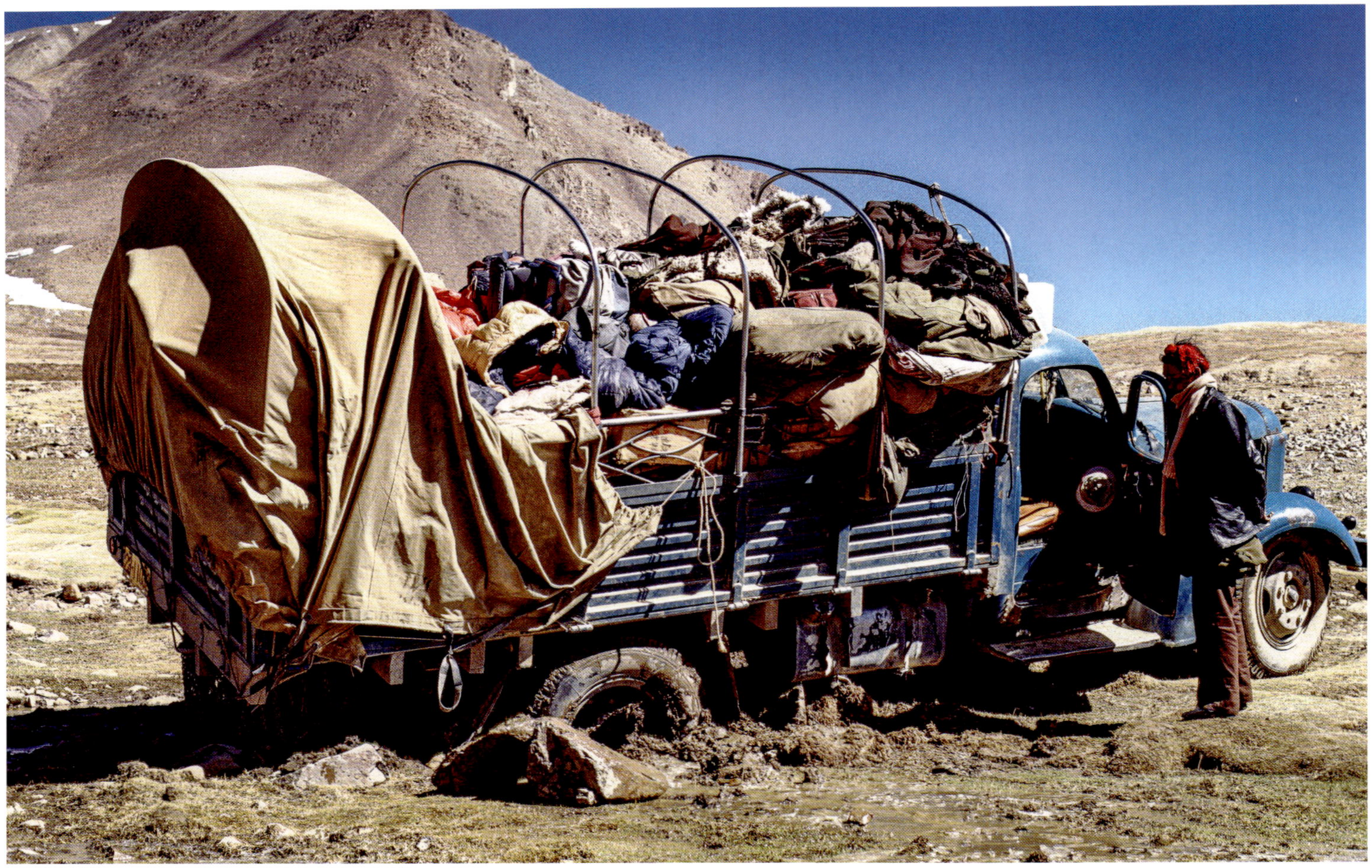

Stuck in the mud again, western Tibet, 1986

scouting truck stops, we found a driver who agreed to take us on the expected week-long trip to the town of Ali, near Mount Kailas. We joined a group of Tibetan men riding in the back of the truck and thinking that they were Buddhists on a pilgrimage but soon found out that this unholy group of traders was hoping to sell religious items to nomads and pilgrims along the way. Bill and I fought with them constantly for space in the back of the truck and waited with irritation while they hawked their wares at every spot where even a few nomads were gathered. At one stop we met another driver who agreed to take us in his roomier vehicle, but, as we grabbed our duffel bags to move them, our original driver, who had not yet been paid, punched me and knocked my glasses off and held a dagger to Bill's throat. With reluctance we re-boarded the back of our original truck and resigned ourselves to spending the remainder of the long road trip there.

Over the next week and a half we shared with our unruly fellow passengers the daily trials of getting stuck in the mud, unloading the truck to lighten it, and then reloading it again after getting back on the road. Some nights we slept in our sleeping bags on the ground alongside the truck and drank hot tea and soup from a pot heated with a blow torch since there was no available firewood. Other nights we slept in rough truck stops with primitive restaurants serving oily noodles. Westerners traveling in this area were rare in those days, and we sometimes attracted a small crowd. At one place two old women peeked inside our room and bowed, as if it were a shrine. We gave them photographs of the Dalai Lama, which they shyly but eagerly accepted, tapped them to their foreheads, and then tucked them deep inside their jackets before backing out of the room.

In a small place called Dumtso, our overnight stop coincided with a horse race festival. Handsome nomads with long, black hair braided with red yarn competed in the races and eagerly bought the beads, butter churns, and sneakers offered by our traveling companions. A Chinese veterinarian we met there told us a sad story in simple English of how he had ended up in Dumtso and then surprised us with a generous treat. He led us to a room next to his office, unlocked a big padlock, and informed us we could take as many of the dried goat legs stacked up there as we wanted to supplement our meager food supplies. We had exhausted most of the canned food we brought from Lhasa and now were on a diet of tsampa mixed with hot tea or water so we took several legs and enjoyed every bite.

The overland trip to Ali took longer than expected because the southern road was washed out in sections, which meant we had to drive north to the only other east-west highway in the region and then work our way south again. When we finally made it to the town of Ali (now also called Shiquanhe), we found no accommodations available in that frontier outpost, so we sneaked into the unfinished rooms of a hotel under construction and used water from a nearby stream for washing. Late one night Chinese authorities in trench coats arrived and didn't say anything but drove us out by turning our overhead lights off and on and drilling in the next room. Our next refuge was the floor of a room rented by LanDing, a young Chinese school teacher we had met in town. Mice scampered over us all night, and the room was stifling. When we tried to open the only window, the frame and glass fell out, leaving a big hole in the wall.

We spent four days in Ali working to obtain a government permit for onward travel and searching for a ride to Mount Kailas. People told us that Mr. Lee was the official responsible for issuing travel permits, so every morning we checked into a dusty government office and waited patiently for him to appear while a taciturn Chinese woman cleaned the room around us. On the fourth day we walked in and saw her sitting behind the big wooden desk with a pile of papers, which she signed and stamped and pushed across the desk to us, this time with a sly smile.

Our plan was to travel to Mount Kailas via the ancient cities of Tholing and Tsaparang in the former Kingdom of Guge. Guge is little known today, but it grew in importance after the dissolution of Lhasa as the capital of Tibet during the ninth century CE. By the eleventh century, it was the center of Buddhist culture, with Tholing and Tsaparang alternately serving as the capital. The large religious complexes in the two cities were filled with magnificent murals and statues. The religious and military influence of the kingdom was felt by inhabitants throughout this vast region for several centuries, but by the seventeenth century a series of events—including religious strife, invasion by the neighboring kingdom of Ladakh, and increasing land desertification—led to Guge's downfall. Earlier visitors, including Father António de Andrade, the Portuguese Jesuit priest who visited there in the early 1600s, had described lush fields and irrigated gardens, but today the area is dry and windblown. For the past four centuries the once-grand Kingdom of Guge has been largely unpopulated and forgotten.

We left Ali by jeep early in the morning before the sun came up and arrived at Tholing, now called Zanda, by evening. The monastery complex there, built around 1,000 CE, was once the most important seat of learning in the region. The building we visited was stripped of exterior details, but it was still imposing with its tall mud walls. Inside there was little light, but, squinting, I could see vivid wall murals depicting what I recognized as Tibetan Buddhist religious scenes. On one wall I could see the faint outline of a destroyed Buddha figure. The ground below was scattered with small Buddha amulets molded in terra cotta. I pocketed one and took it back to our room at the local hotel but awoke in the middle of night after a dream told me to return it. I slipped out without waking Bill and ran down the deserted street to the monastery, where I laid it back down with the others.

Without a map or guide we set off walking the next day to find Tsaparang, reputed to have even grander temples with murals and exquisite statues somehow still intact in spite of the destruction unleashed by China's Cultural Revolution, which reached even this remote corner. Guiseppe Tucci (1894–1984), the Italian professor and author, visited the area during the 1930s and reported on the remaining but precarious frescoes in his book, *The Temples of Western Tibet and Their Artistic Symbolism* (1989). During the late 1940s, the German-born monk, Lama Anagarika Govinda (1898–1985), and his wife, Li Gotami Govinda (1906–1988), made a pilgrimage to Tholing and Tsaparang and spent three months documenting the interiors of the temples there through photographs and sketches. The resulting two-volume *Tibet in Pictures: A Journey into the Past* (1979) by Li Gotami provides an important photographic account of the treasures found there.

Walking across the wide valley separating the two former cities, we passed hundreds of chortens made of mud and plaster that blended into the surrounding landscape of sandstone walls molded by years of wind and desiccation. We searched for Tsaparang for five hours but were unable to find a trace. The heat was brutal, and the only available water was from irrigation ditches. Reluctantly, we turned around to get back to the town before dark. Today, there is a road from Zanda to Tsaparang, and it takes about an hour to travel between them by car.

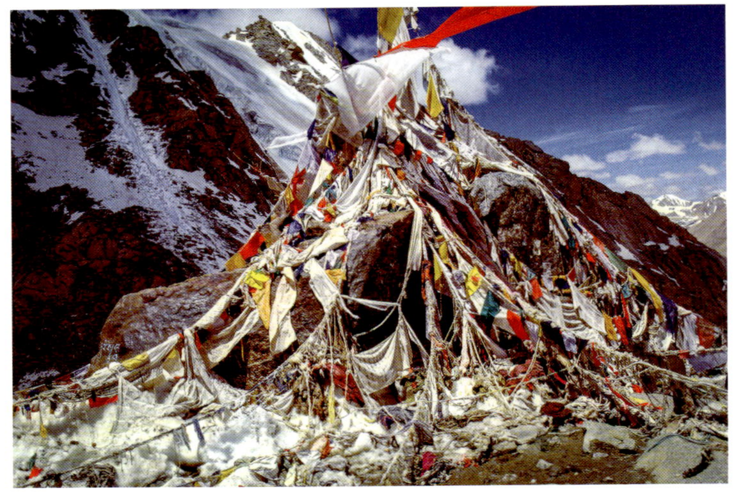

Dolma La, the high point on the pilgrimage around Mount Kailas, western Tibet, 1986

We left Zanda by jeep early the next morning and traveled for several hours before catching the first views of Mount Kailas. The sacred mountain is a 21,778-foot high, pyramid-shaped peak with horizontal ridges of snow that rises singularly on the high Tibetan plateau. Arriving at the base of Mount Kailas, we set up our tent outside the small pilgrim outpost of Darchen. At 6:00 the next morning, we awoke to the sound of people walking by, so we quickly got dressed, took down our tent, packed our things, and joined groups of Tibetan pilgrims, known as *yatris*, on the *parikrama*, the walk around the mountain. Around the first bend of the trail, we encountered Bon pilgrims walking down from the left side of the peak in contrast to Tibetan

Buddhists, who always pass clockwise on the left when they walk past a holy shrine or monument and who walk clockwise around the mountain.

Today, most tours around Mount Kailas take three days to complete the thirty-mile circuit, but we assumed everyone did it in one day, as most Tibetan pilgrims do. It was a long day, and we could not have completed the walk in one day without the support of the pilgrims we met along the way, often greeting us with a smile and the Tibetan salutation "Tashi Delek" or sticking out their tongue, another way of saying hello. At mid-day, as we approached the pass on a steep section of trail, a group of pilgrims stopped to share their food with us and load my backpack on one of their horses. Without this encouragement I could not have made it over Dolma La, a high mountain pass of more than 18,000 feet. Crossing Dolma La is a symbolic transition from one life to another. The top was filled with hundreds of prayer flags and articles of clothing placed there by pilgrims. We paused to admire the scene before hurrying down the steep backside, knowing we had many more hours of walking ahead of us. Throughout the afternoon we passed through beautiful valleys, barely noticing the scenery because we were so tired. Few people remained on the trail, and even the old Tibetan ladies were passing us. We plodded on and watched the sun set on the mountain at 8:30. An hour later, we stumbled into Darchen and were pleased to learn that completing the pilgrimage around Mount Kailas in one day confers extra merit.

The next day we drove by jeep to nearby Lake Manasarovar, a freshwater lake, which, like Mount Kailas, has been a holy pilgrimage site for hundreds, maybe thousands of years. Hindus believe that bathing in the lake cleanses the sins of a lifetime and that those who drink from it will go to the abode of Shiva after they die. The lake water was very cold, and we hesitated to immerse ourselves completely as Hindus do, so instead we poured water over our heads and drank the holy lake water for good luck as we watched the Tibetans do.

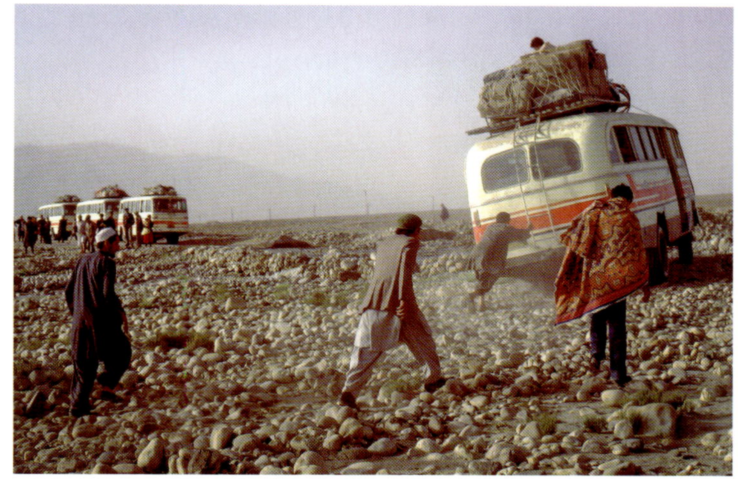

Our broken-down bus on the road between Kashgar and the Khunjerab Pass at the border of China and Pakistan, 1986

From Mount Kailas and Lake Manasarovar we returned to Ali to pick up our school-teacher friend and search for a ride north to the city of Kashgar in northwest China. LanDing was on a circuitous route to his family's home somewhere in central China. He proudly insisted on bringing his new television with him as proof of his new prosperity. With his help we found a truck with a driver from the Muslim Turkic ethnic minority group known as Uyghurs on his way to Kashgar. Uyghurs are now facing terrible persecution from the Chinese government, but, in those days, we saw few Chinese officials and witnessed no overt discrimination. Even today, it is difficult to know the full extent of the oppression of the Uyghurs due to Chinese government limits on journalists, but it is estimated that more than a million Uyghurs have been detained in camps that are very disingenuously called "vocational training camps."

Bill and LanDing rode in the back of the truck while I sat in the cab with the affable driver who gave me dried fruit to eat as an antidote to altitude sickness. We needed it. As we headed north through China's Xinjiang Uyghur Autonomous Region our truck crossed countless high passes that left us breathless even just sitting. We drove on across a moonlike tundra. Near the truck stop of Mazar, on our way down from around 16,000 feet toward the Taklamakan Desert plateau, we noticed smoke at the back of the truck. Our driver pulled over and yanked off the left back brake shoe before roaring off downhill. At 1:00 am we arrived at a checkpost where a group of sleepy young policemen in their underwear took in stride our fantastic story about why we were traveling through a restricted area in the back of a truck. At the next checkpoint we hid under a tarp in case that group of local authorities was not so gullible.

After four days on the road we pulled into Kashgar, the legendary market city on the Silk Route. It still had the appearance of a remote outpost. Only a sprinkling of colonial-era buildings amongst the gray adobe walls gave a clue to the city's former importance in the "Great Game," the political battle of intrigue for control of Central Asia between Britain and Russia from 1830 to 1907. We went by horsecart to our hotel at the Chini Bagh, a complex that housed the British Consulate until 1947. The stately two-story main house with Raj-era carved wooden details was still standing but was off limits to backpackers like us. Our accommodations were in the back of the complex in much simpler buildings that looked like former stables. Kashgar had just opened to foreign tourists in 1983, so we were still oddities. Bill and I splurged on a four-person room for some privacy, but without a functioning door lock we were barraged day and night by yogurt sellers, money changers, and curious fellow guests who also followed me into the women's bathroom. After barricading our door from the inside, we looked up and saw faces staring through the glass transom above the door and later found lawn chairs stacked up outside.

We spent our days in Kashgar walking the streets and wandering through the city's exotic market that was filled with produce and products from all over Central Asia. We ate hand-stretched noodles, lamb kebobs cut right off the carcass and grilled on a tiny brazier, and ice cream scooped from a mound on top of a big chunk of ice. In the late afternoon we would place a dinner order with the proprietor of a giant wok that was seasoned with lamb fat and filled with uncooked rice and chunks of lamb, carrots, onions, and spices. A few hours later, we would return to the wok and eat our platefuls of this delicious Central Asian specialty known as *plov*.

The dusty streets of Kashgar served as a backdrop to the colorful people we encountered: an old bearded bookseller in his long blue-and-white-striped robe and embroidered skullcap seated next to his stacks of books, the barber with an ancient chair and mirror set up under a tree alongside the road, vegetable sellers with antique hanging scales sitting amongst piles of cabbages and heads of cauliflower, butchers stationed in front of goat haunches hanging on iron hooks. The most frenetic activity at the market centered around the arrival of wagons filled with the small, green-and-yellow Hami melons; people would storm the wagon with money in hand to get as many as they could of this local product renowned for its sweetness.

We hated to leave Kashgar, but we had to get to Rawalpindi, Pakistan, in one week to meet a group we were leading for a trek to K2 Base Camp. After once again searching grimy truck stops, we hooked up with a group of Pakistani merchants who had hired a bus to get home with their stock of new Chinese bicycles. The next morning at 5:30 we all boarded the bus and began our long trip down the Karakoram Highway through the countryside in our decrepit vehicle. At 11:30 p.m. the bus hit a huge pothole, and we could hear the sharp sound of bolts loosening on the wheels. With that, all the passengers, including us, piled out, grabbed blankets or sleeping bags, and instantly went to sleep on the ground. Repairs were made the next morning, and we rode uneventfully for another day, passing the impressive peaks of Kongur and Muztagh Ata just to the east of the highway. In the evening of the second day, at the border between and China and Pakistan, we became some of the first tourists to cross the Khunjerab Pass, the highest point on the Karakoram Highway (elevation 15,397 feet) and the highest paved international border crossing in the world, which had just opened to foreign travelers earlier in the year.

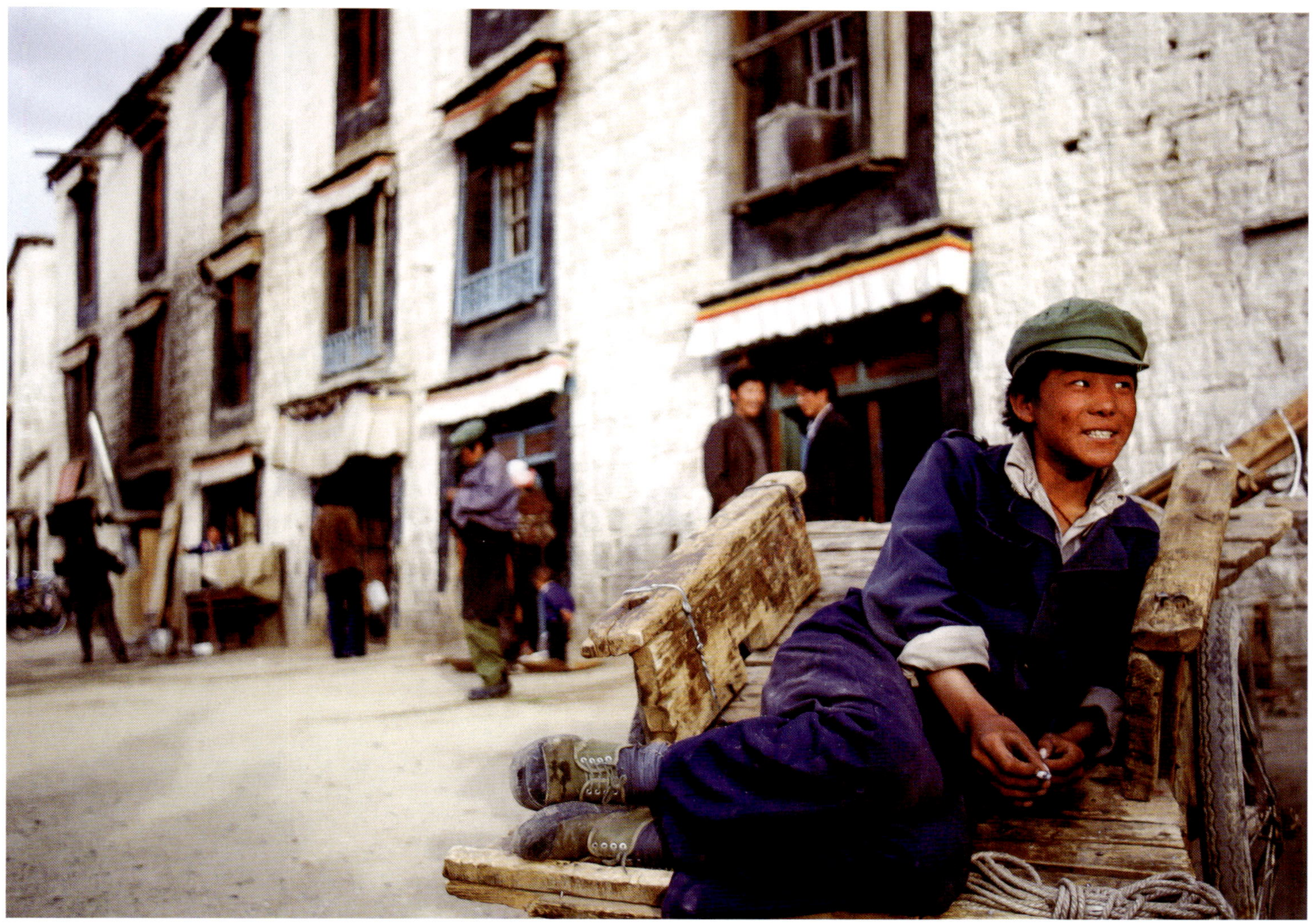

Lhasa, Tibet, 1986

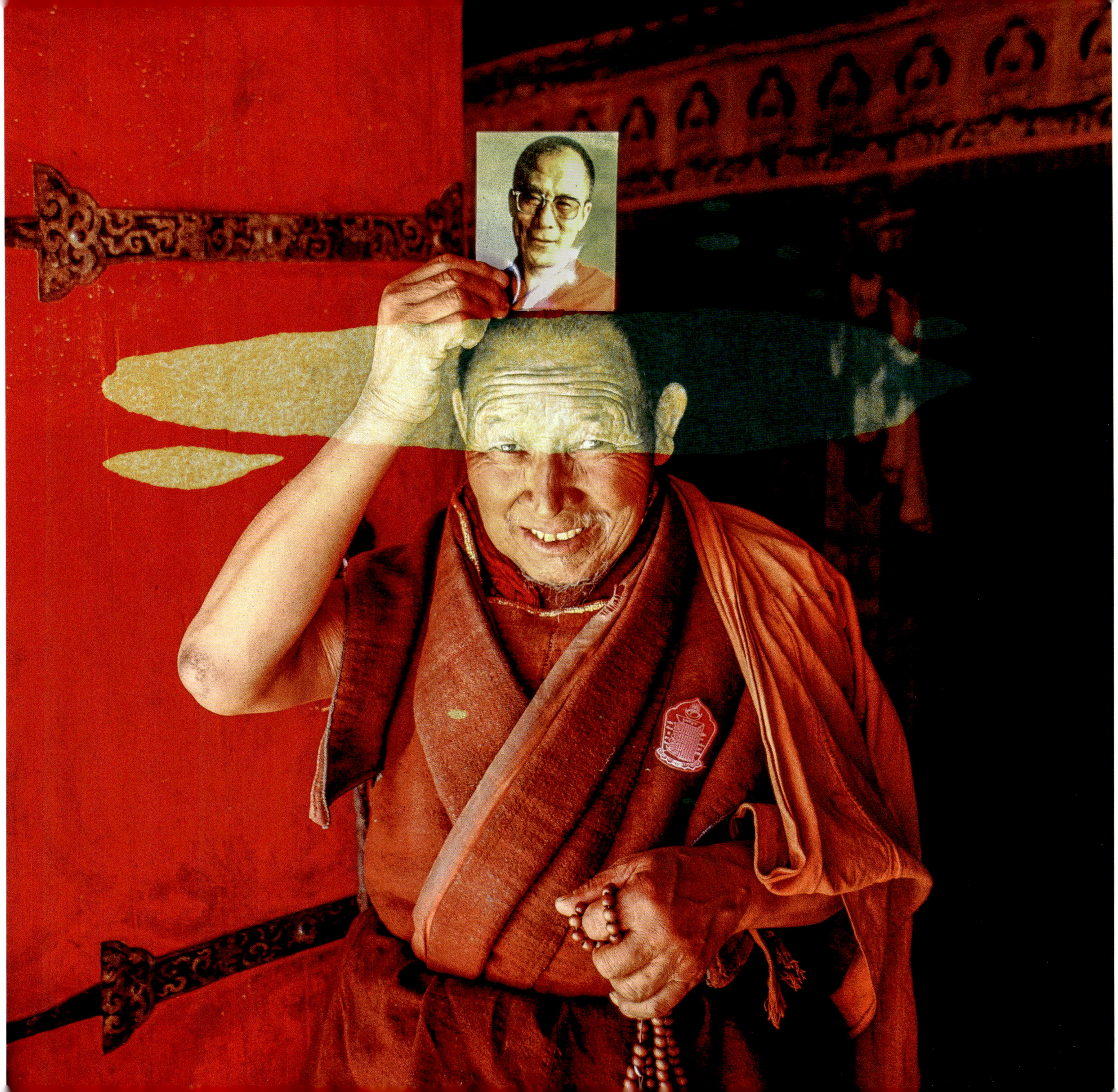

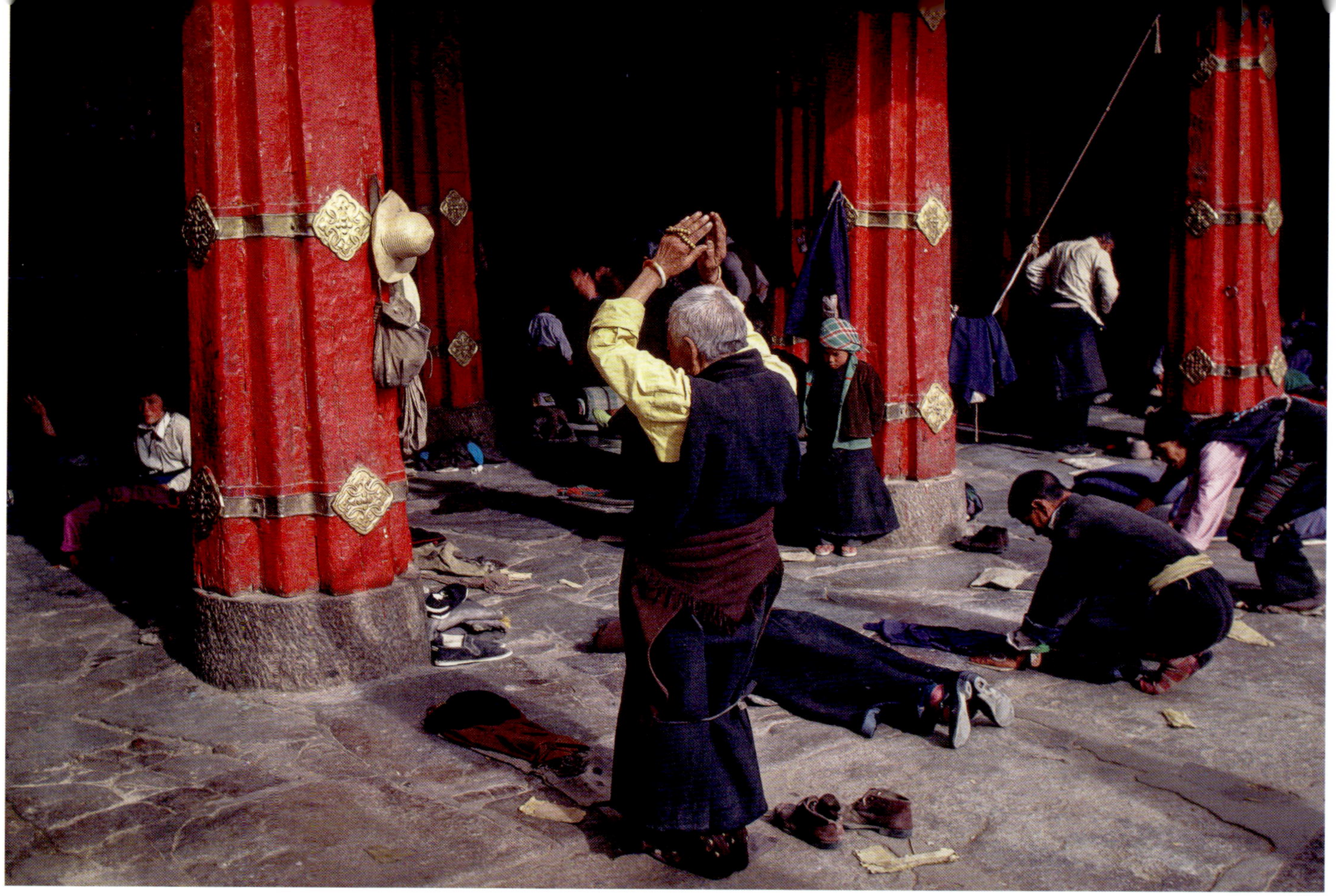

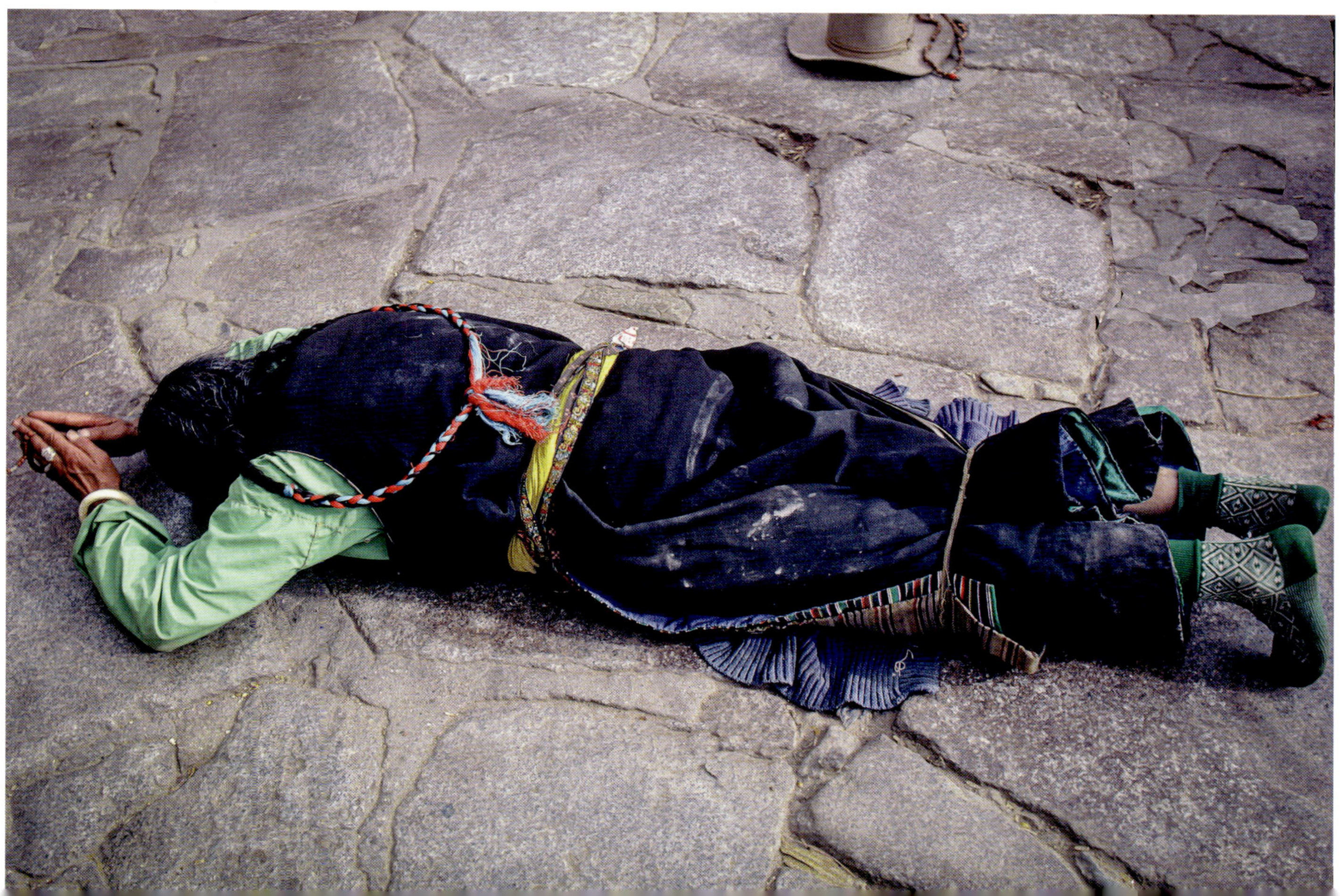

Above: Tibetans praying inside the Jokhang Temple, Lhasa, Tibet, 1986

Left: Tibetan prostrating inside the Jokhang Temple, Lhasa, Tibet, 1986

Opposite: Tibetan Buddhist monk venerating a photograph of His Holiness the Dalai Lama, Lhasa, Tibet, 1986

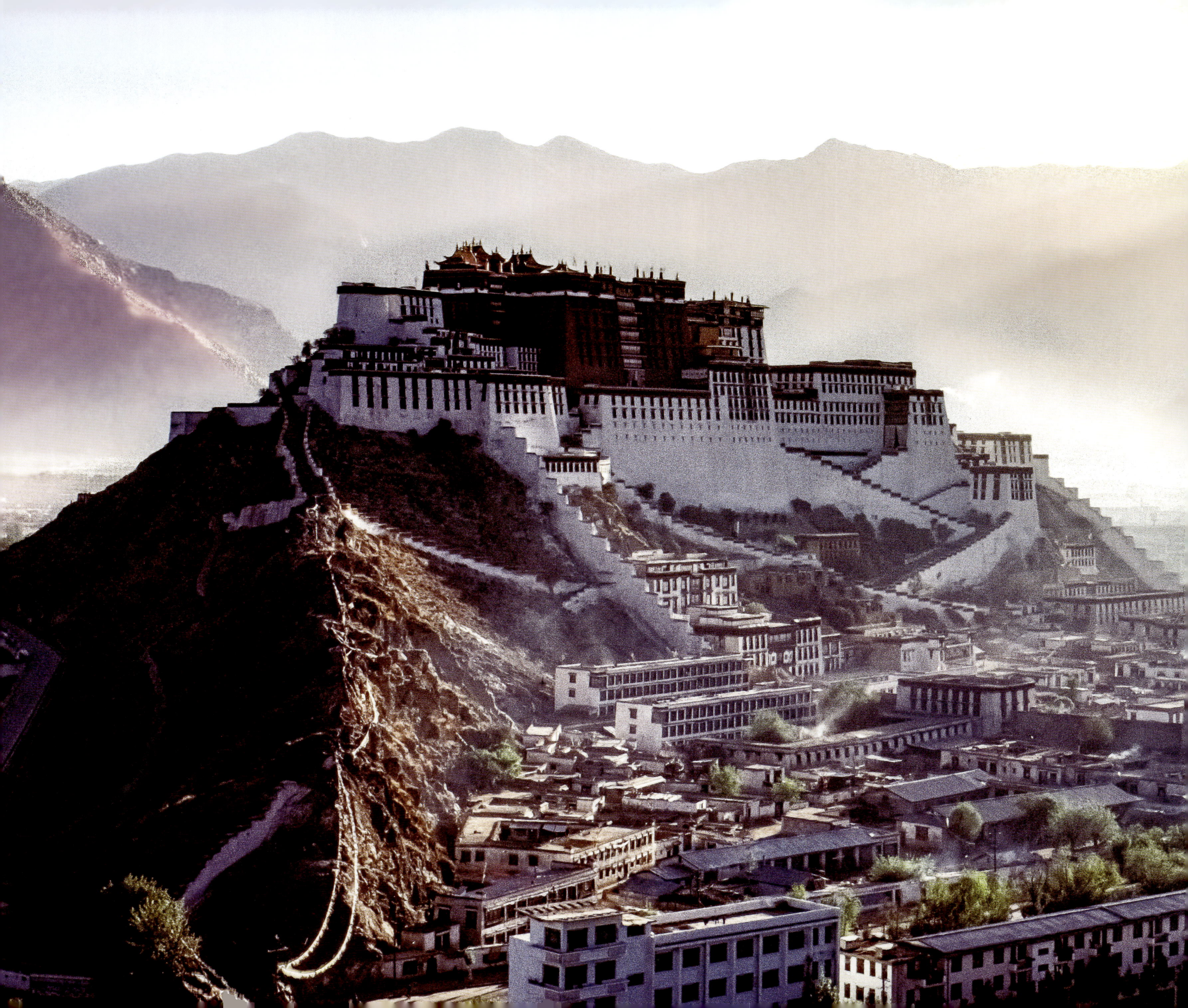

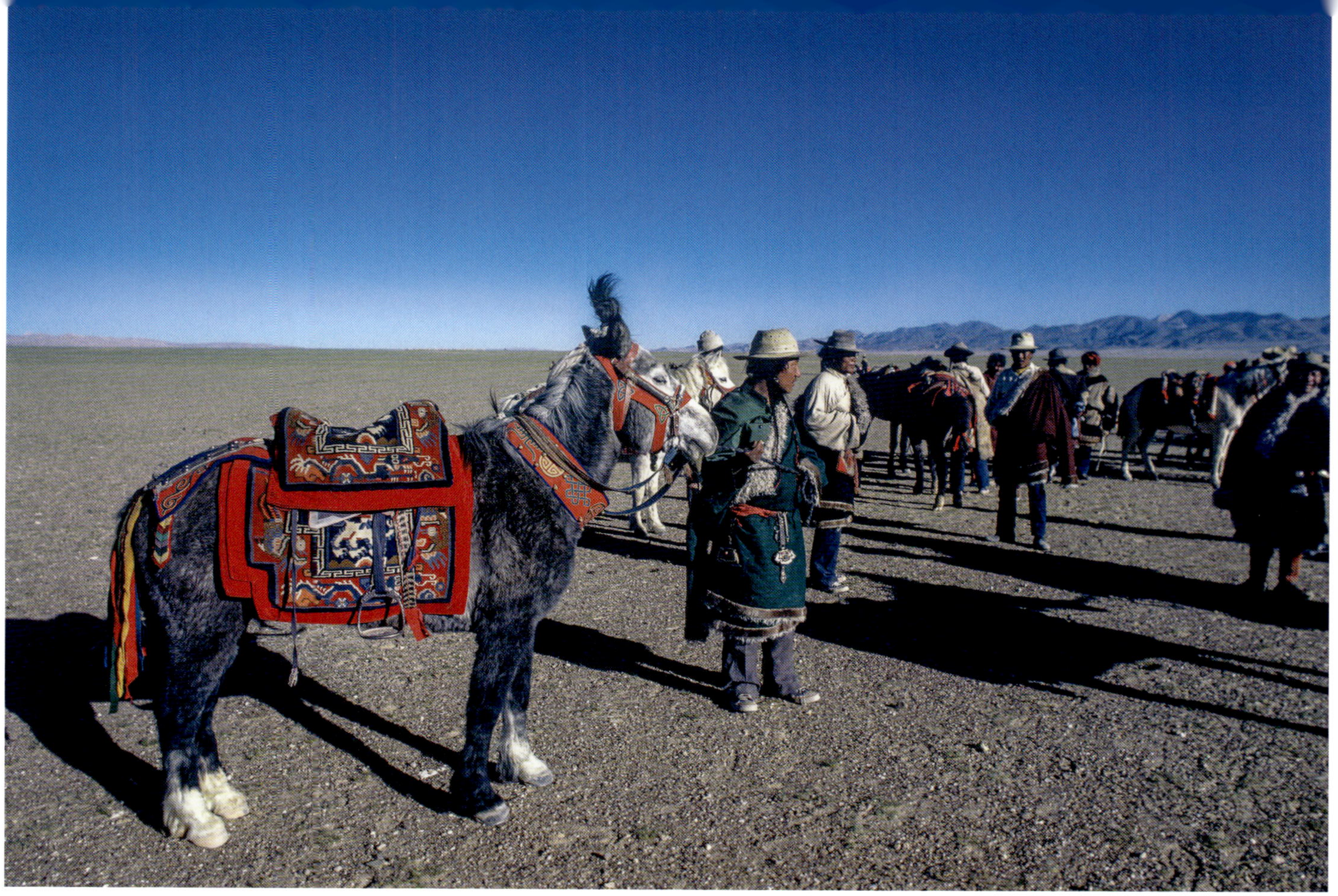

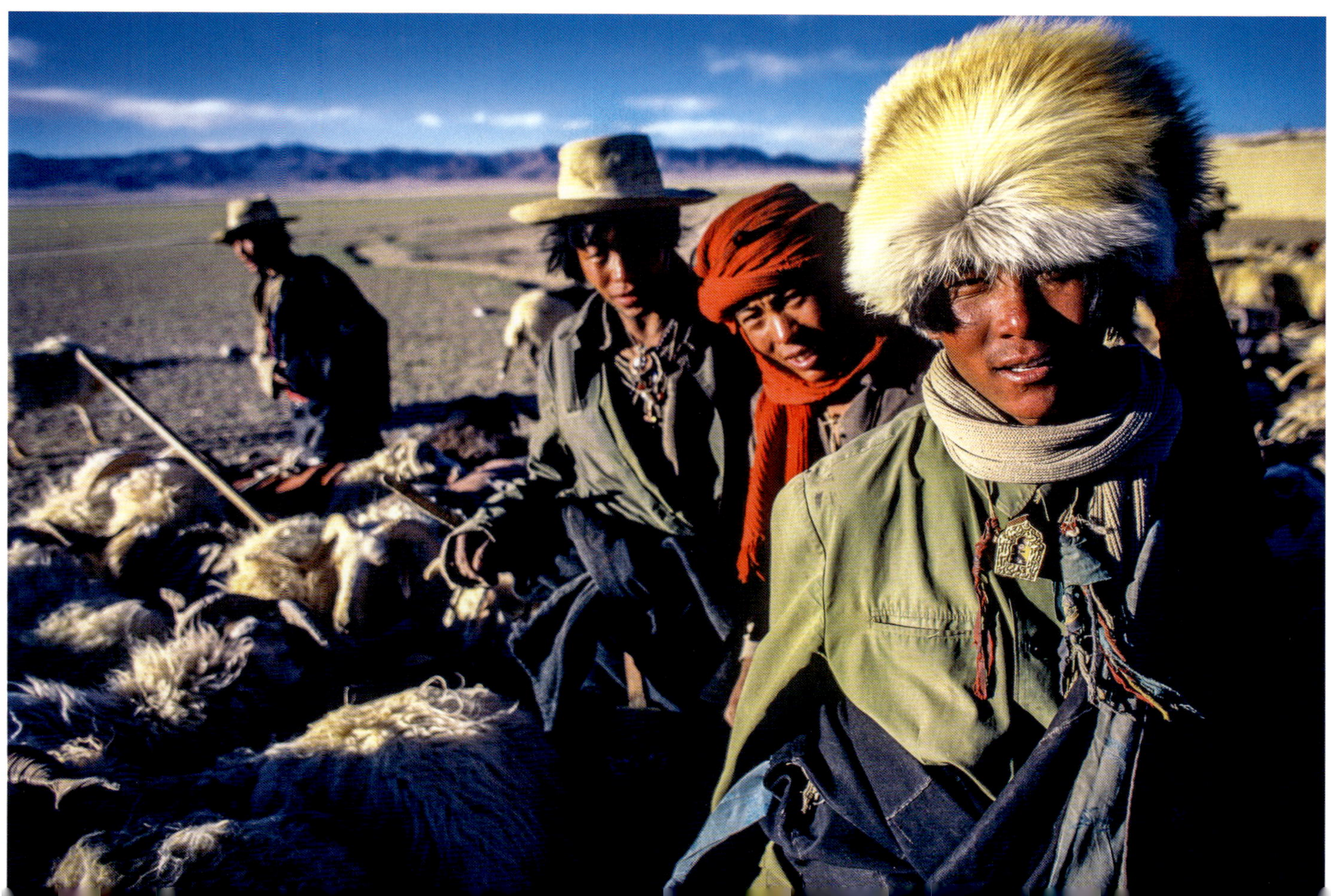

Above: Tibetan nomads near Gêrtsê, western Tibet, 1986

Left: Nomads, western Tibet, 1986

Opposite: Potala Palace, home of the Dalai Lamas for centuries, Lhasa, Tibet, 1986

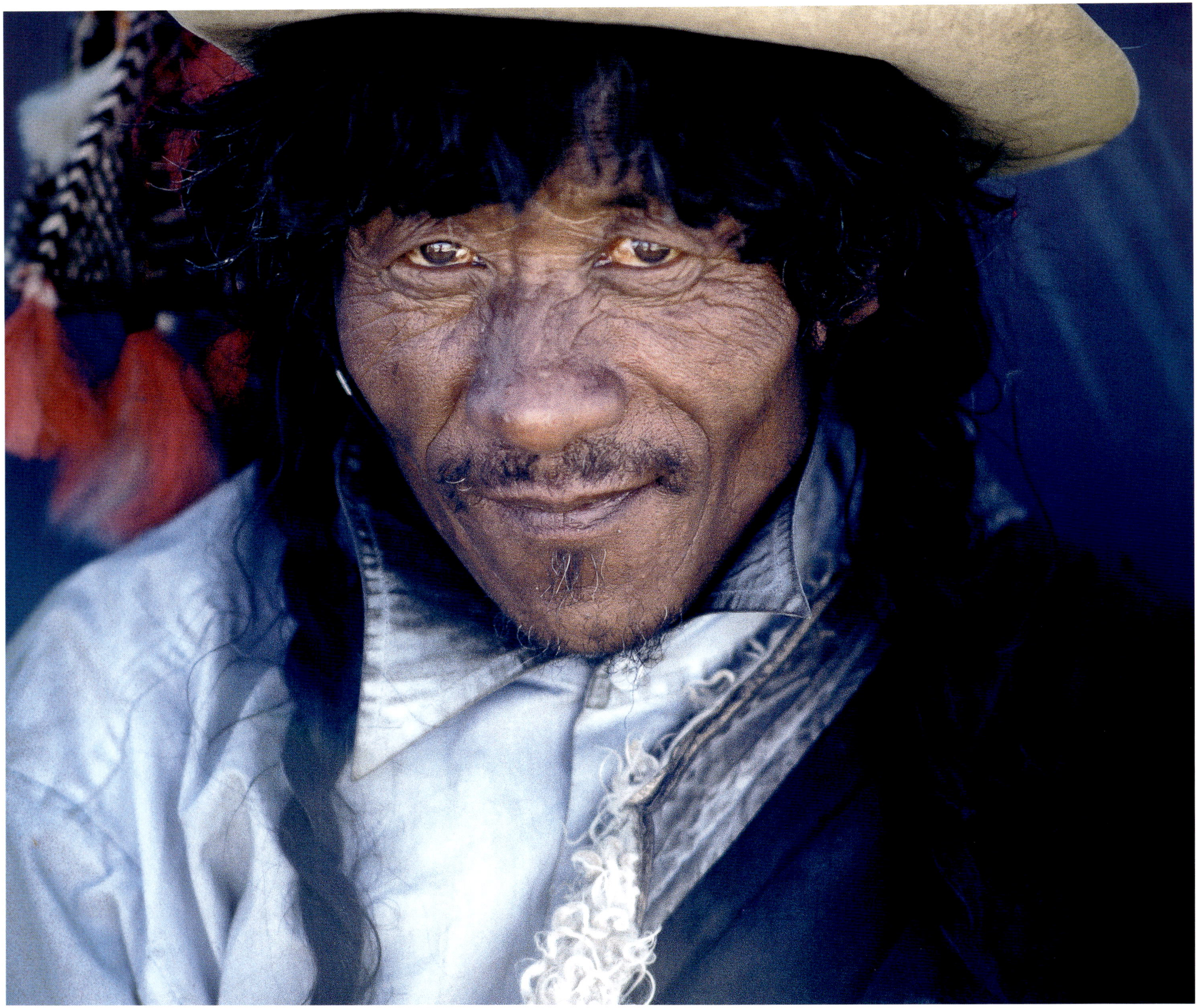

Tibetan nomad, western Tibet, 1986

Opposite: Ruins of an ancient monastery between Tholing and Tsaparang, western Tibet, 1986

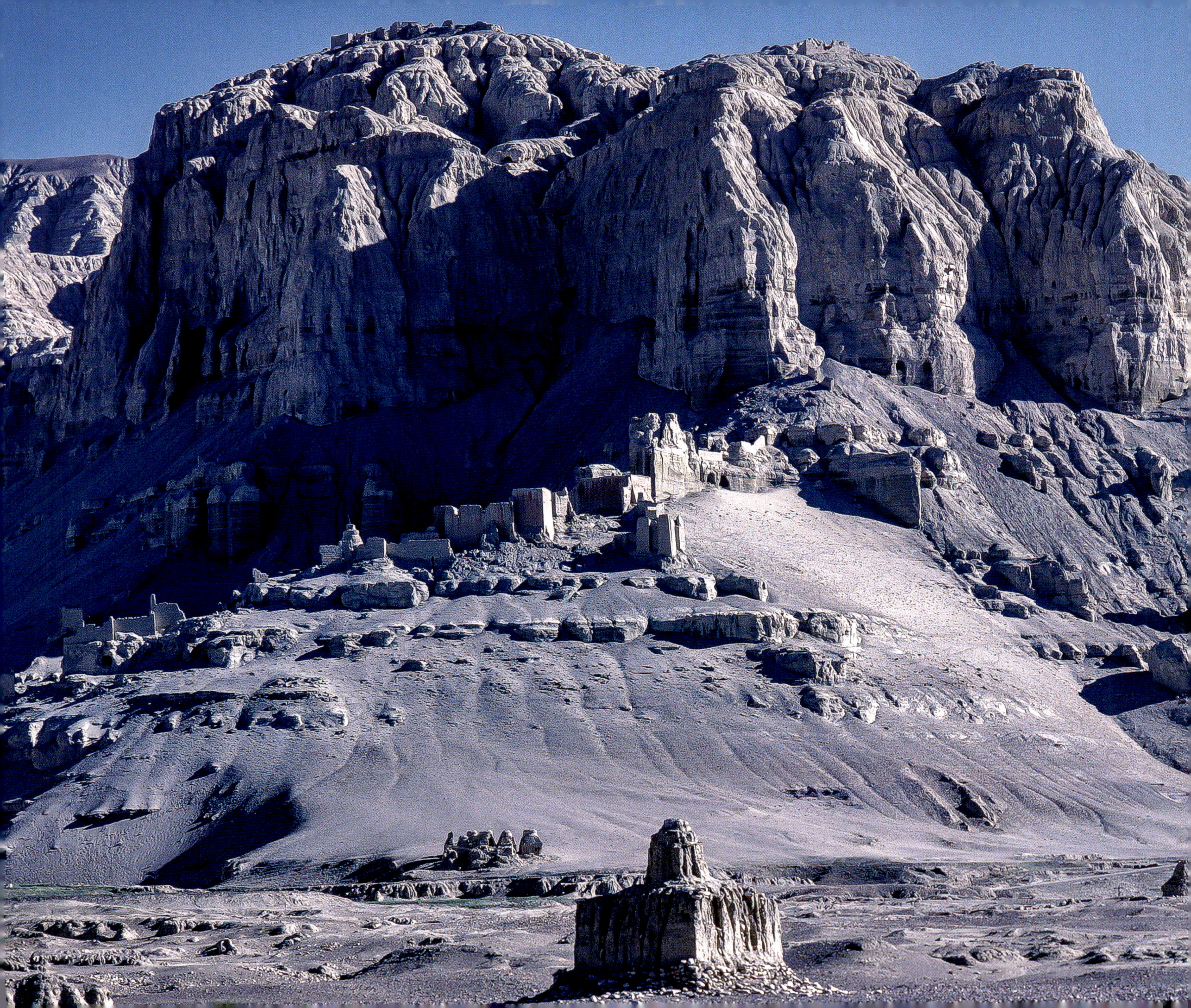

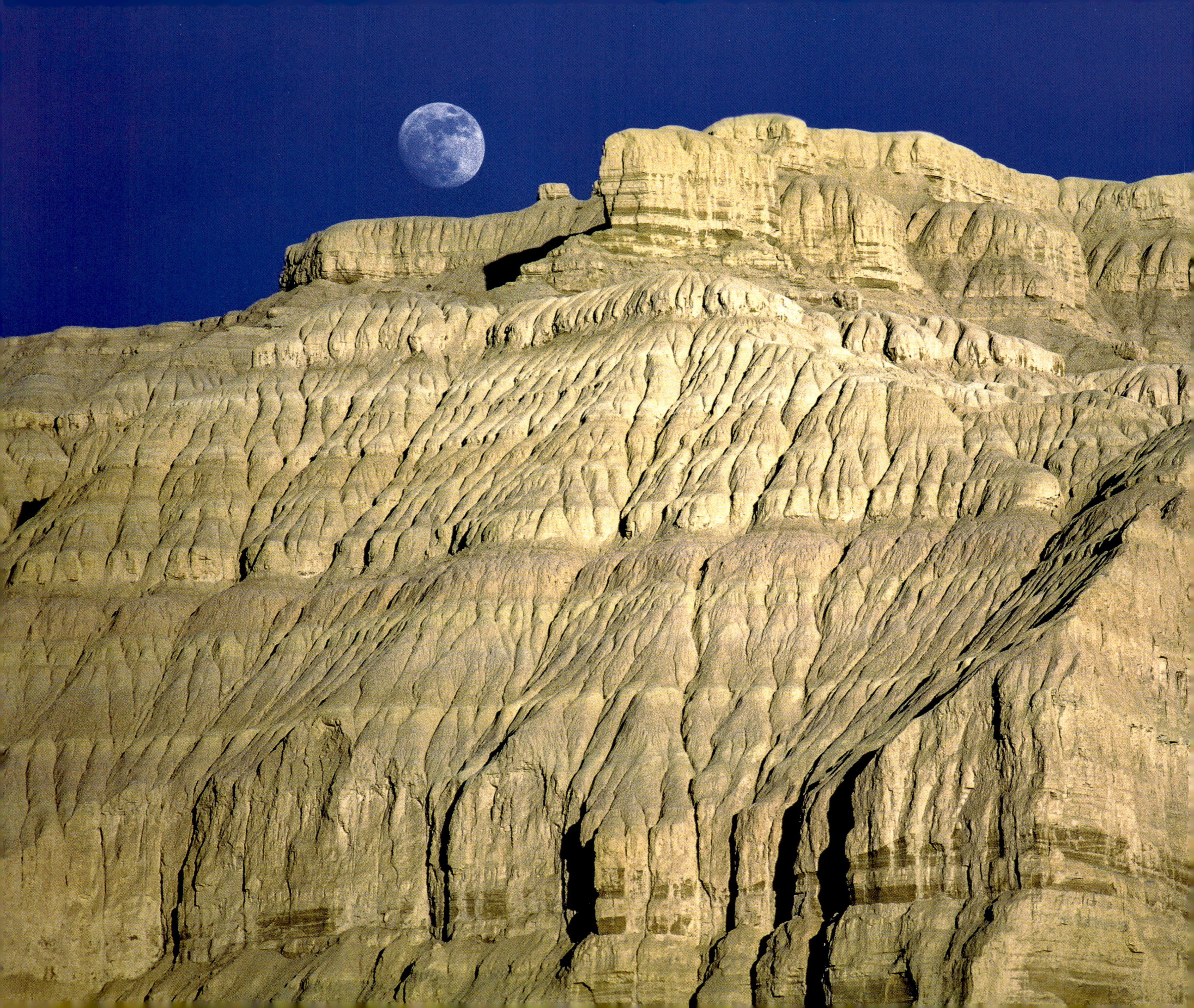

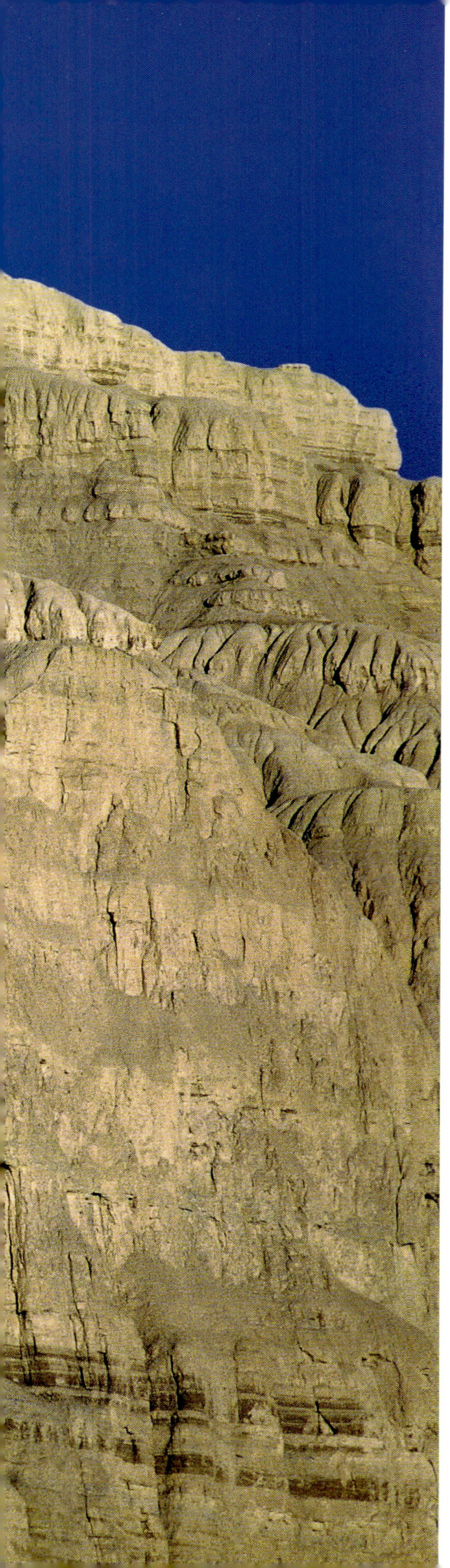

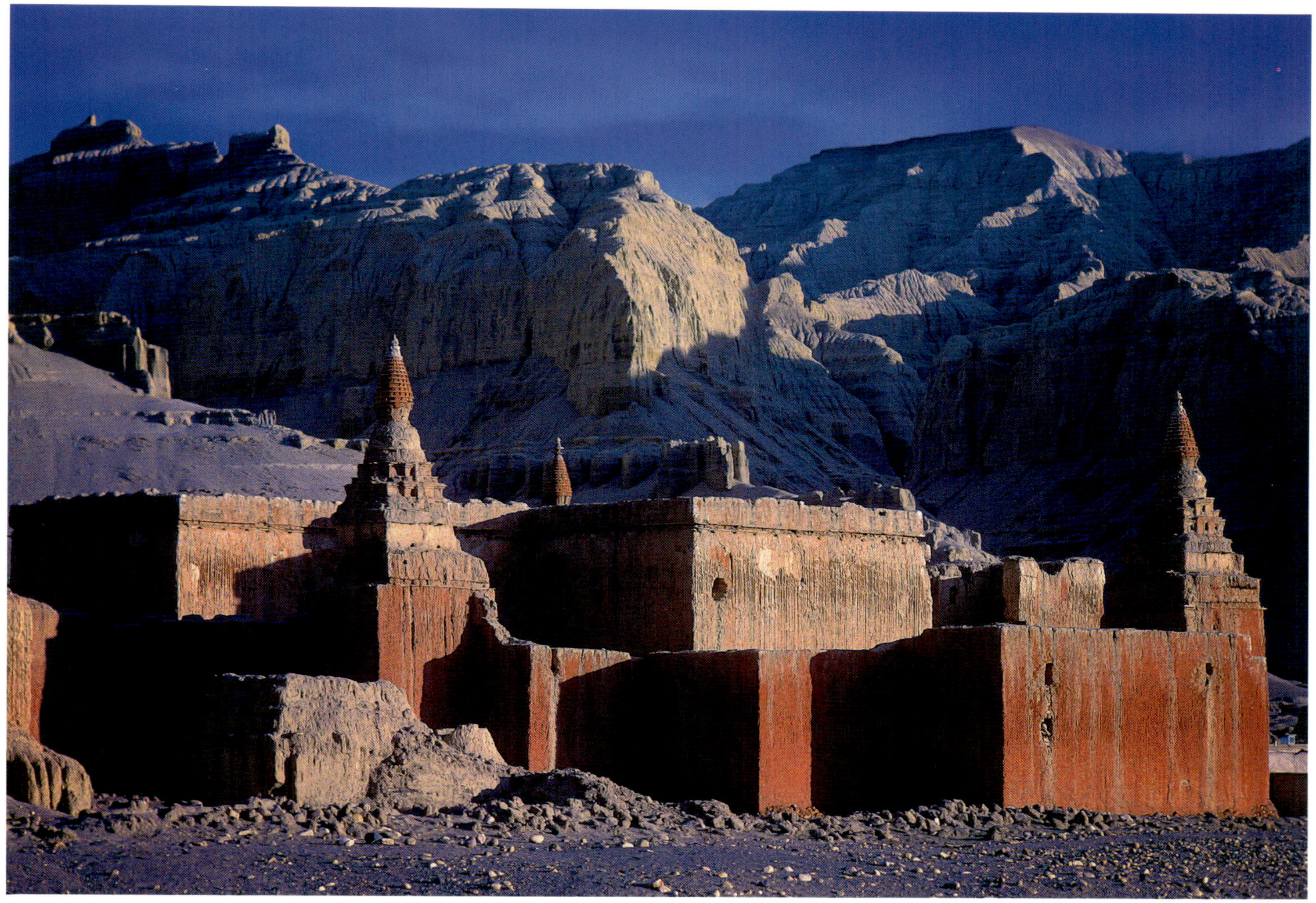

The Golden Temple of Tholing, western Tibet, 1986

Moonrise over high peaks in the Tholing Valley of the ancient kingdom of Guge, western Tibet, 1986

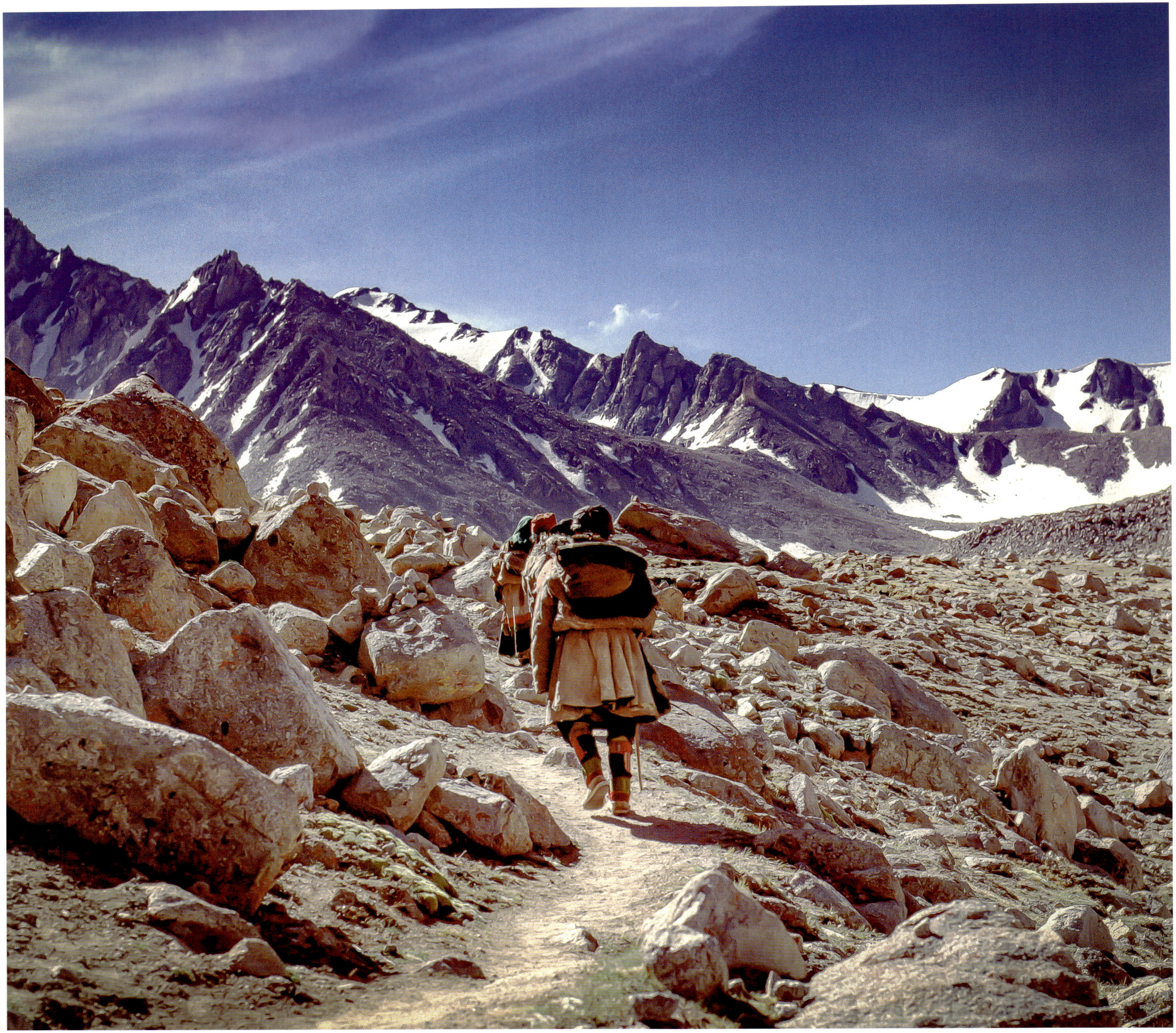

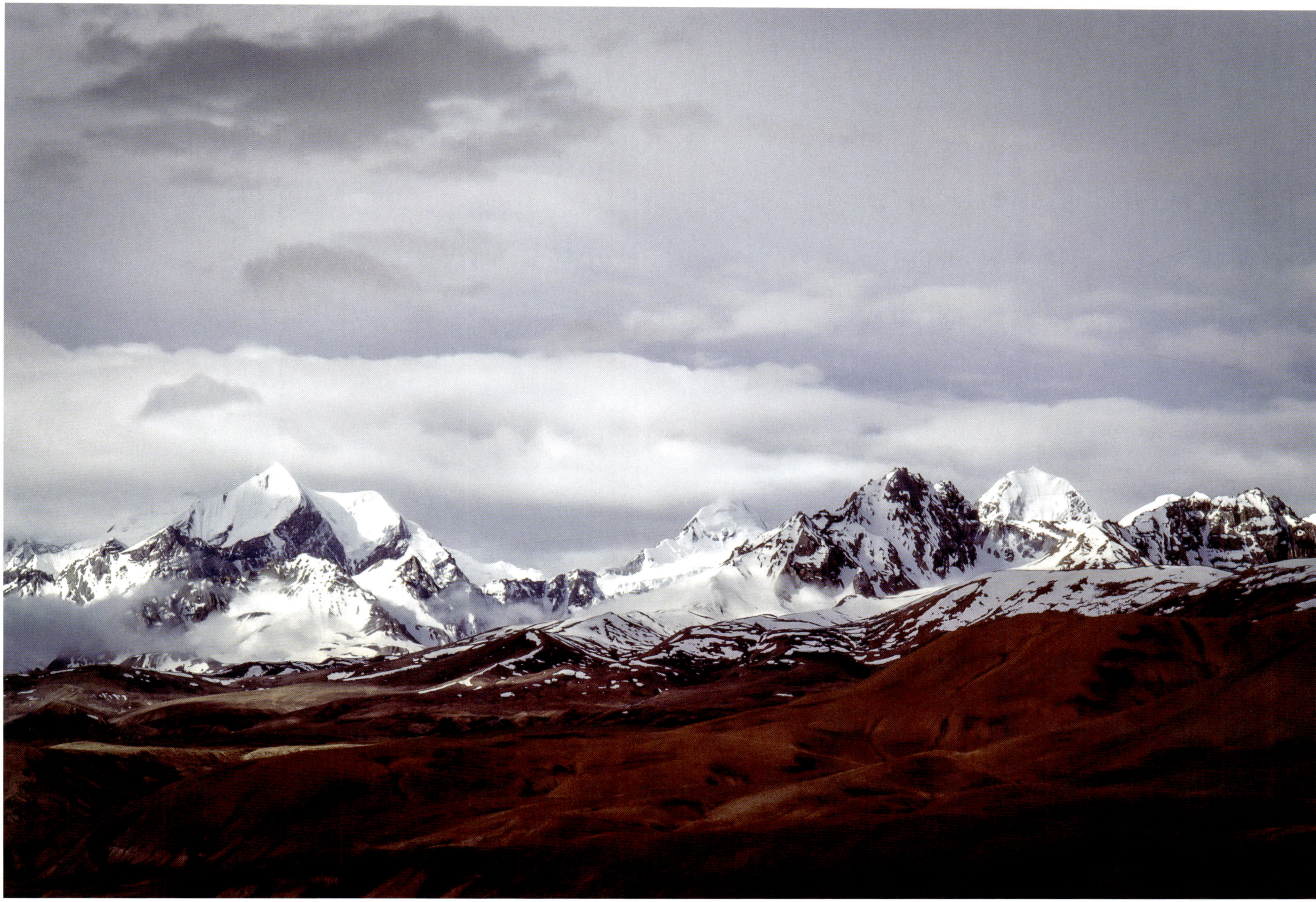

The road between Ali and Kashgar in far western China, 1986

Opposite: Tibetan pilgrims circumambulating sacred Mount Kailas, western Tibet, 1986

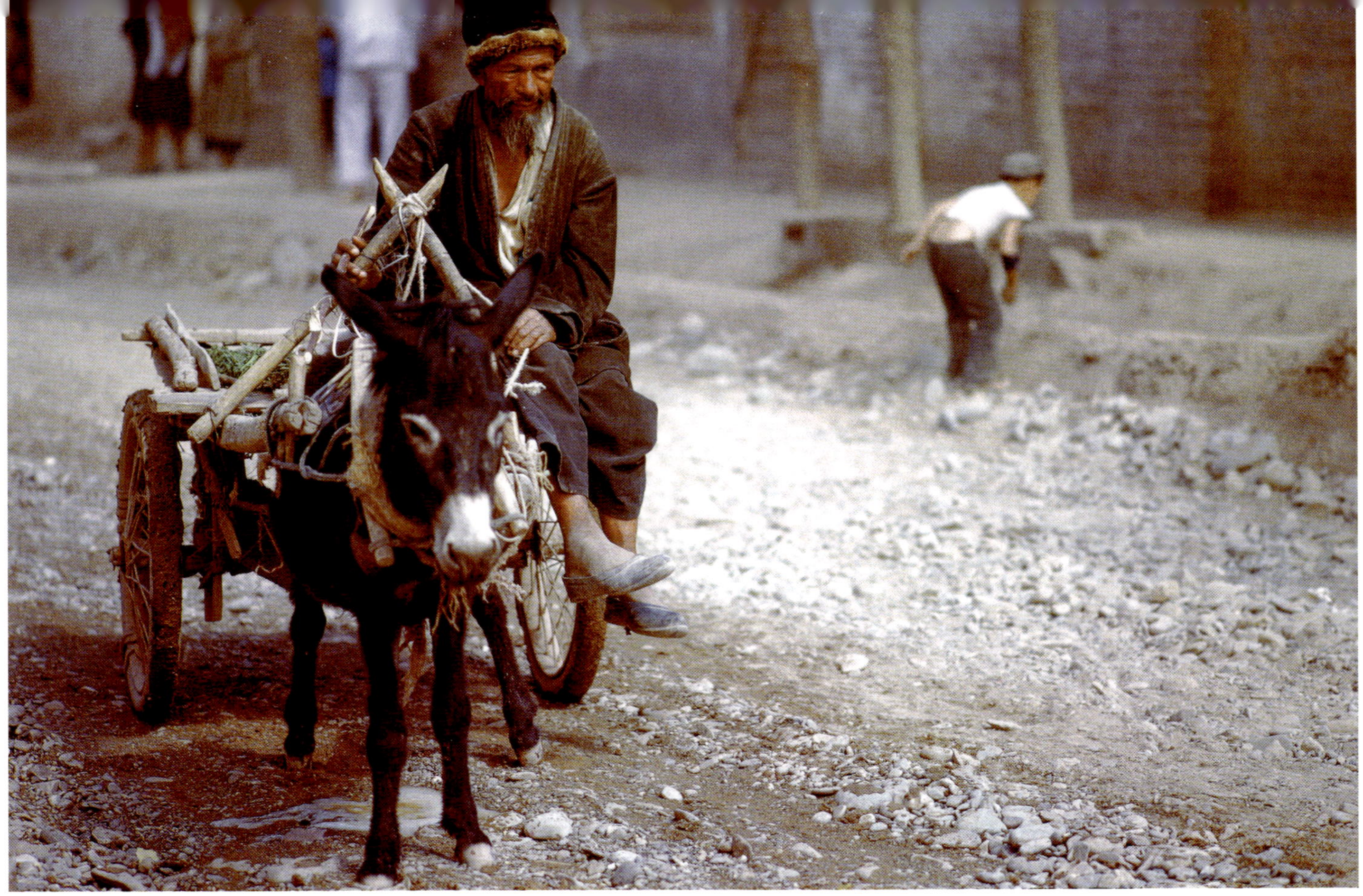

Kashgar, Xinjiang,
China, 1986

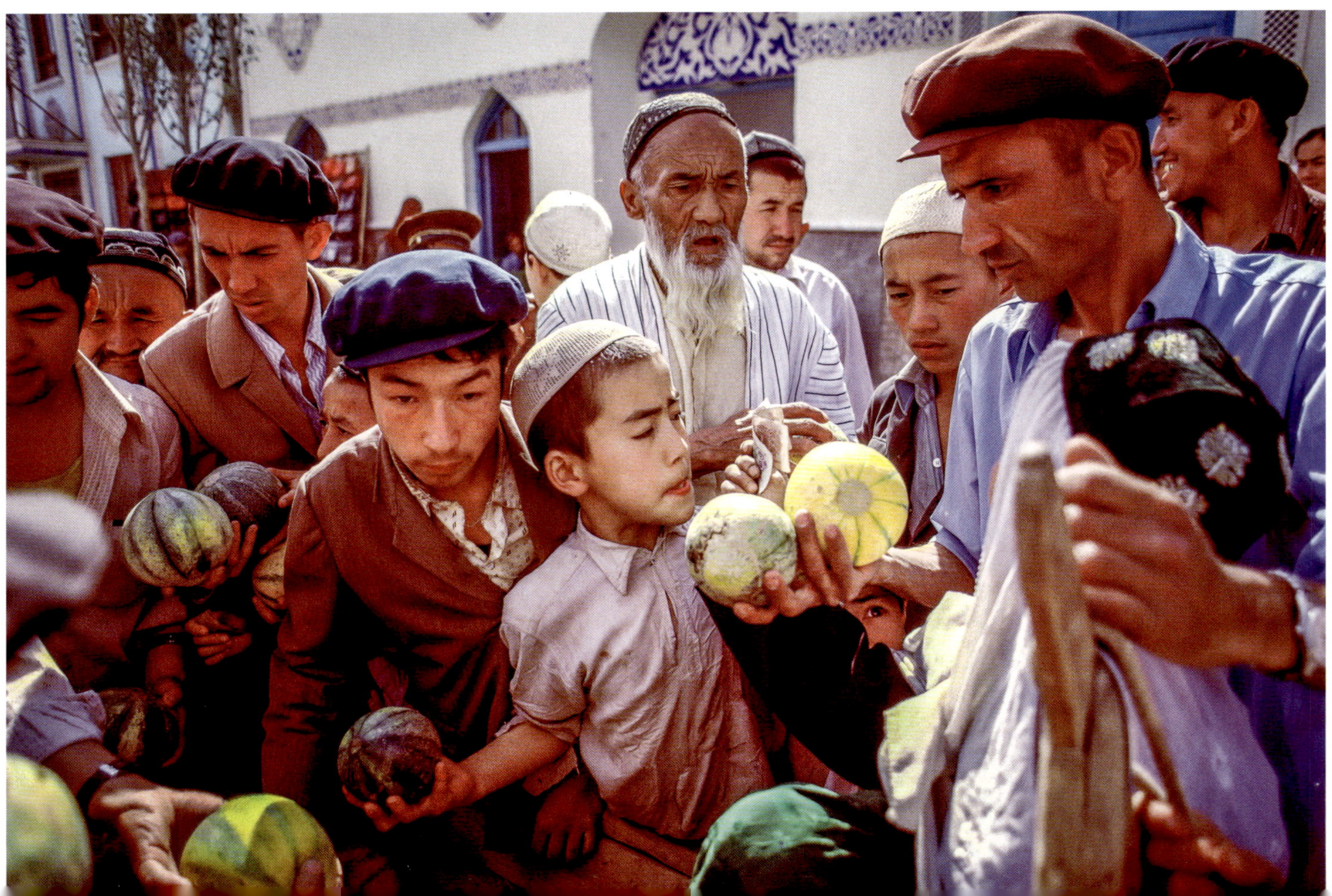

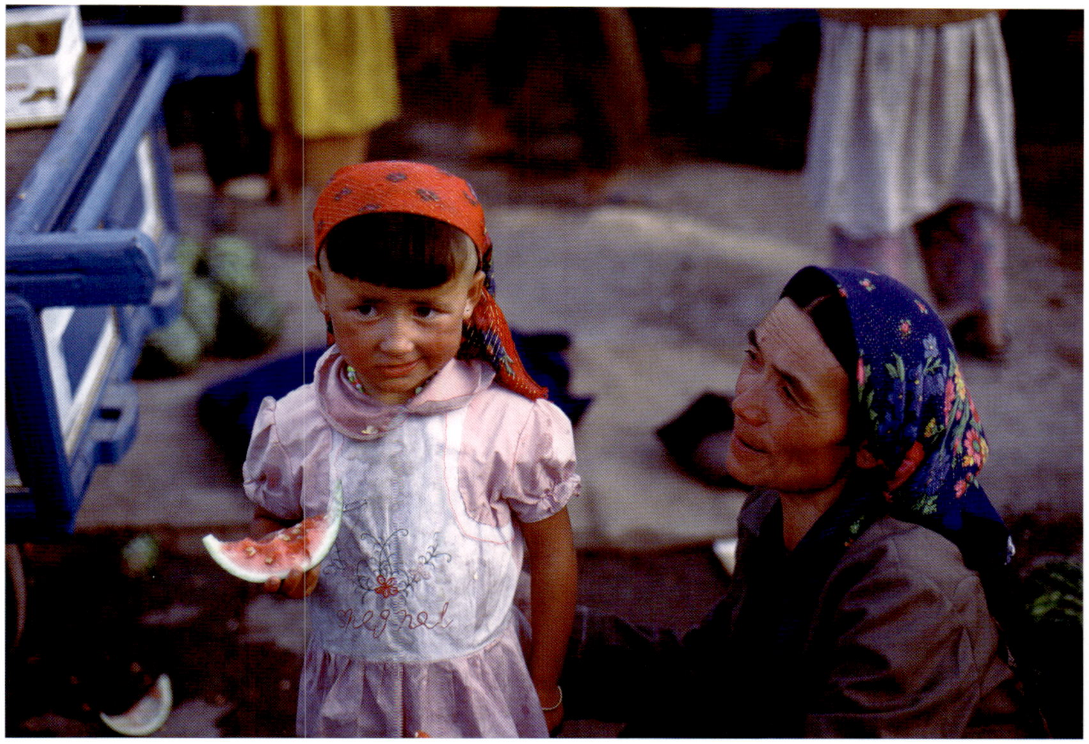

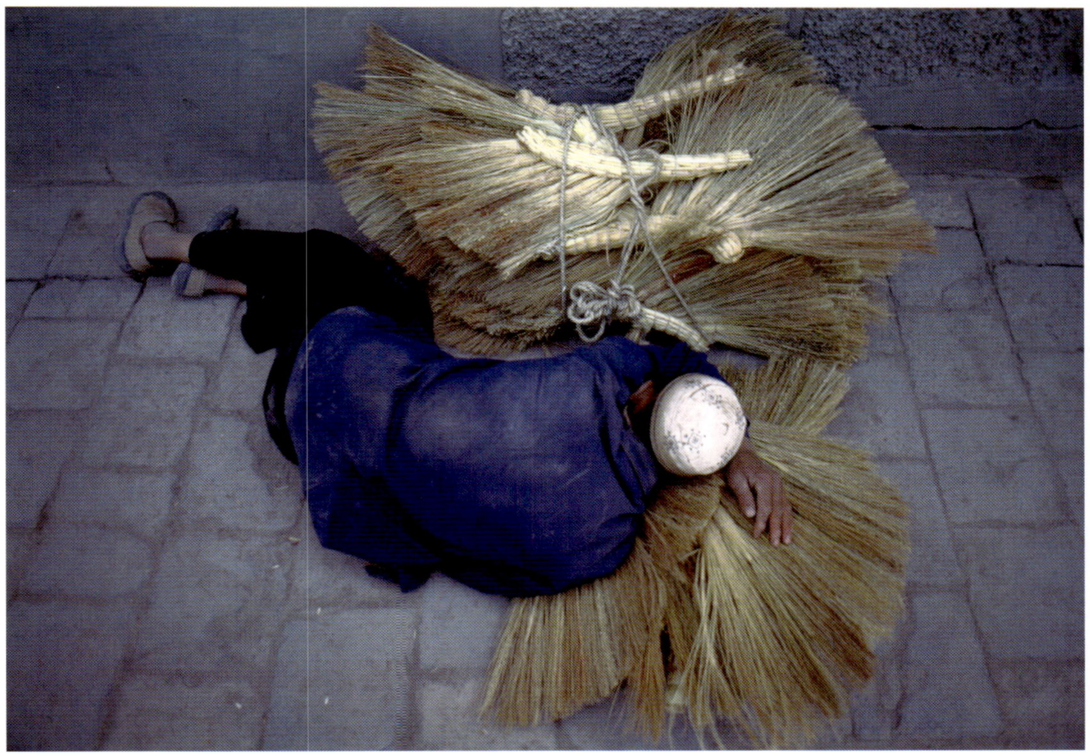

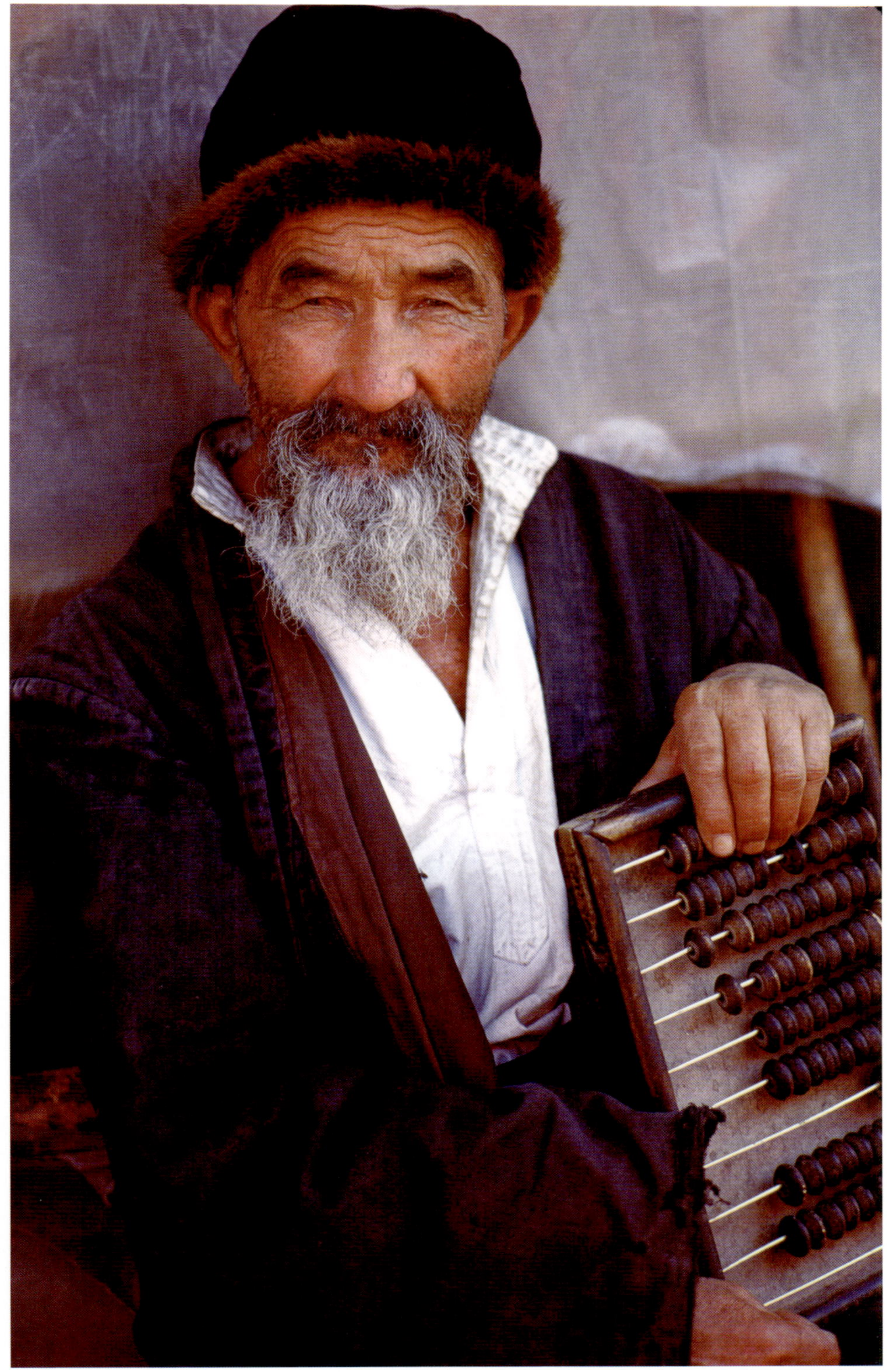

Kashgar, Xinjiang, China, 1986

Following spread: Muztagh Ata, on the road from Kashgar to the border of Pakistan, 1986

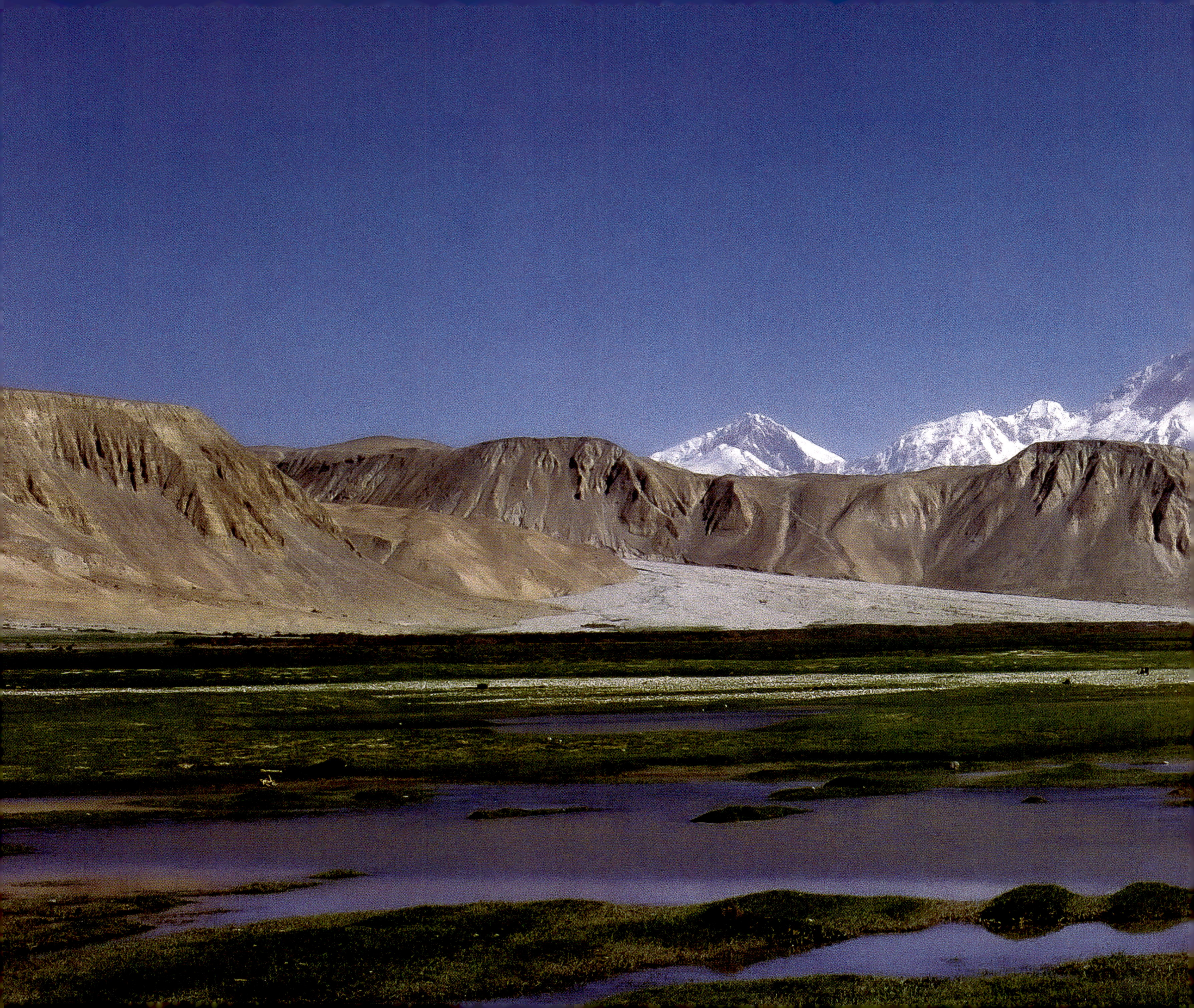

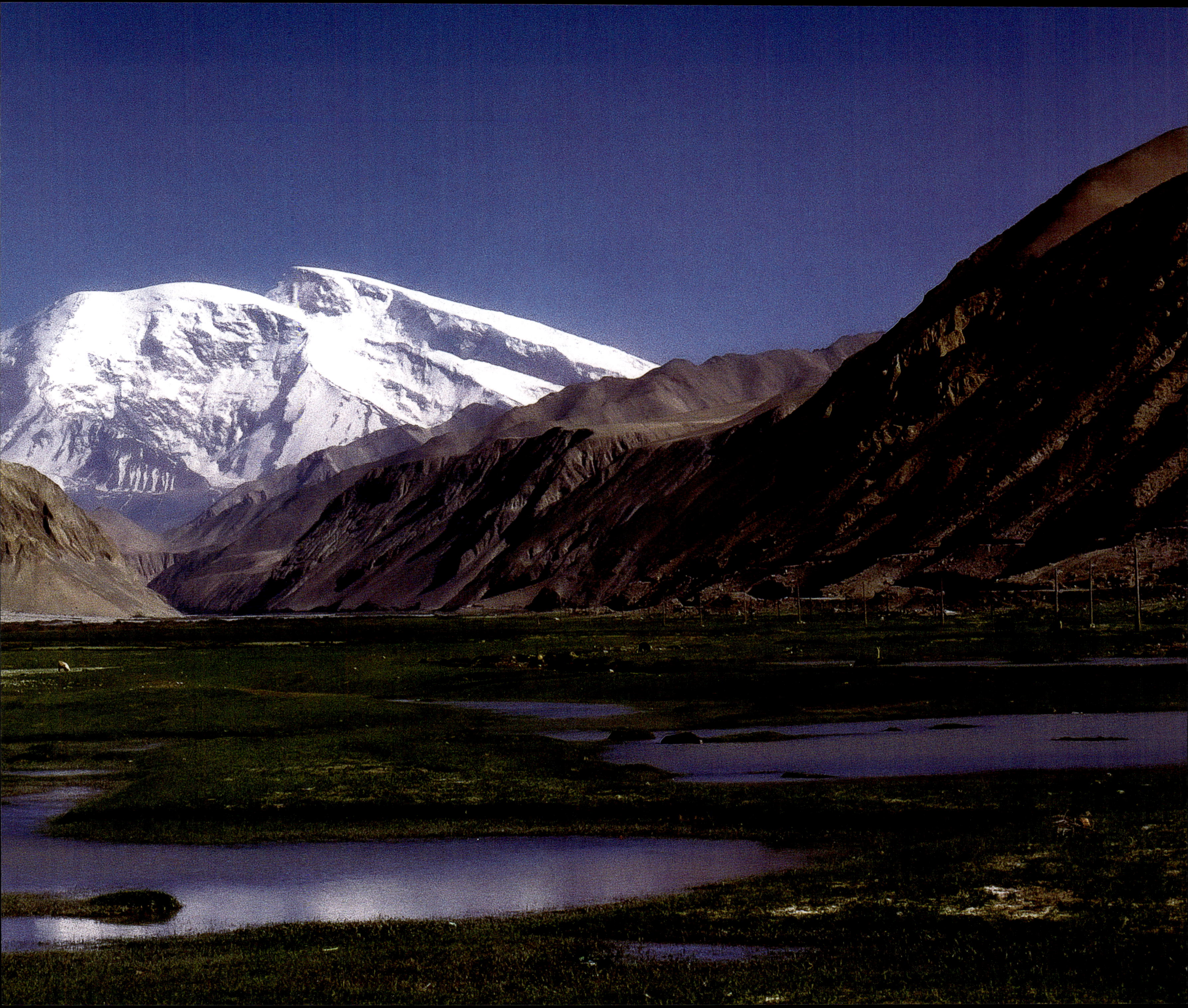

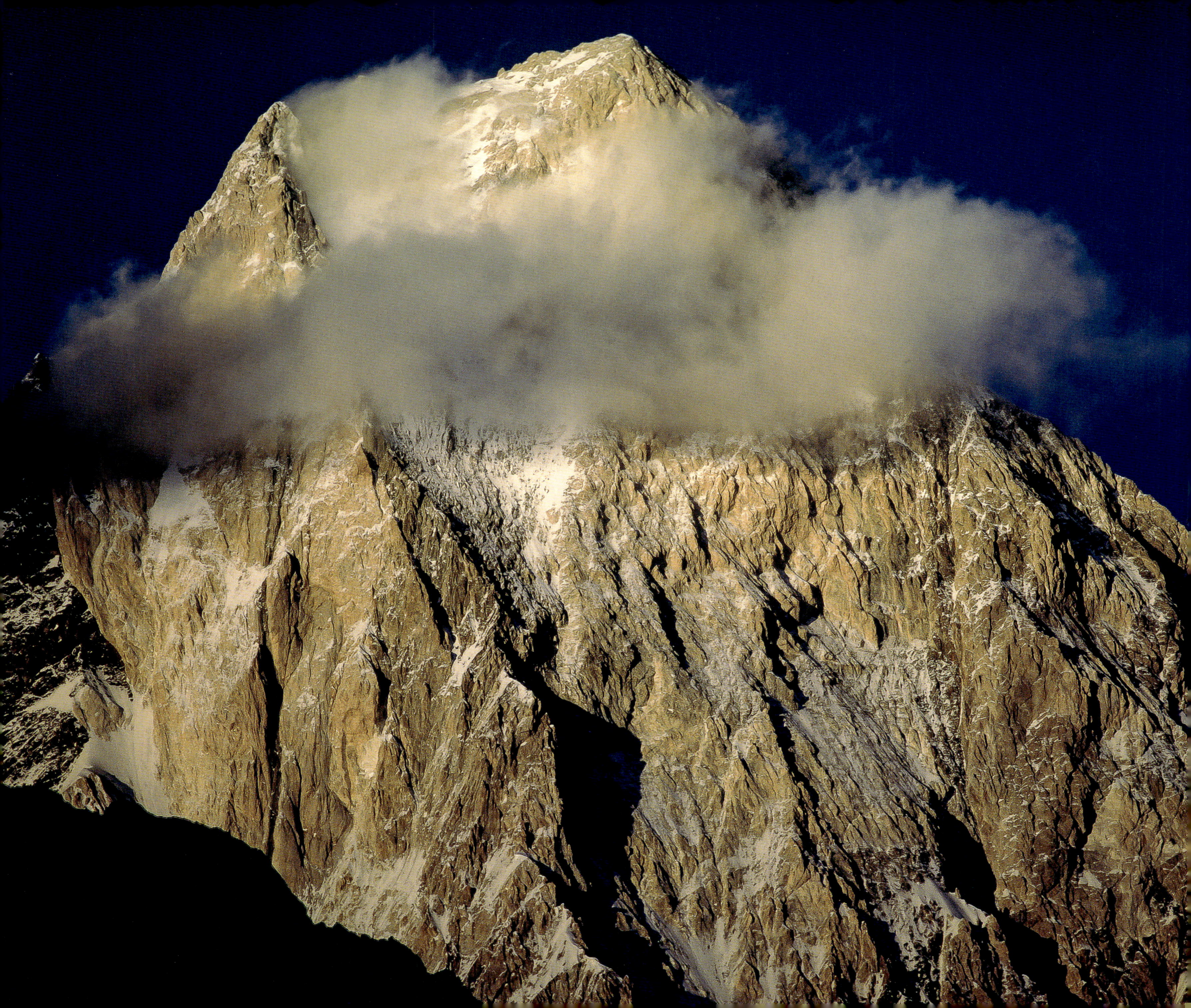

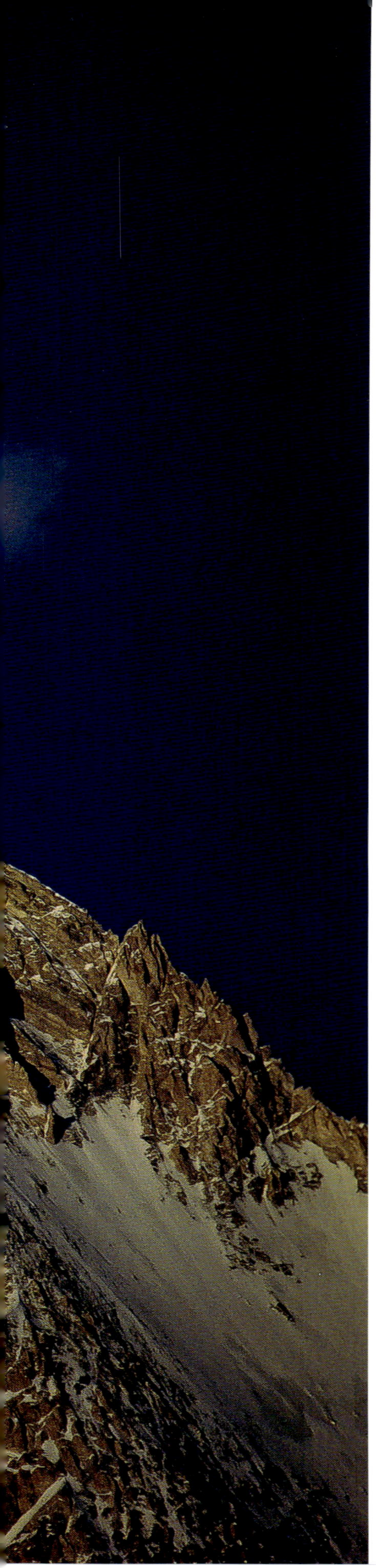

Crossing Rivers and Glaciers to K2, Pakistan

JULY–AUGUST 1986

K2 Base Camp was buzzing with excitement about the possibility of climbers from all over the world summiting the peak in the coming days. We arrived on August 2 after thirteen days of hiking, stayed for two days, and only found out later that an unexpected snowstorm and avalanche ended in tragedy there after we left: Five mountain climbers from the United Kingdom, Austria, and Poland died on the mountain between August 3 and 10. In total, twenty-seven people summited K2 that season, but thirteen died in the process, seven on their descent.

It is no wonder that K2, the second-highest mountain in the world (elevation 28,251 feet), is known as the "Savage Mountain" for its record as a deadly climbing peak. Attempts to climb K2 started as early as 1902, and over the years several teams reached impressive heights until 1954, when Italians Achille Compagnoni and Lino Lacedelli became the first people to reach the summit. They did it via the Abruzzi Ridge, the route on the southeastern side used by many subsequent expeditions. The area including K2 was closed to foreigners between 1961 and 1974, but in the years since then the proportion of climbers losing their lives on K2 compared to climbing attempts there has been significantly higher than on Mount Everest.

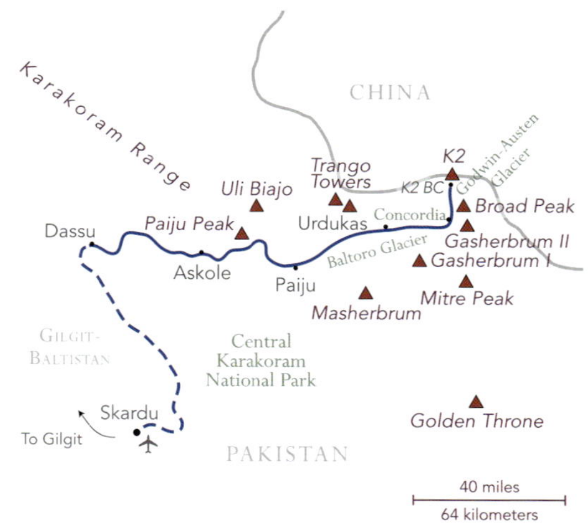

Our trek to the South Base Camp of K2 began in mid-July with a group of ten friends, including Ram, who flew in from Nepal. We took a small plane from Islamabad to the town of Skardu in northern Pakistan and then boarded jeeps to take us to the village of Dassu, where we hired a large contingent of porters from Baltistan, the area south of K2. Baltis are believed to be of ancient Tibetan descent, although they are now Muslim. Like countless mountaineering and trekking expeditions before us, we relied on their services to carry our bags and provisions on their backs since no pack animals can navigate the trails. Ikram Beg, our capable

K2, the world's second-highest peak, Pakistan, 1986

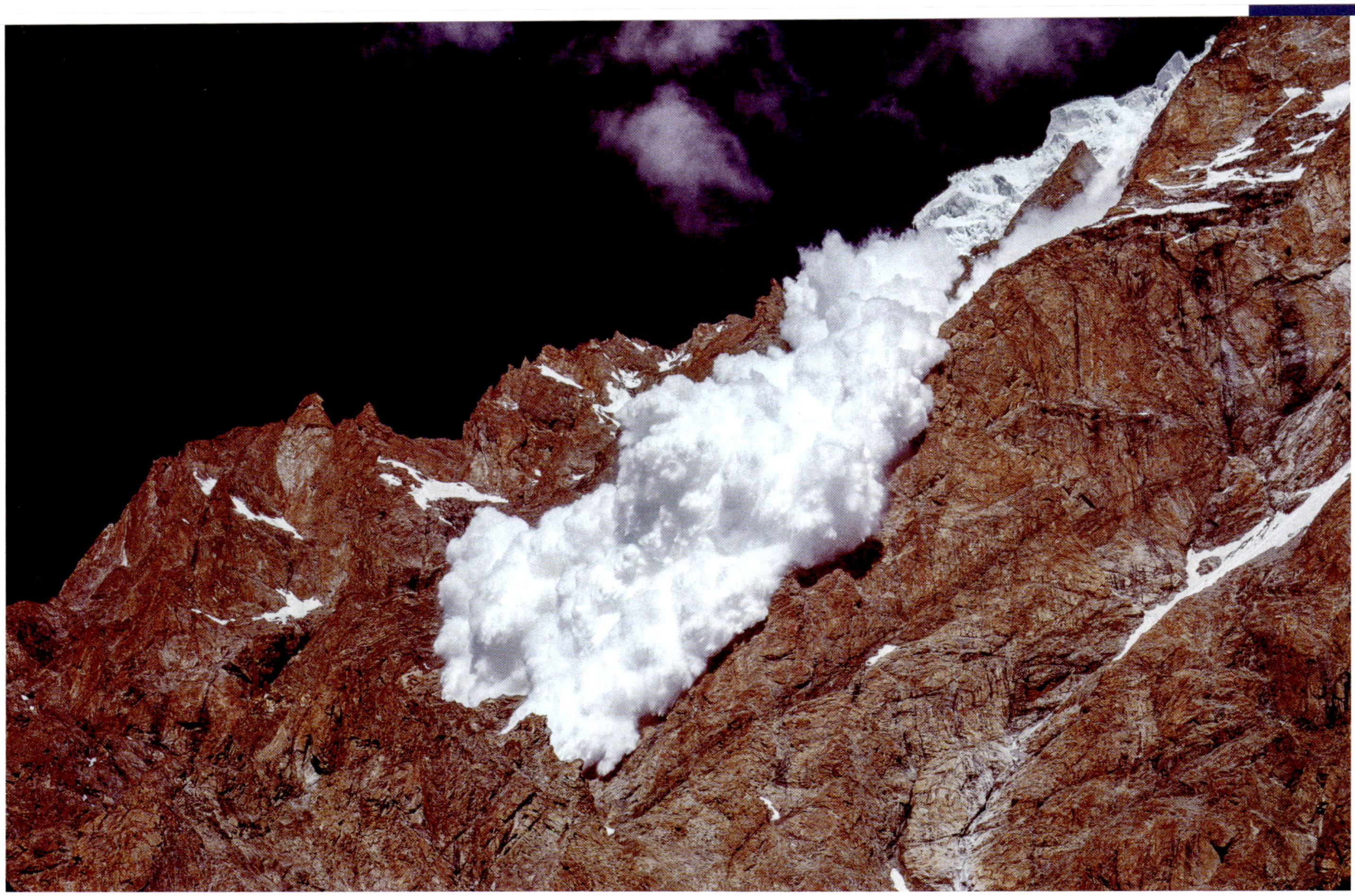

This page and opposite:
Avalanche above K2 Base Camp, Pakistan, 1986

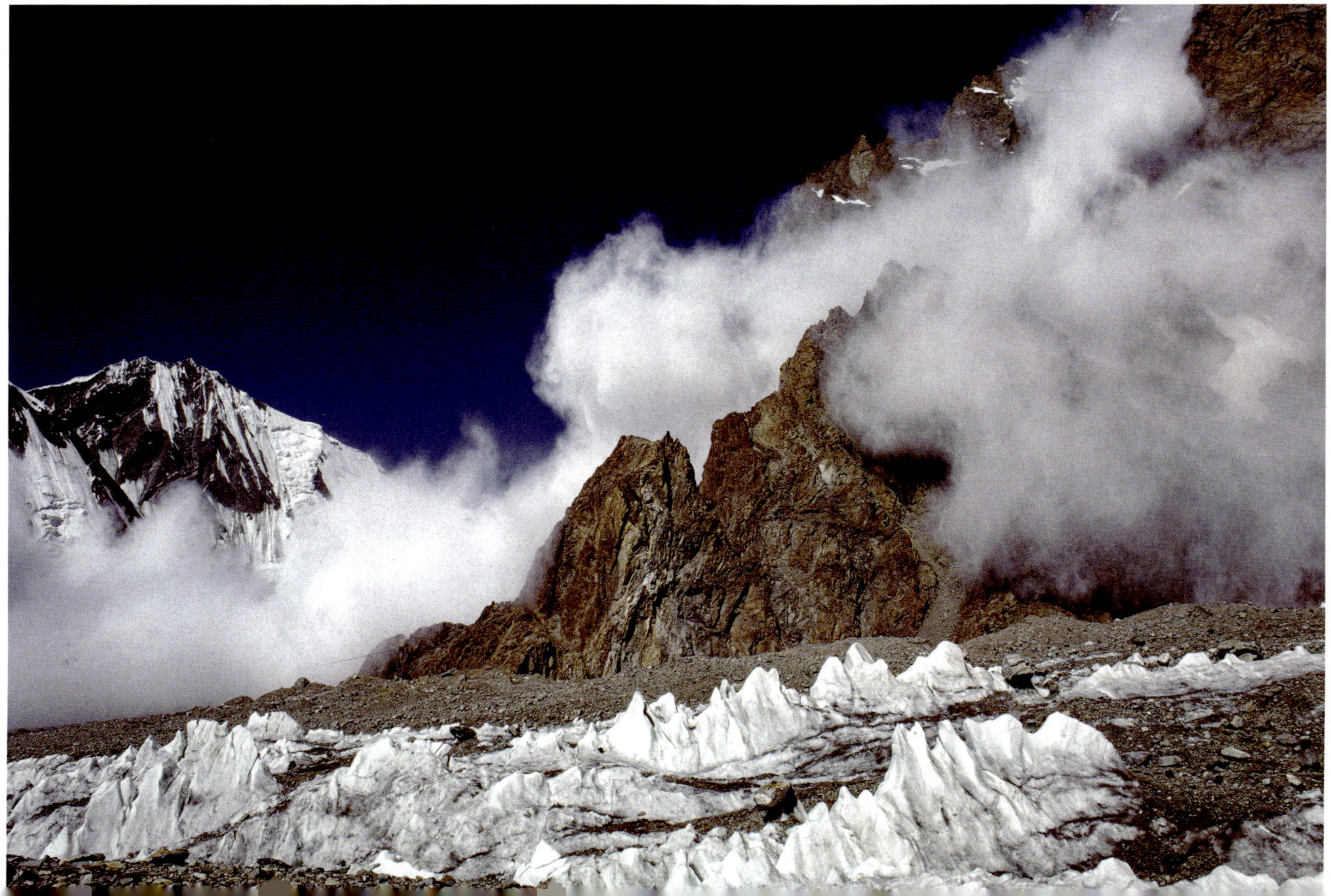

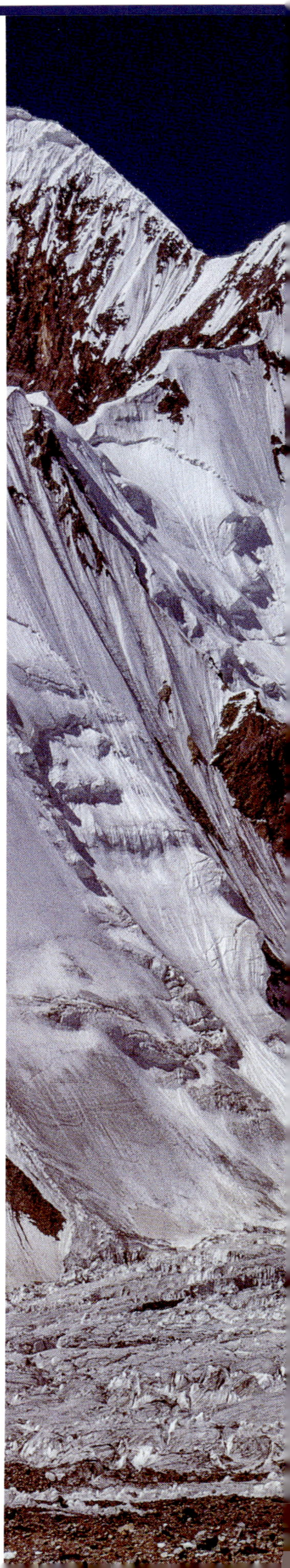

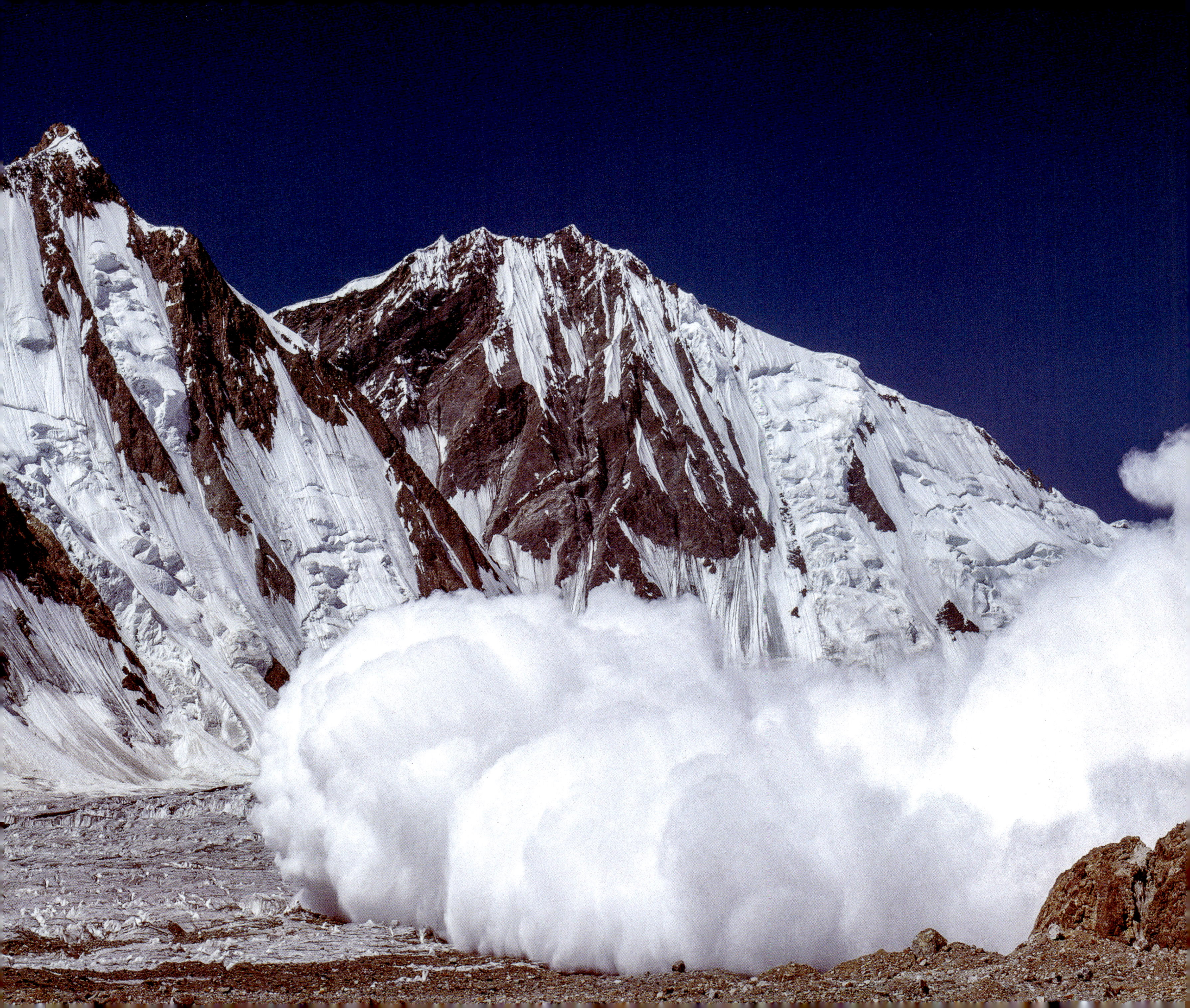

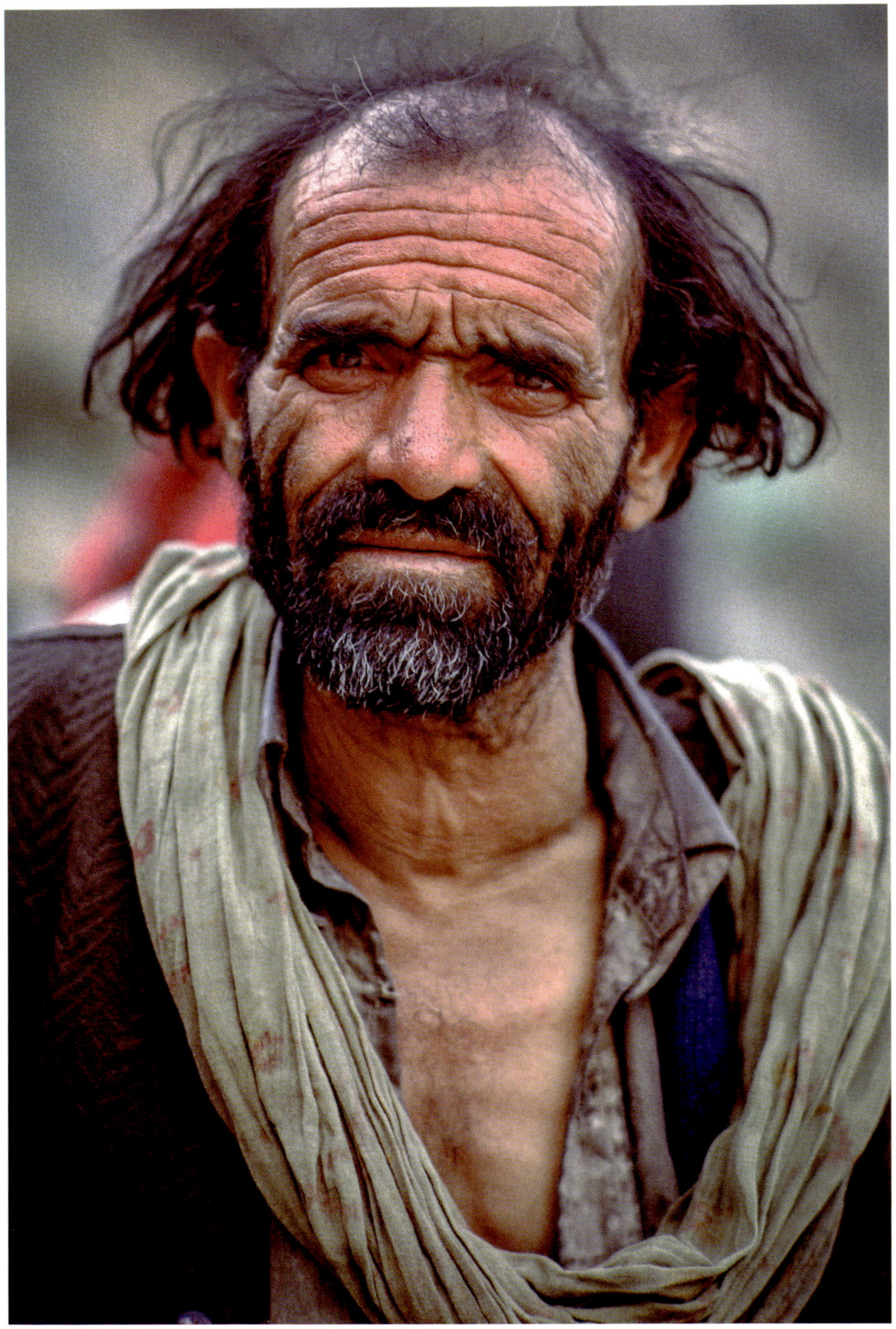

Porters on the trail to K2, Pakistan, 1986

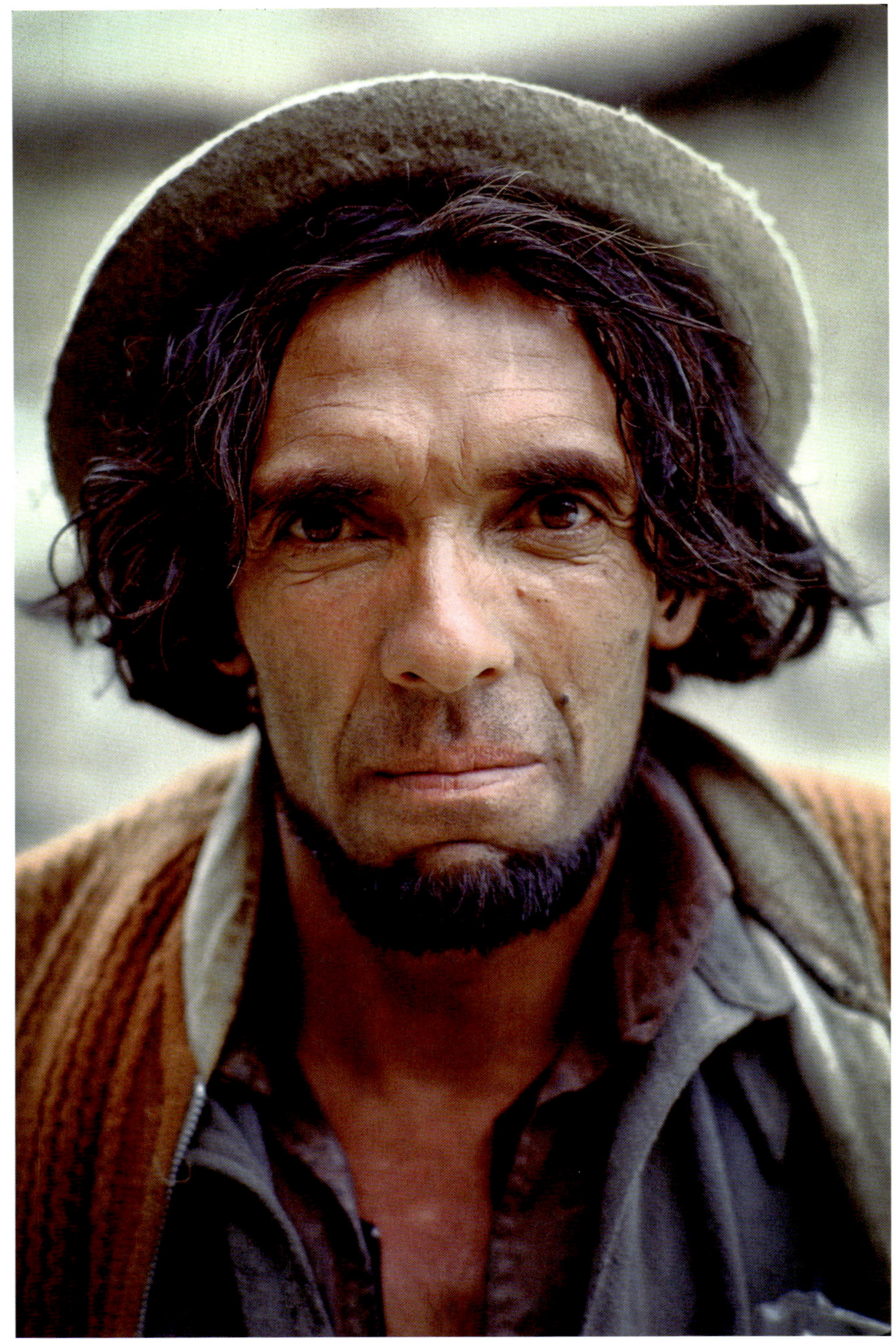

Opposite: Porters at Concordia, the confluence of the Baltoro and Godwin-Austen Glaciers, Karakoram Range, Pakistan, 1986

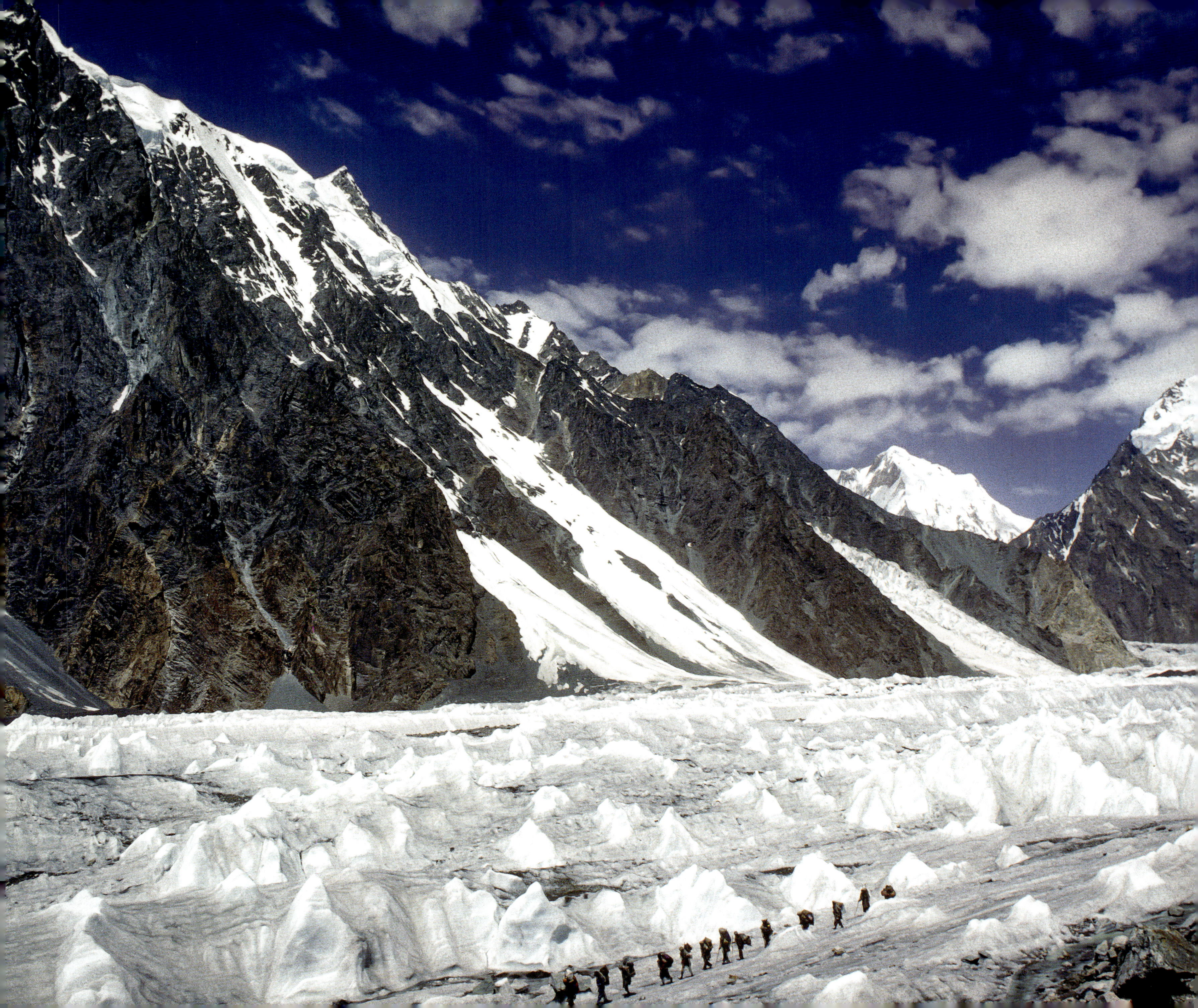

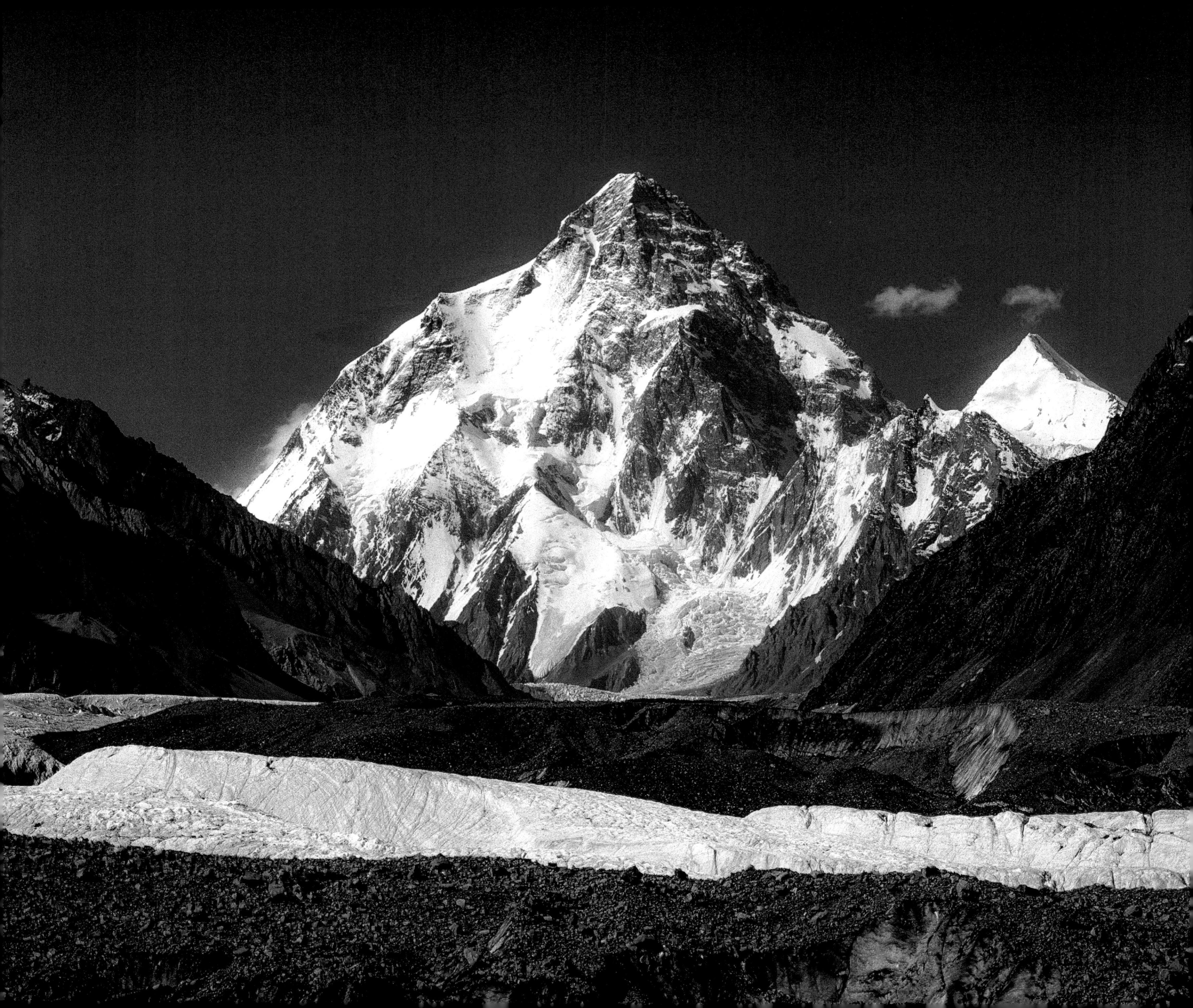

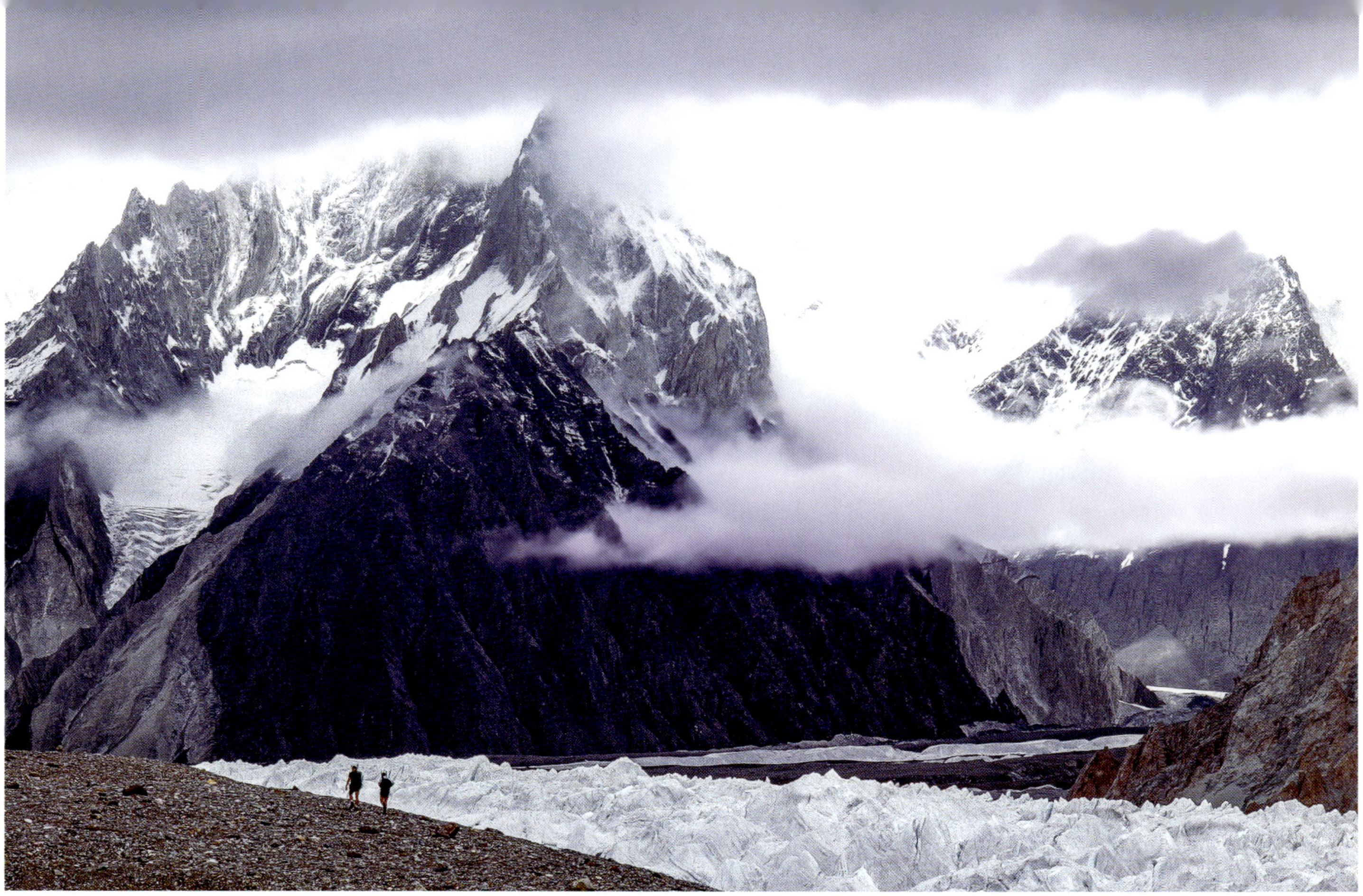

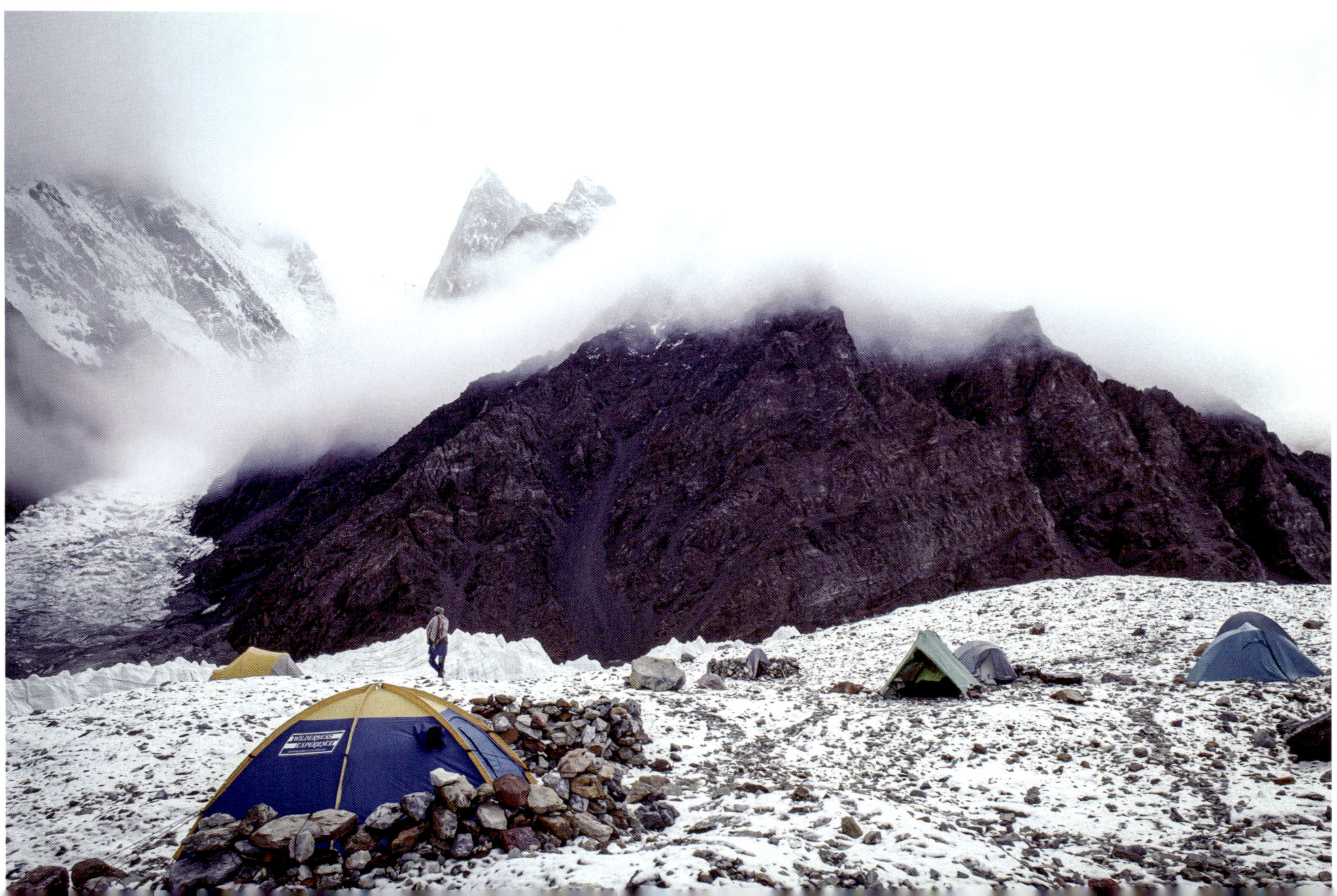

Above: On the trail to Gasherbrum I above the Baltoro Glacier, Pakistan, 1986

Left: Gasherbrum I (Hidden Peak) camp, Pakistan, 1986

Opposite: K2 from Concordia, Pakistan, 1986

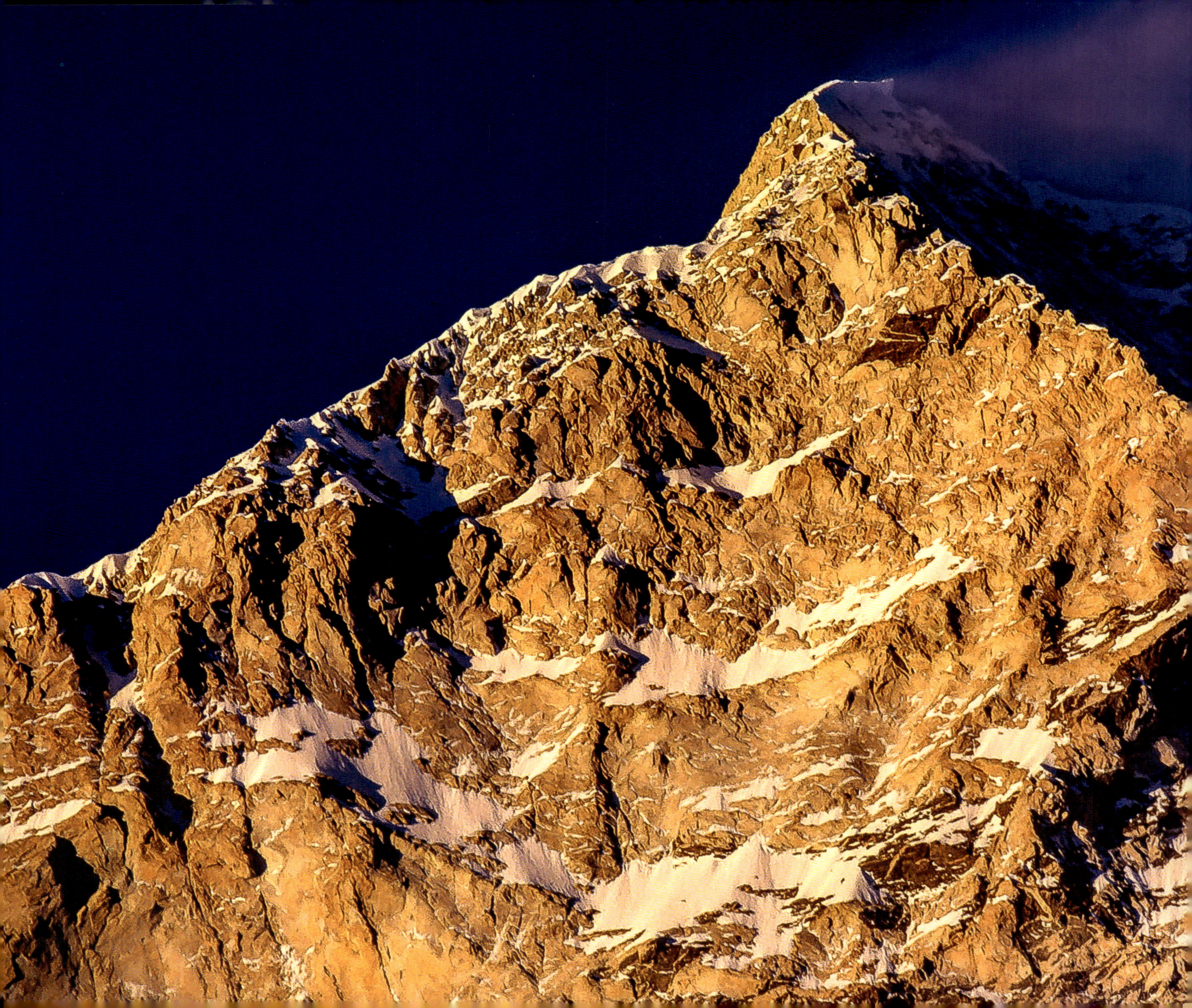

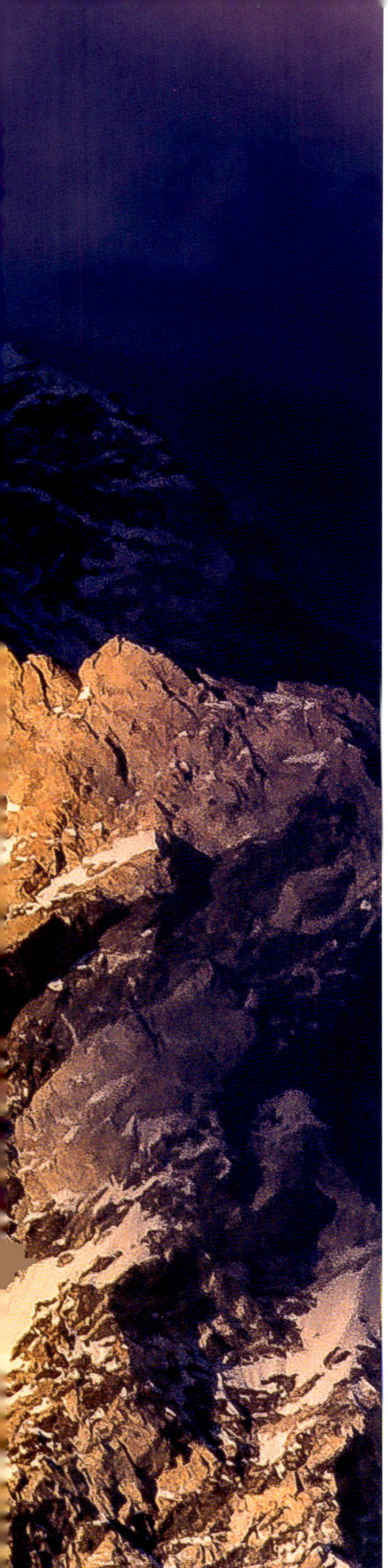

Up the Arun Valley to Makalu, Nepal

NOVEMBER 1986

Mount Makalu was an unknown destination to Bill and me—we knew it as the fifth-highest mountain in the world (elevation 27,766 feet) and one of only fourteen 8,000-meter peaks. A three-week round-trip walk to the Makalu Base Camp gave us a deep appreciation for the beauty and diversity of the Arun River Valley that leads to Makalu, a region that ranges from tropical to alpine in just about 100 miles.

Makalu is situated only around twelve miles southeast of Mount Everest. Although a trail connects Makalu to the Everest region, it is a technical track that requires crossing several high passes, including the notoriously difficult Sherpani Col. Eric Shipton, Edmund Hillary, and others took this route when they walked into this area, which Shipton called the "unexplored region east of Khumbu" in June 1952. They had no idea if the Barun and Arun Valleys below Makalu were inhabited and were very happy to find small villages and trails to follow on their way out by these valleys. Makalu was first climbed successfully in 1955 by a French and Sherpa team that put an amazing nine climbers on the summit over several days.

Our trekking route started in the town of Dhankuta, south of Makalu, and followed first the Arun Valley and then Barun Valley northward for two weeks to reach the Makalu

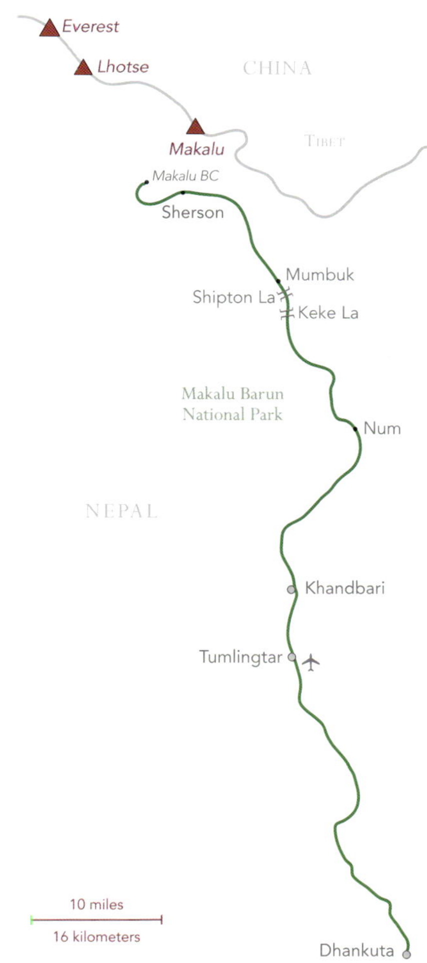

Makalu, from Makalu Base Camp, Nepal, 1986

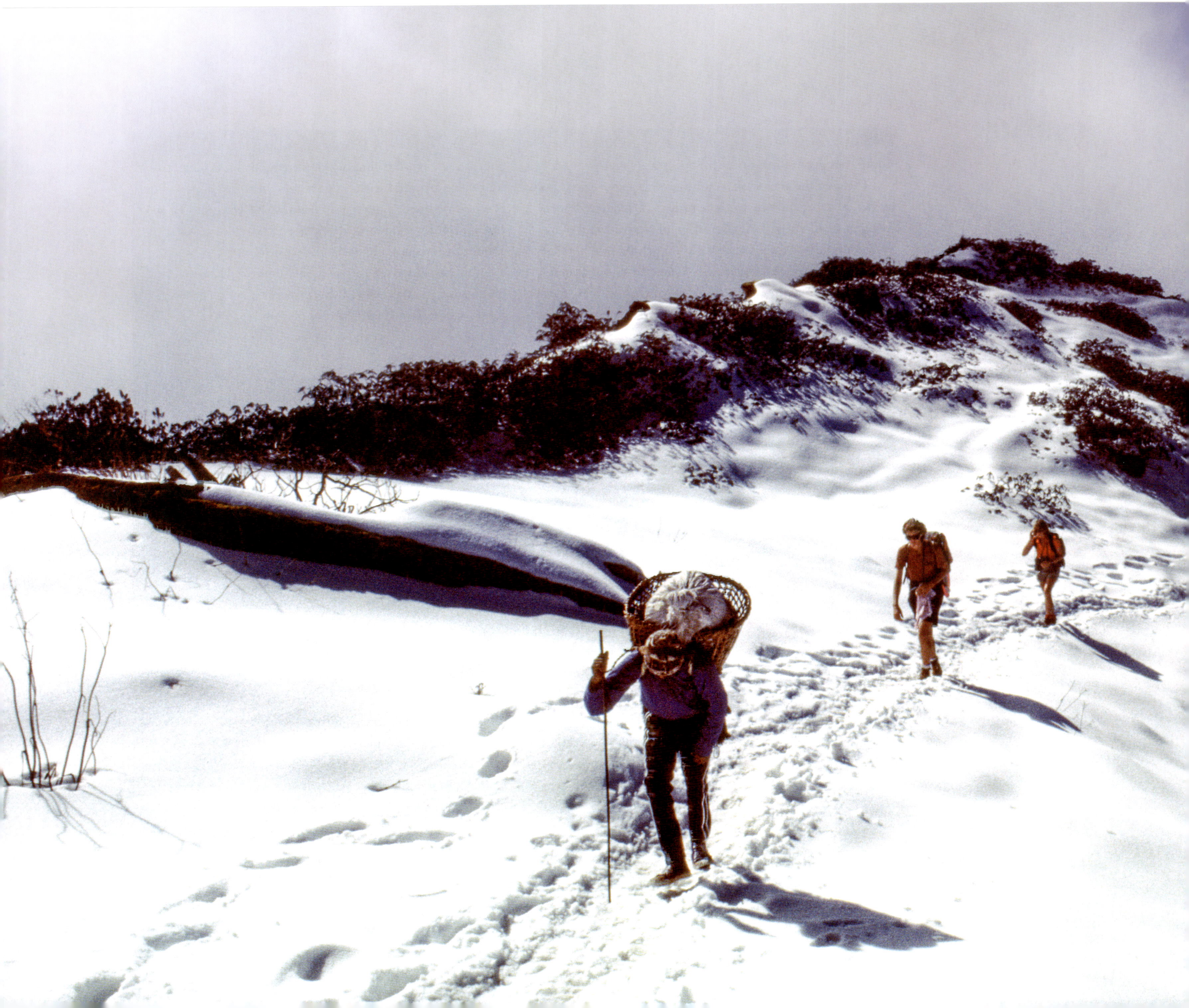

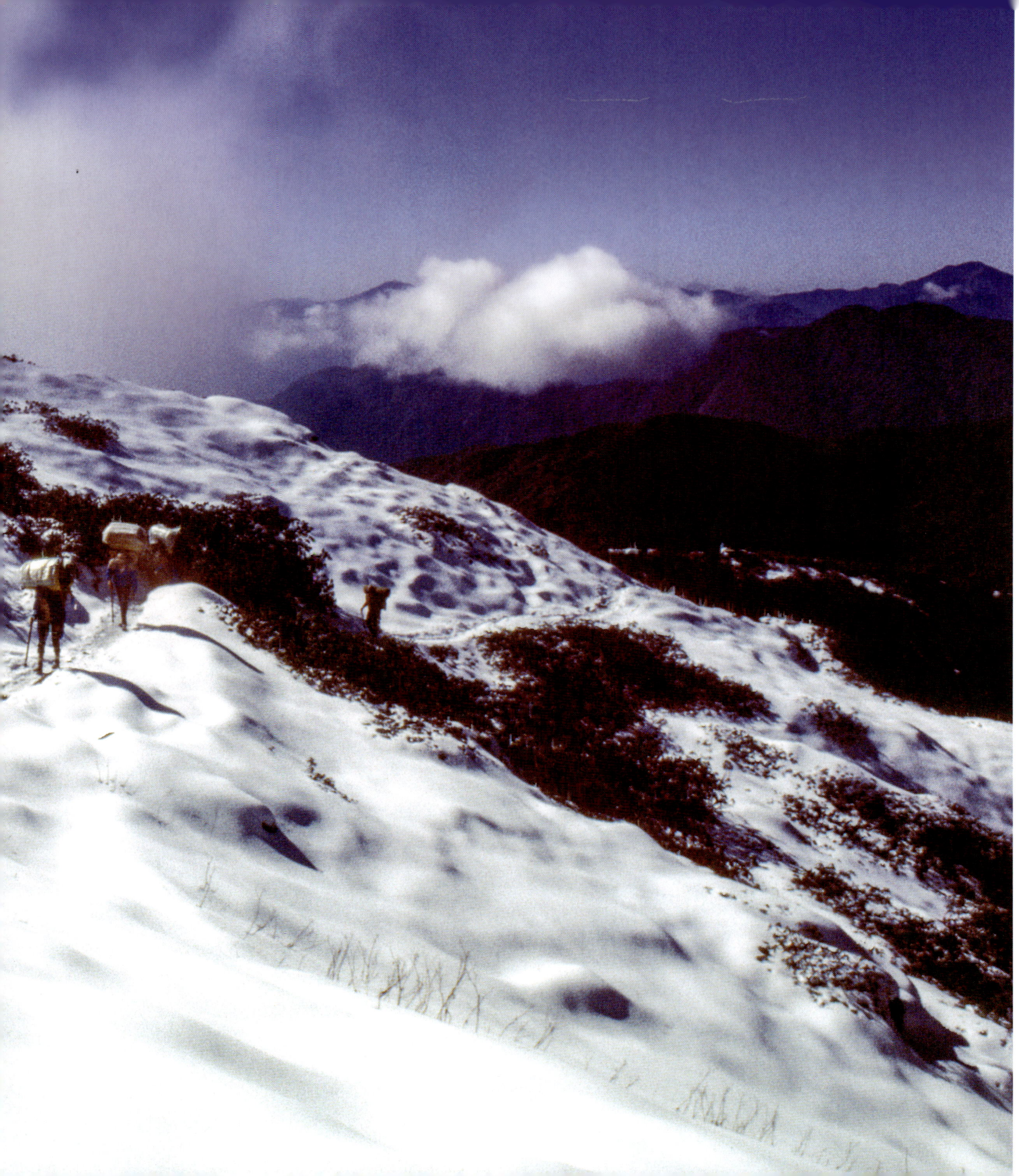

The descent from Shipton La, Nepal, 1986

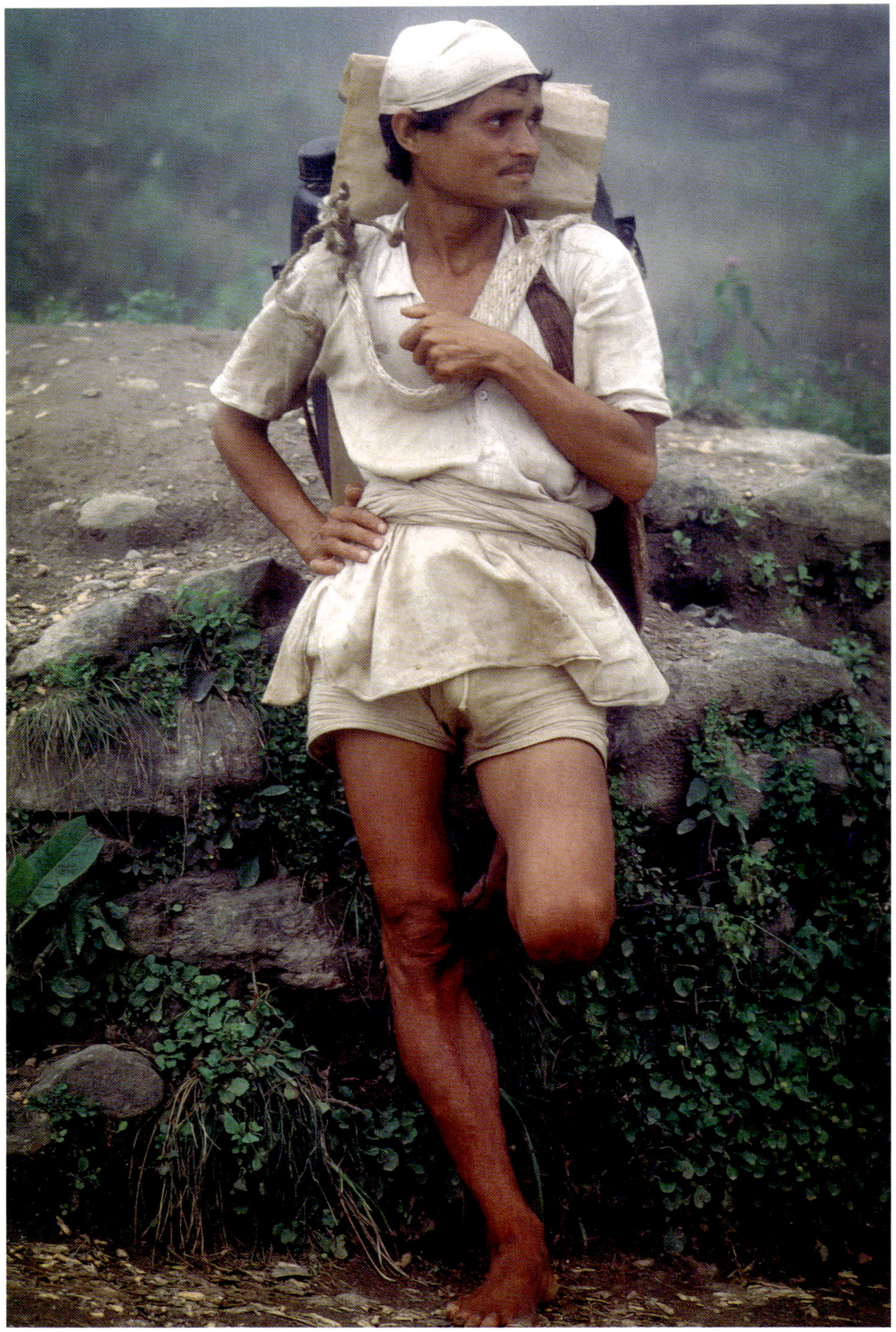

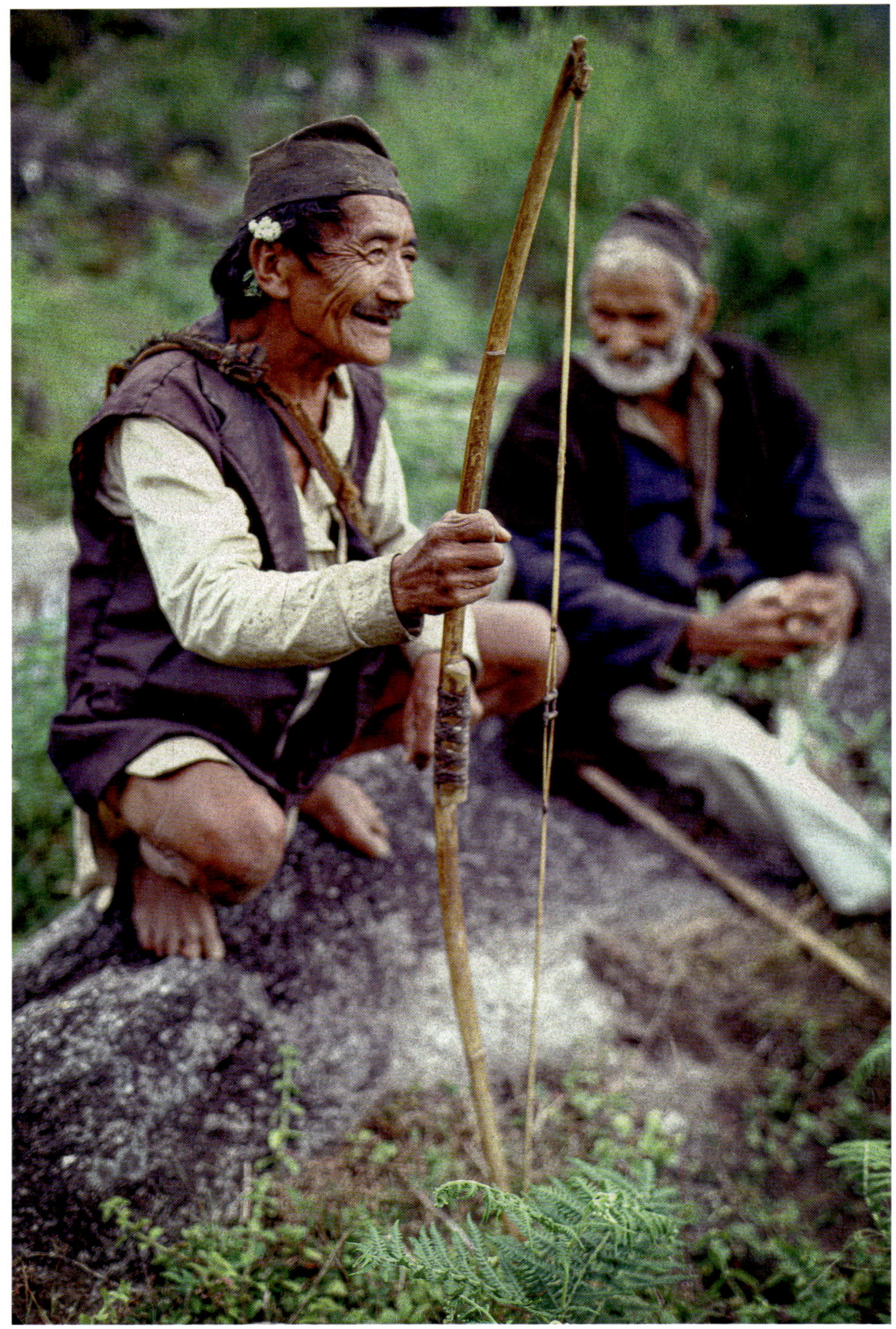

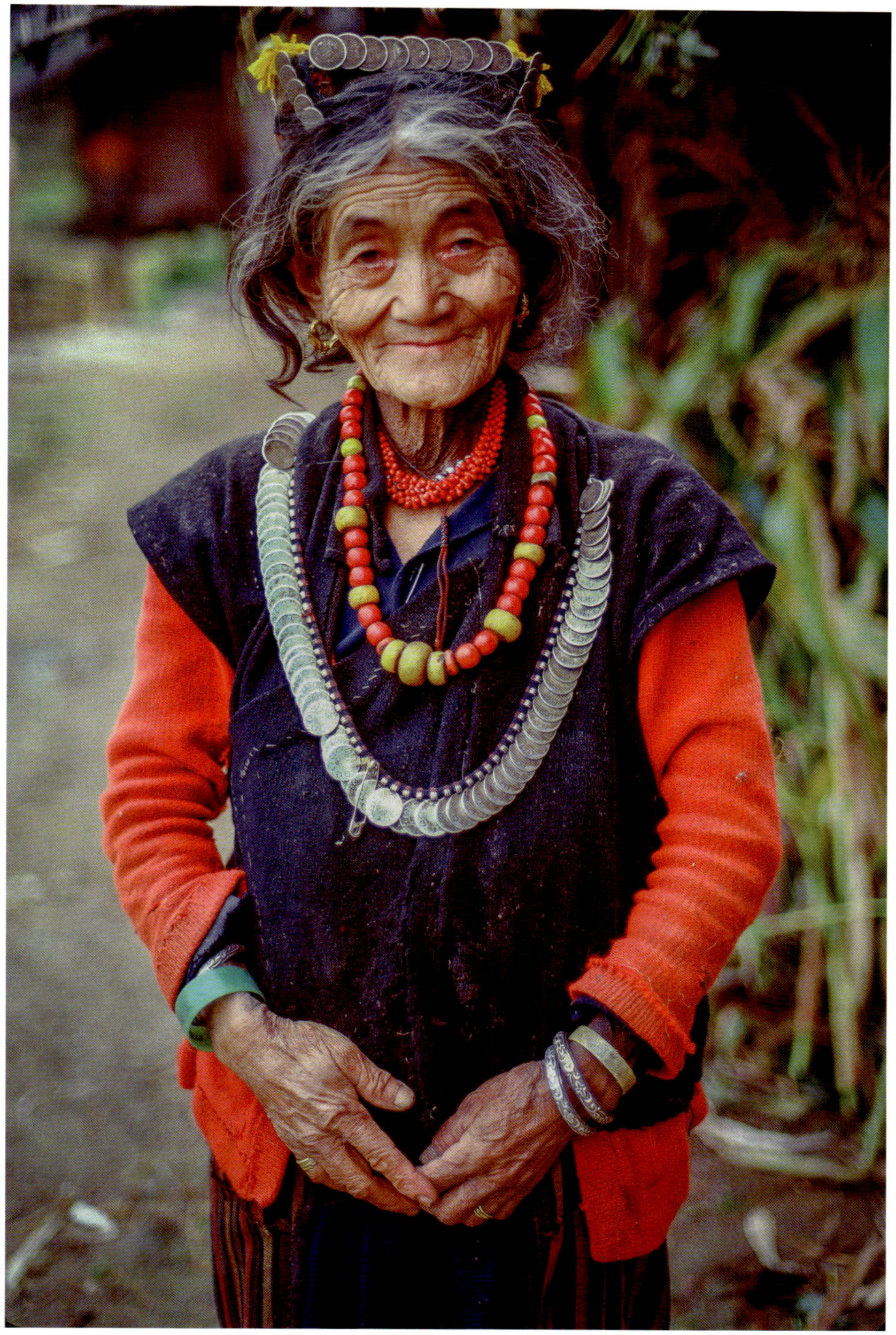

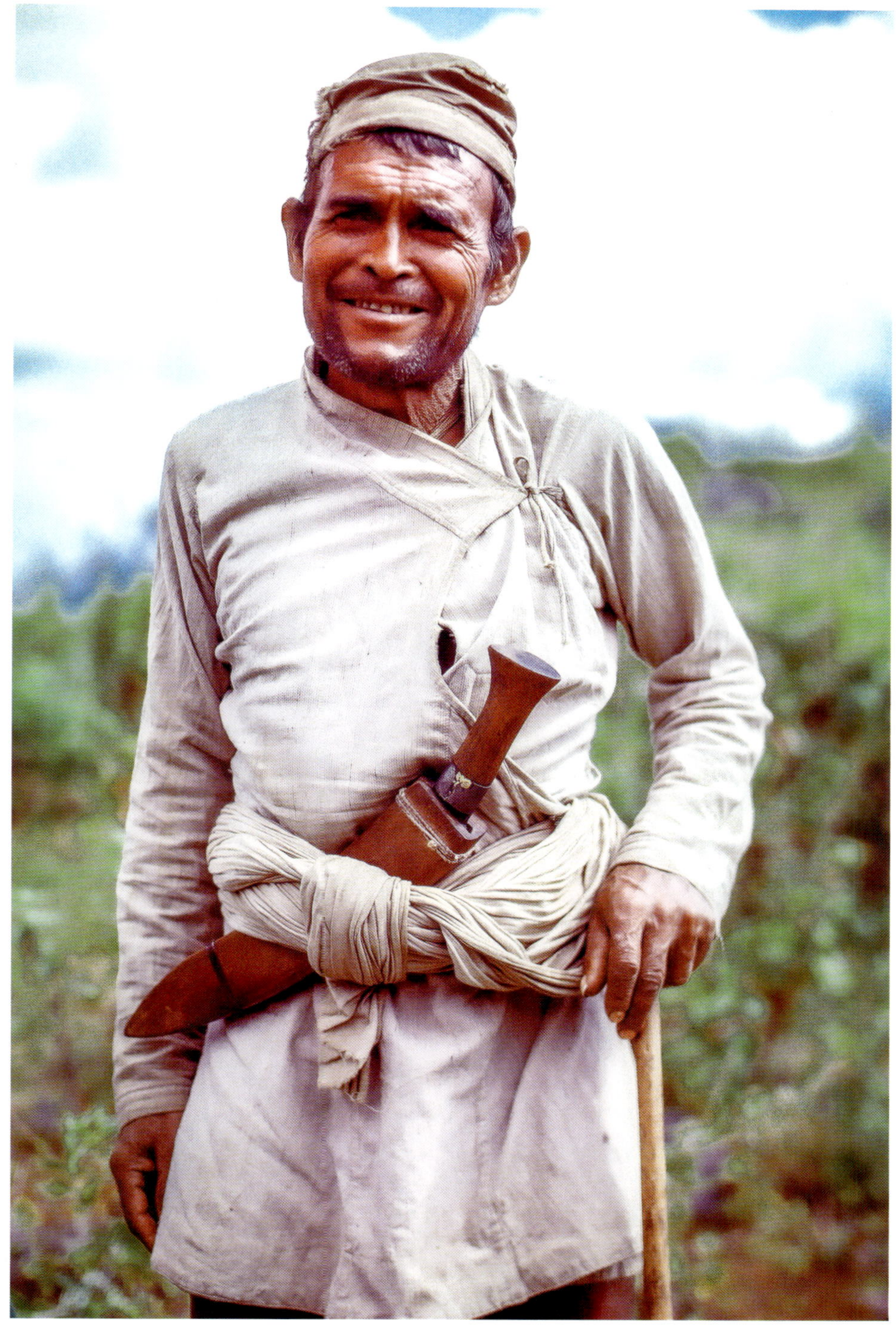

This page and opposite: Nepalis in the Arun Valley enroute to Makalu, 1986
Following spread: Chamlang from Makalu Base Camp, Nepal, 1986

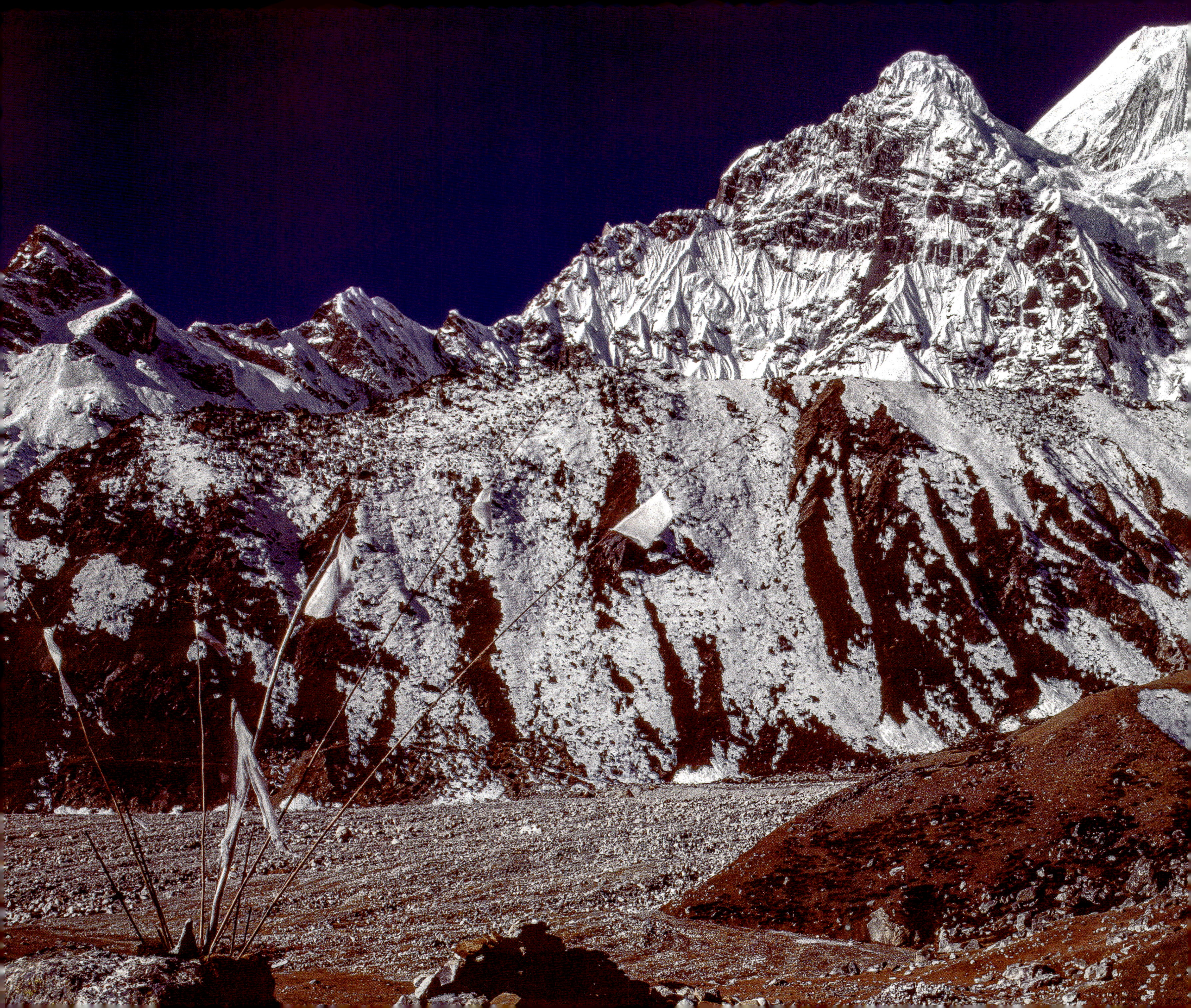

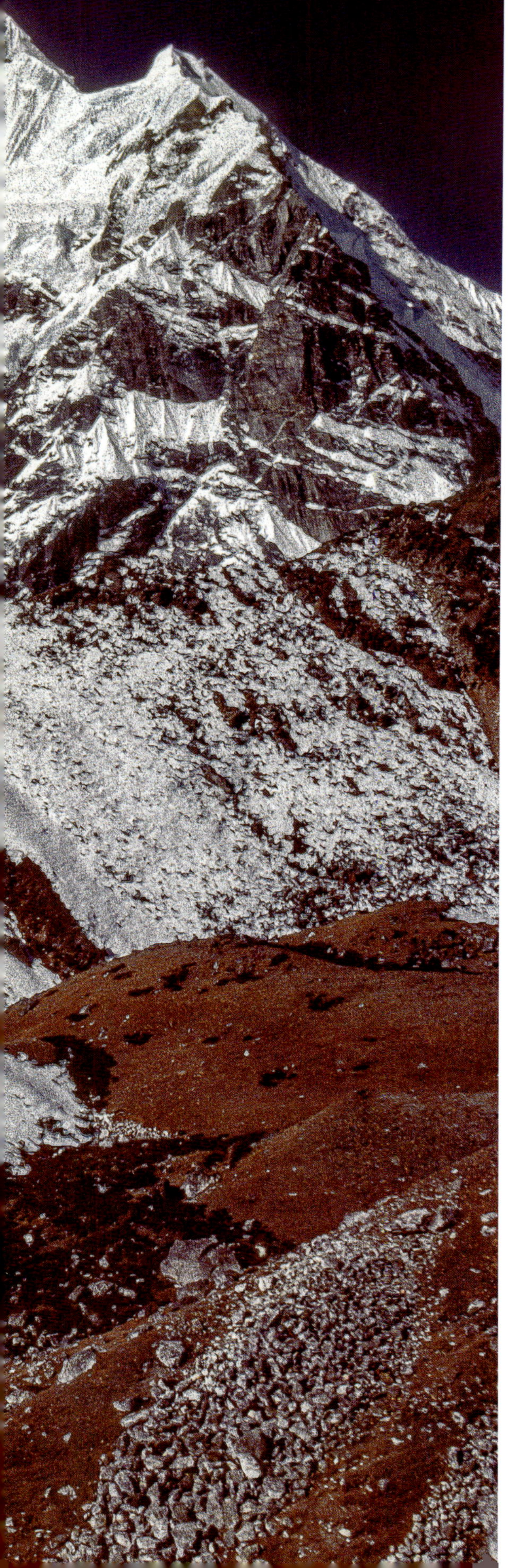
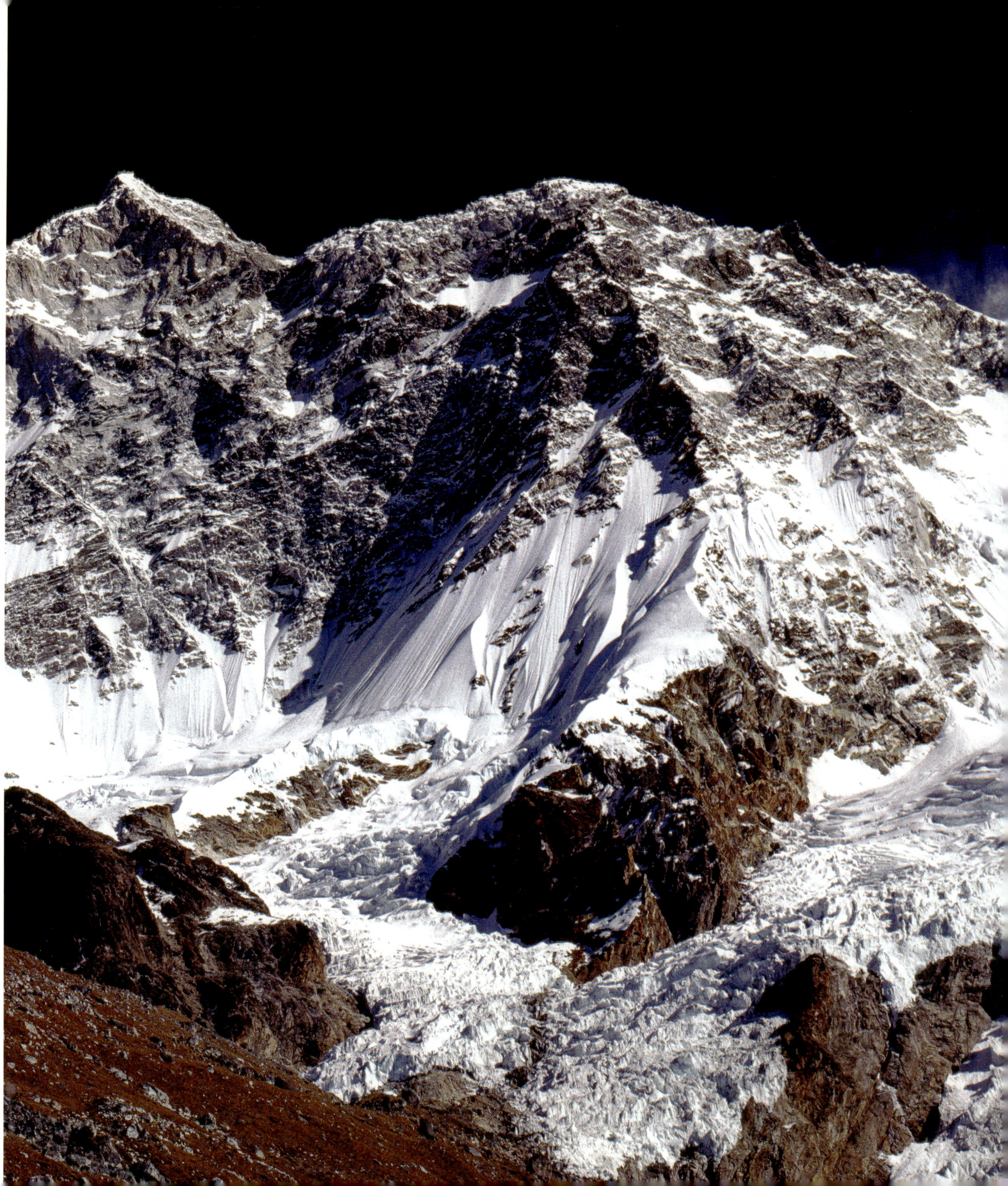

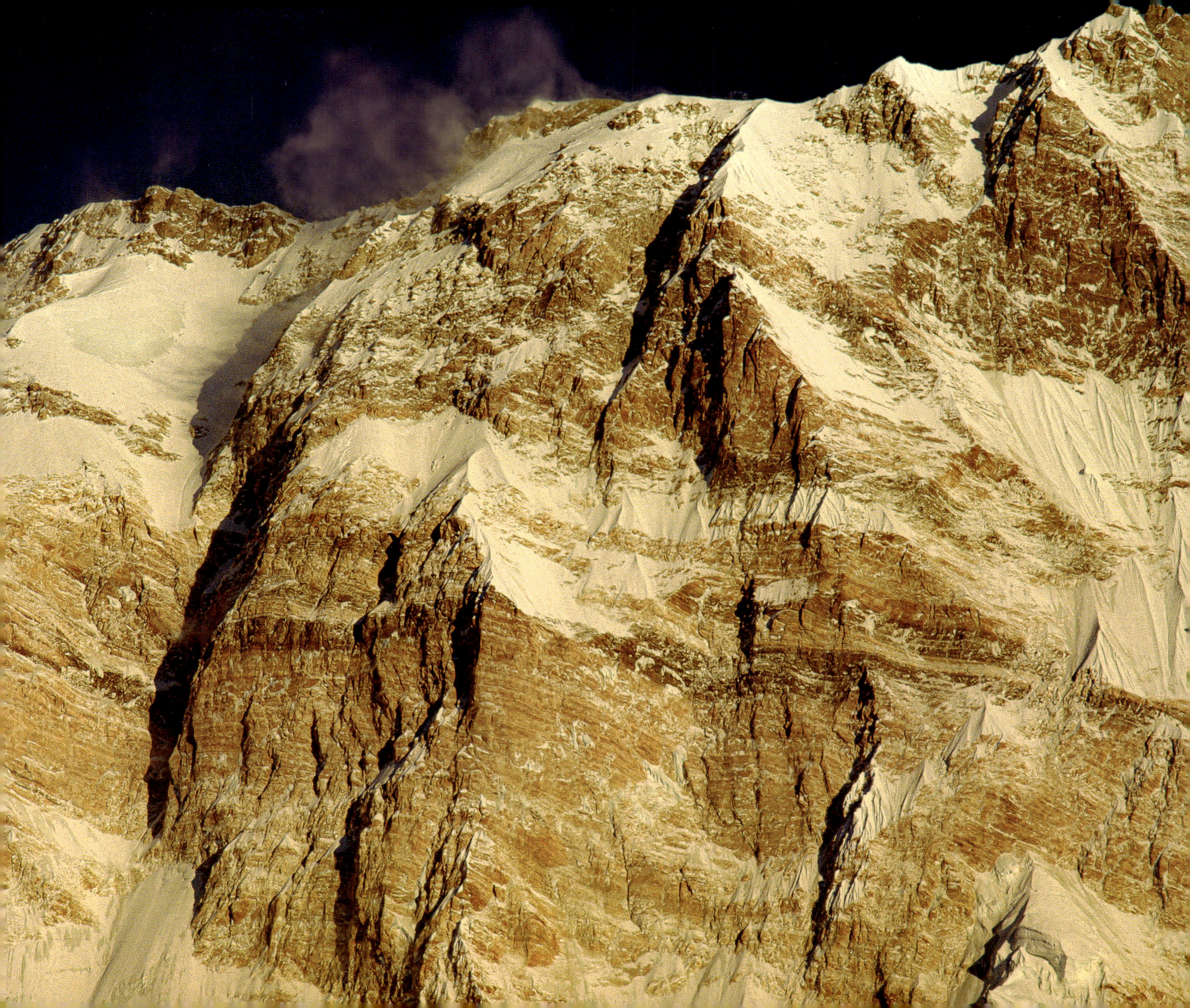

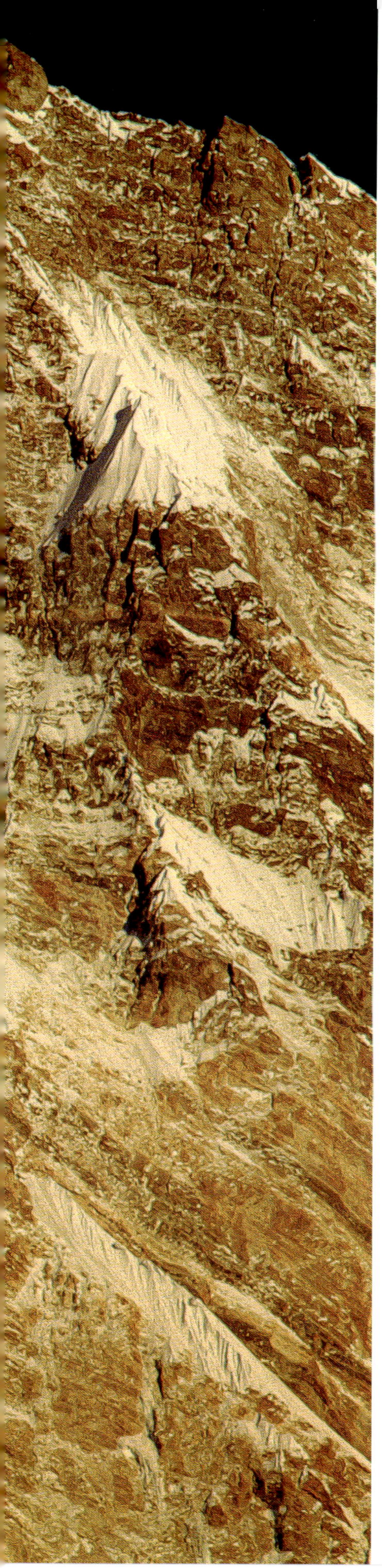

Annapurna, from Annapurna Base Camp, 1986

Into the Annapurna Sanctuary, Nepal

NOVEMBER 1986

Bill and I crawled out of our tent in the cold and dark at 6:00 a.m. to wait for the first rays of sun to hit the mountains surrounding us. At 6:30 sunlight from behind Machapuchare, a distinctively shaped mountain to our east, struck the top of Annapurna I and then lit up the western ridge of Annapurna South. In less than an hour the frigid white expanse surrounding us was glowing in a warm amber color. We were camped in the Annapurna Sanctuary, one of the best places on Earth for magnificent mountain views. The aptly-named sanctuary is a huge natural amphitheater defined by nine peaks more than 20,000 feet in elevation. It is easy to understand why the sanctuary and many of the mountains surrounding it are considered sacred.

The tallest, Annapurna I, is the tenth-highest mountain in the world (elevation 26,545 feet). Annapurna was initially climbed by way of its northern face in June 1950 by Maurice Herzog and Louis Lachenal, members of a French expedition. It was the first of the 8,000-meter peaks to be climbed, a highly celebrated achievement in France and around the world. Today, it is hard to imagine how difficult and unpredictable it was to complete the first ascent of a mountain in those early days of Himalayan mountaineering. When Herzog's team set out to climb either Dhaulagiri or Annapurna, the region was largely unexplored, and the few available

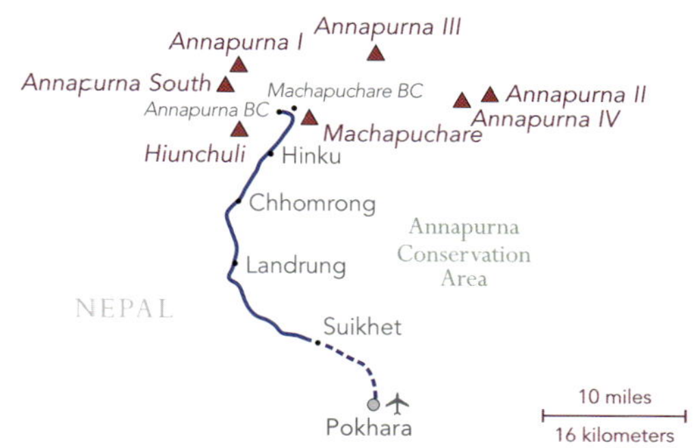

maps were unreliable. It was difficult even to find Annapurna since it was hidden behind other peaks. Herzog's riveting account of the climb in his book, *Annapurna* (1951), describes how he and Lachenal faced a harrowing descent down the mountain, complete with the loss of gloves, long, precarious slides, and a fall into a crevasse. Eventually, they had to be carried out on sledges on the backs of Sherpas. Both men suffered frostbite that required the expedition's doctor to amputate their fingers and toes on the long trek out, without anesthesia, in village houses, in the middle of fields, and finally on a train in India, where Herzog watched as a pile of his toes was swept out onto the station platform.

Fortunately, our walk into the Annapurna Sanctuary was relatively straightforward. From the town of Pokhara we took a cab to a small place called Suikhet and started walking.

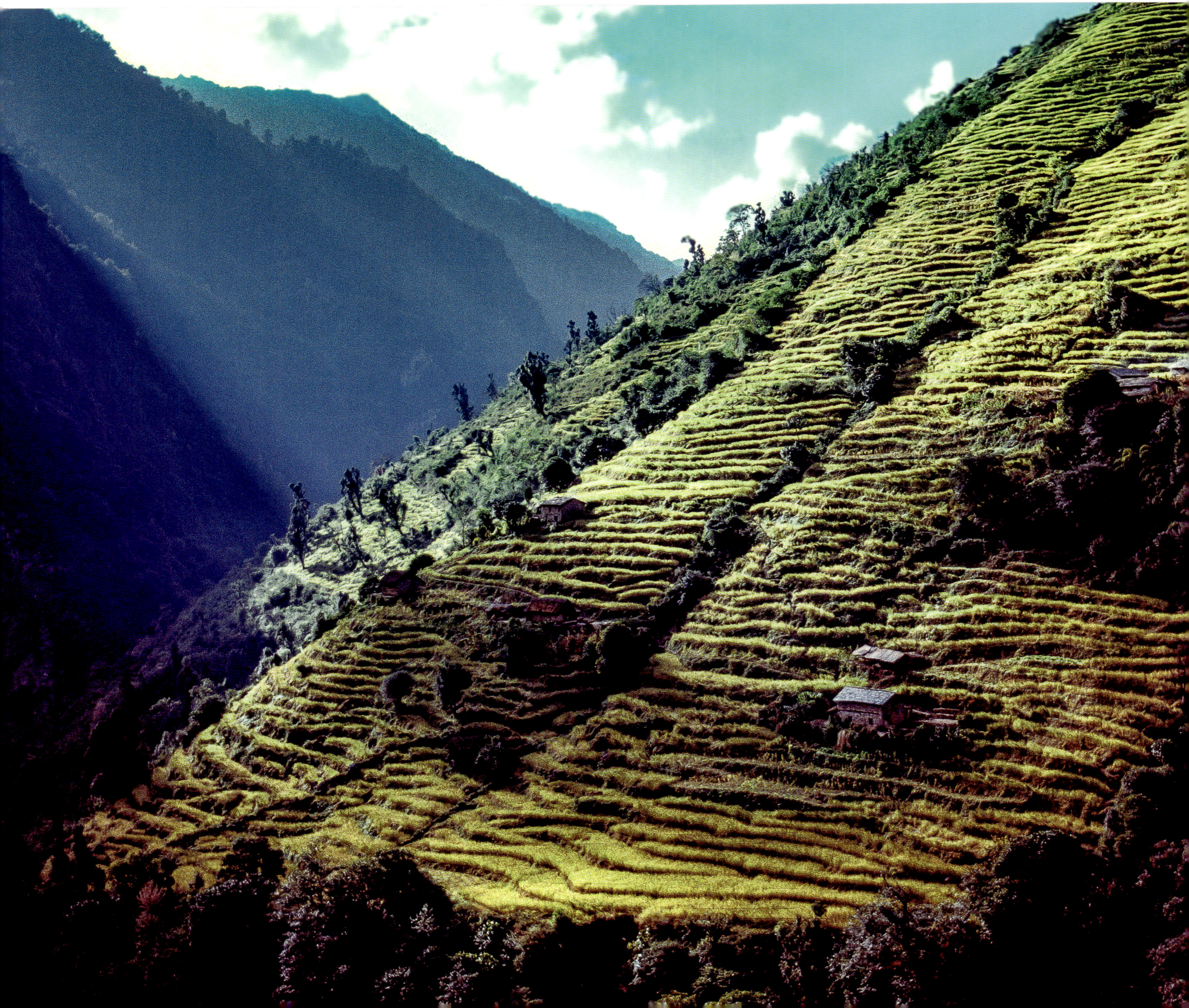

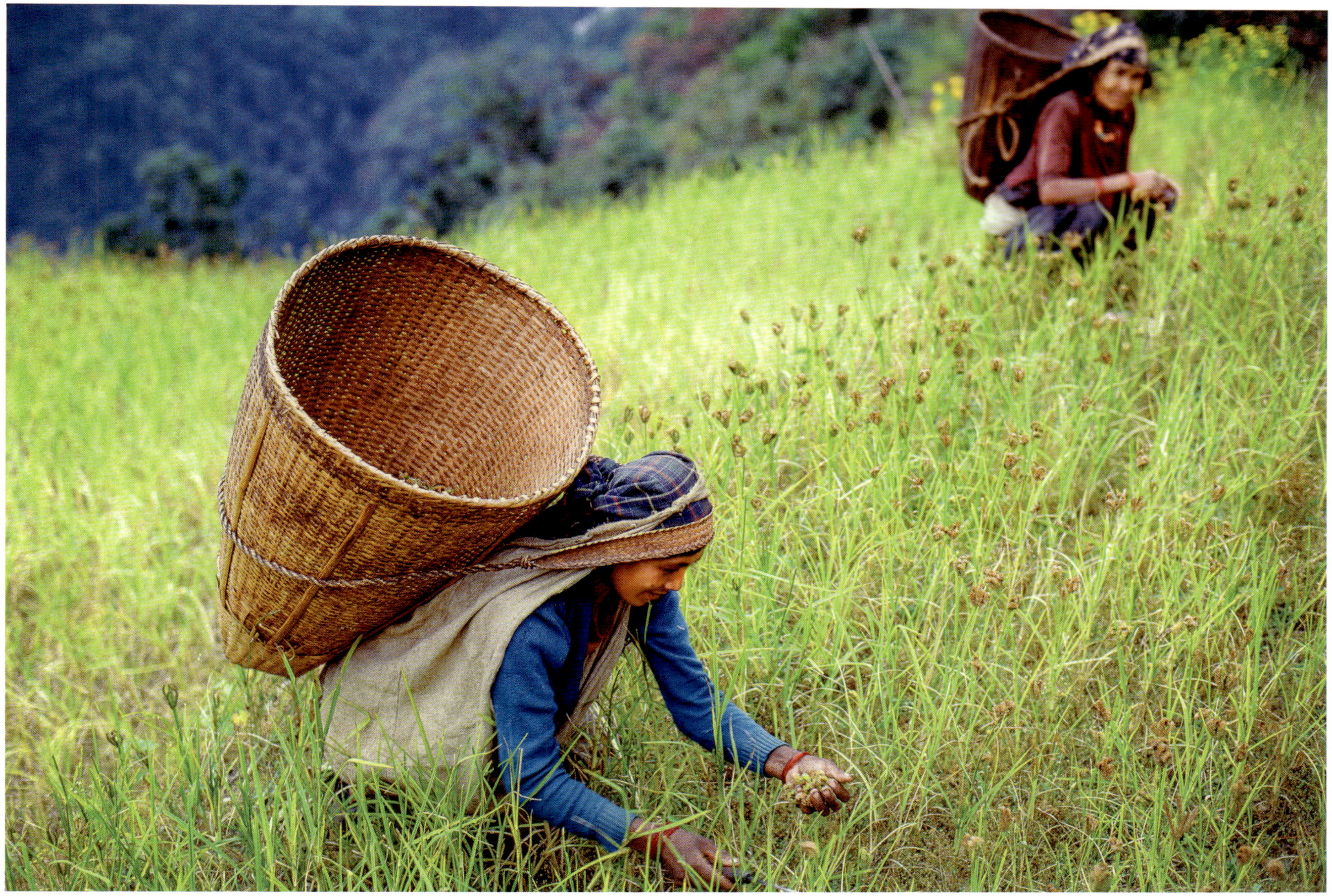

Sorghum fields on the Annapurna Sanctuary trek, Nepal, 1986

Rice terraces near Chhomrong on the Annapurna Sanctuary trek, Nepal, 1986

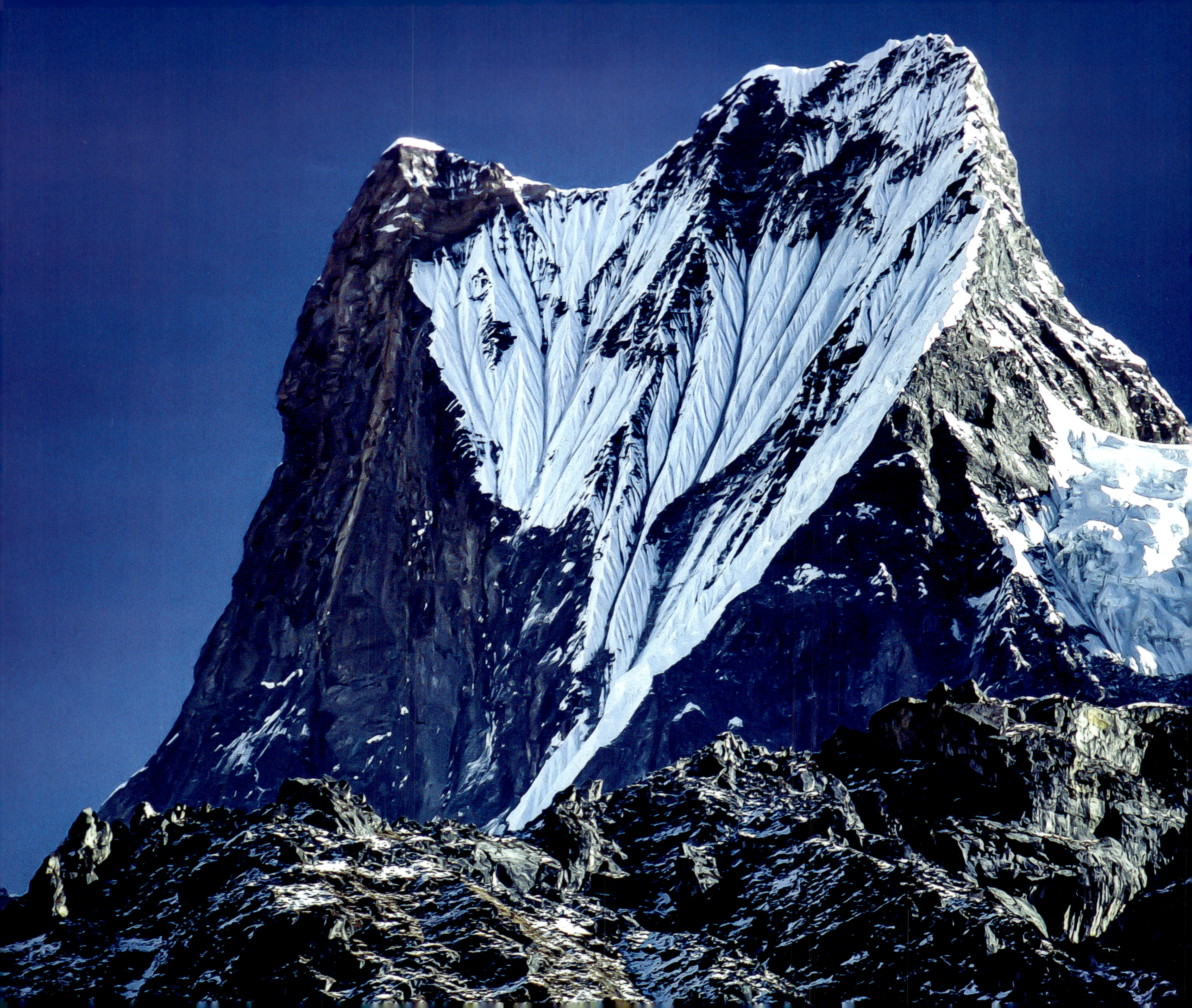

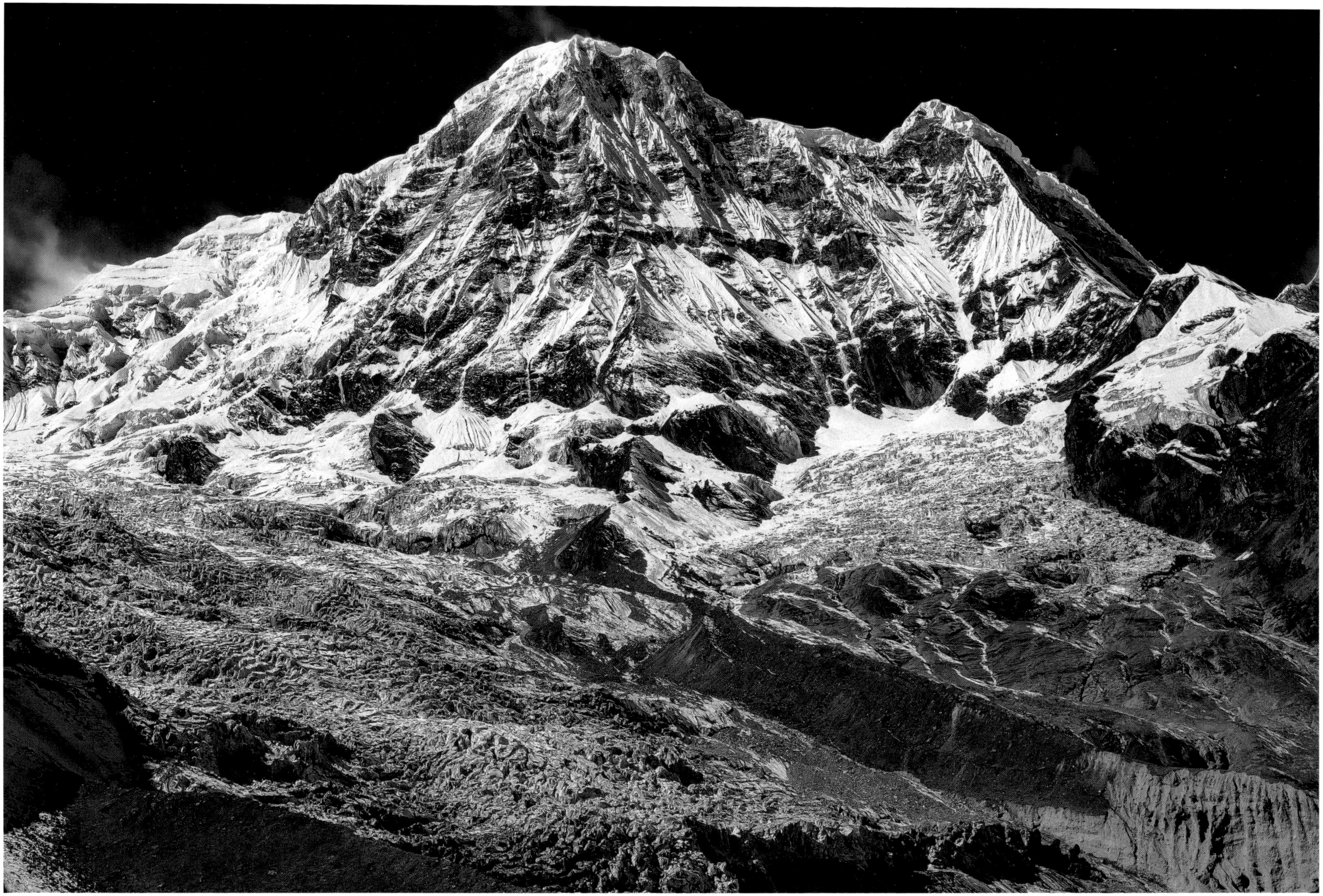

Annapurna South Base Camp, Nepal, 1986

Opposite: Machapuchare, Nepal, 1986

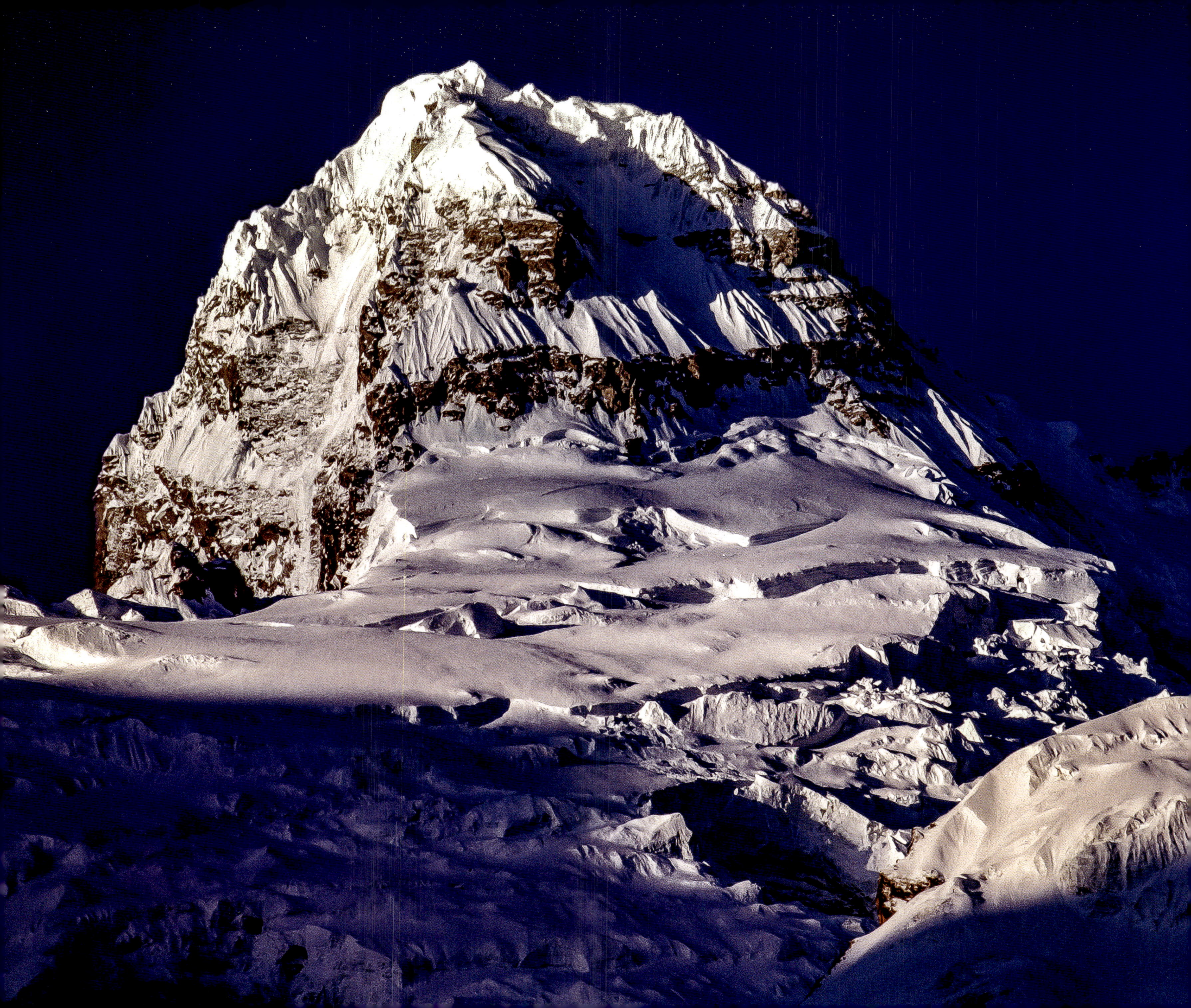

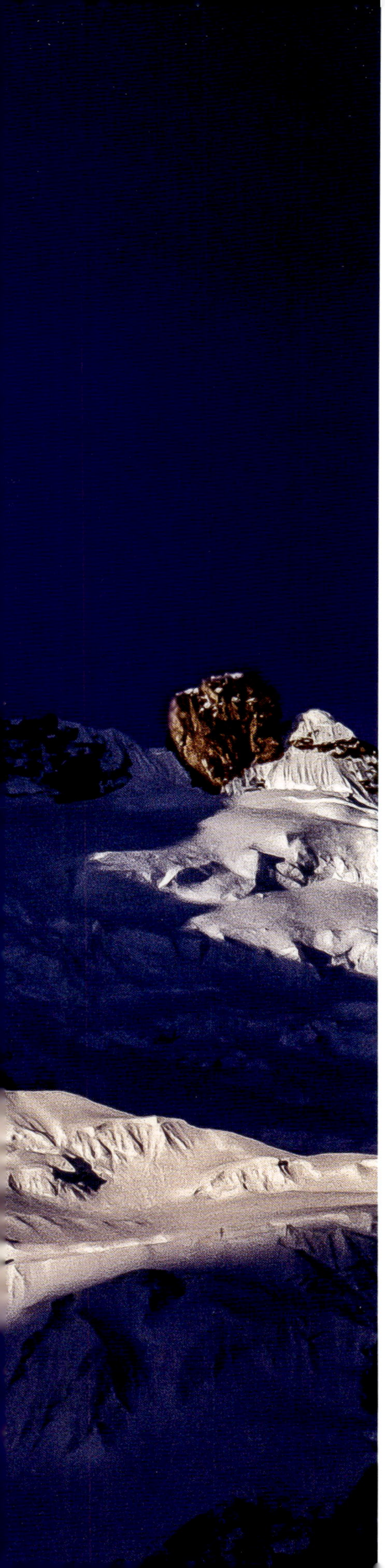

Through the Tsum Valley to the Ganesh Sanctuary, Nepal

MAY 1991

THE TSUM VALLEY IN 1991 was a remote and mysterious place with wary locals and still-cherished tales of Milarepa, the Tibetan scholar who spent months meditating in caves there nearly 900 years ago. Walking up through the valley, our destination was the base of Ganesh I, the tallest of eighteen peaks that make up the Ganesh Himal Range and create a semi-enclosed sanctuary. Ganesh I, also known as Yangra, is a distinctive mountain with the appearance of Ganesh, the elephant-headed Hindu god who is believed to be the remover of obstacles. Bill, Ram, and I had been planning to visit this area for years and were thrilled to set out, not knowing if it would be possible to enter the Tsum Valley since it was still a restricted area. Even then there were few detailed written accounts of travel in the area, the most interesting by David Snellgrove, a Tibetologist who walked across Nepal in 1956 and wrote *Himalayan Pilgrimage: A Study of Tibetan Religion by a Traveller Through Western Tibet* (1988). The Tsum Valley officially opened to tourism in 2008, and some small hotels have been built. Plans are also underway to construct a road through the Tsum Valley as a connection to China, so many changes are inevitable.

Our trek started in the town of Gorkha. For eight days we followed the Buri Gandaki River on a trail we had taken ten years earlier on our way to Manaslu. It was May, and the weather was hot, with long days and warm nights and land-

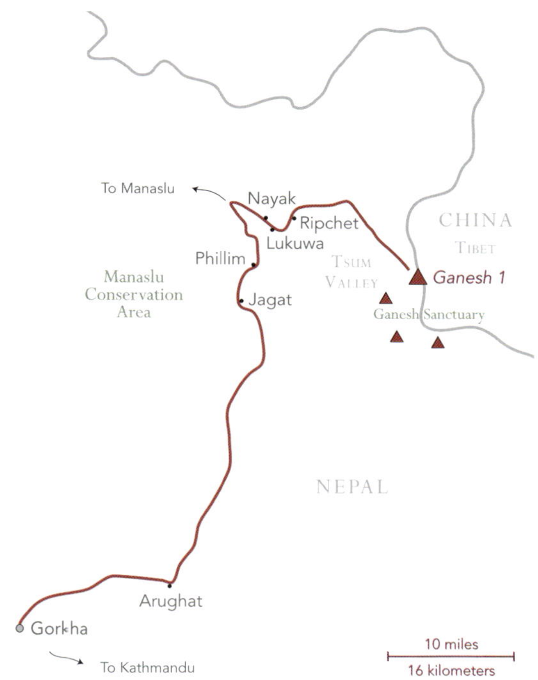

scapes of flowering pink and white rhododendrons covering hillsides in the distance. We walked through small villages populated by the Gurung ethnic group. Near the village of Deng we saw a shaman sitting at a small altar filled with strange objects playing a drum. Just a short distance later, we noticed fleas that stayed with us for the rest of the trip, and we wondered if the shaman was trying to tell us something.

The trail was exposed with steep drop-offs and frequent river crossings on structures that ranged from lashed-together

Ganesh I from Ganesh I Base Camp, Nepal, 1991

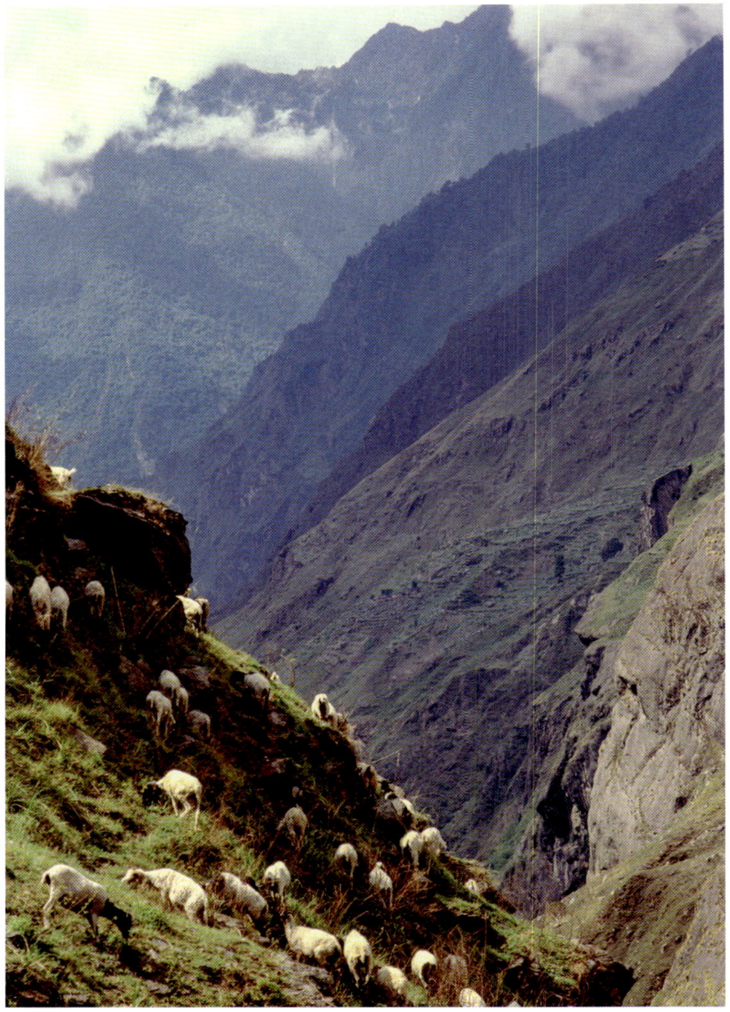

The trail to the Tsum Valley, Nepal, 1991

logs over the narrow sections to long, pedestrian suspension bridges high above the wide sections. We turned to the east at the village of Nyak and hiked down a steep trail to cross the Buri Gandaki once again. Soon, we entered the Tsum Valley and immediately saw the change from the Gurung people of the middle hills to the Tsumbas who live in the higher elevations and embrace a more pure Tibetan culture, as reflected in their faces and dress. We heard music coming from somewhere below us, so we followed it until we reached a small village where a Buddhist lama and other monks in maroon robes with bells and *dorjes* (small ritual scepters) were leading a procession of chanting men and women. The gompa there was just a small white stone and stucco structure marked with a pine pole and prayer flags. Inside were brightly painted Buddha figures rather than the ornate bronze and gold-covered figures we often saw in larger gompas.

The older people we met were friendly but guarded, while teenagers examined us menacingly and hung around our camps all night. As in many of the difficult-to-reach places across the Himalaya, there were no medical facilities, so local people would gather at our tent asking for medicine or treatment for everything ranging from burns to a dehydrated baby. We did the best we could with first-aid cream, band aids, aspirin, and advice.

On the eleventh day the trail turned to the southeast, and we could see the Ganesh Himal Range in the distance. There were no more villages, and the trail became narrower and was often blocked by landslides of loose dirt and rocks that we had to scramble over. It took another day before we reached our camping spot in the Ganesh Sanctuary, a space defined by snowy peaks and filled with pink azaleas and yellow and blue wildflowers. We spent the night in the sanctuary and awoke to mountain views that lasted an hour before the clouds rolled in.

Upon returning to the main trail we met a Tibetan Buddhist Rinpoche (religious leader) who was on his way to inspect gompas in the region. We asked him if we could reach the ancient Mu (Moo) Gompa, several days' walk to the north, not far from the border. He told us that the trails were bad and restricted, so we should try again later. Maybe we will one day.

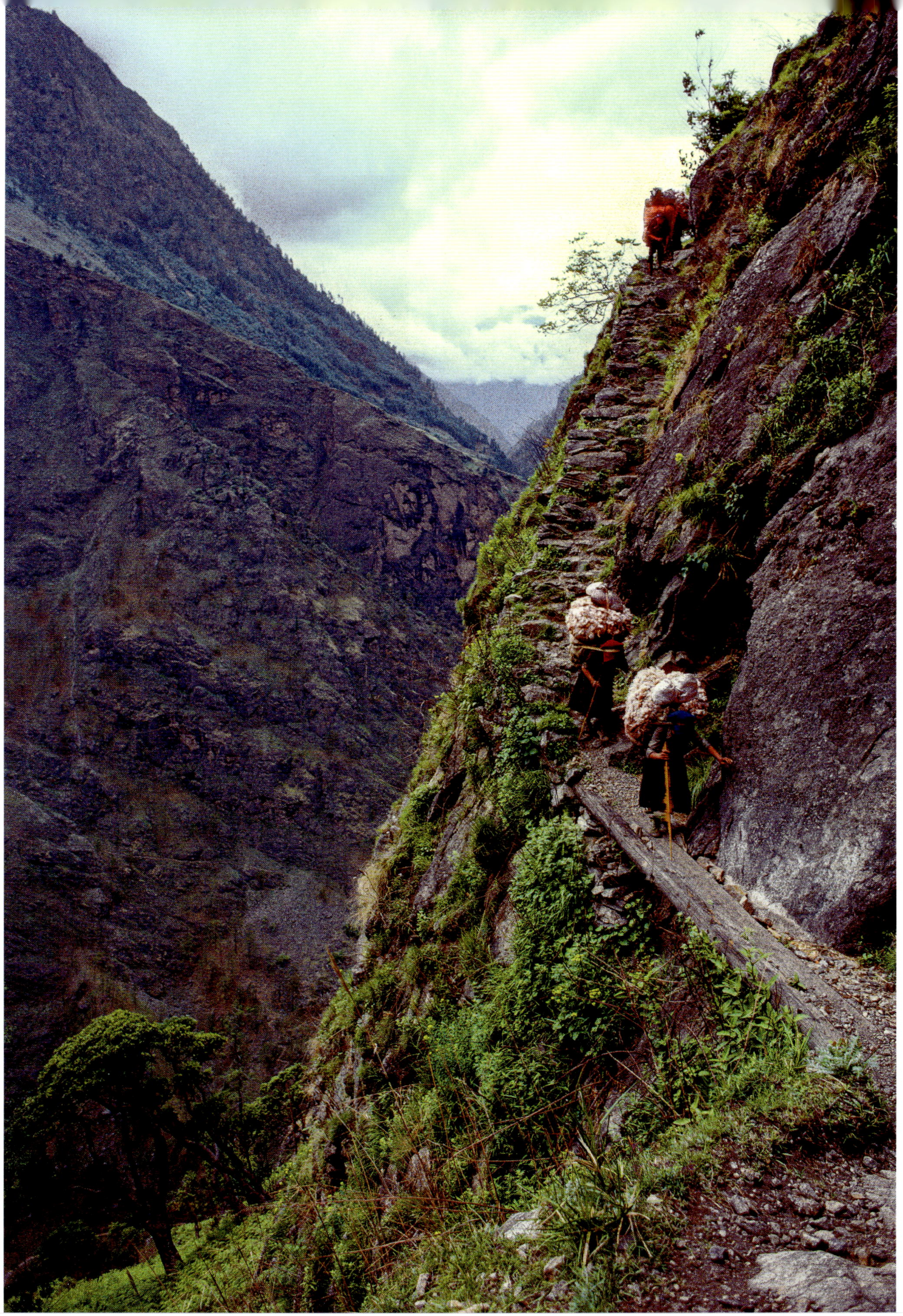

A trail on the route to the Tsum Valley, Nepal, 1991

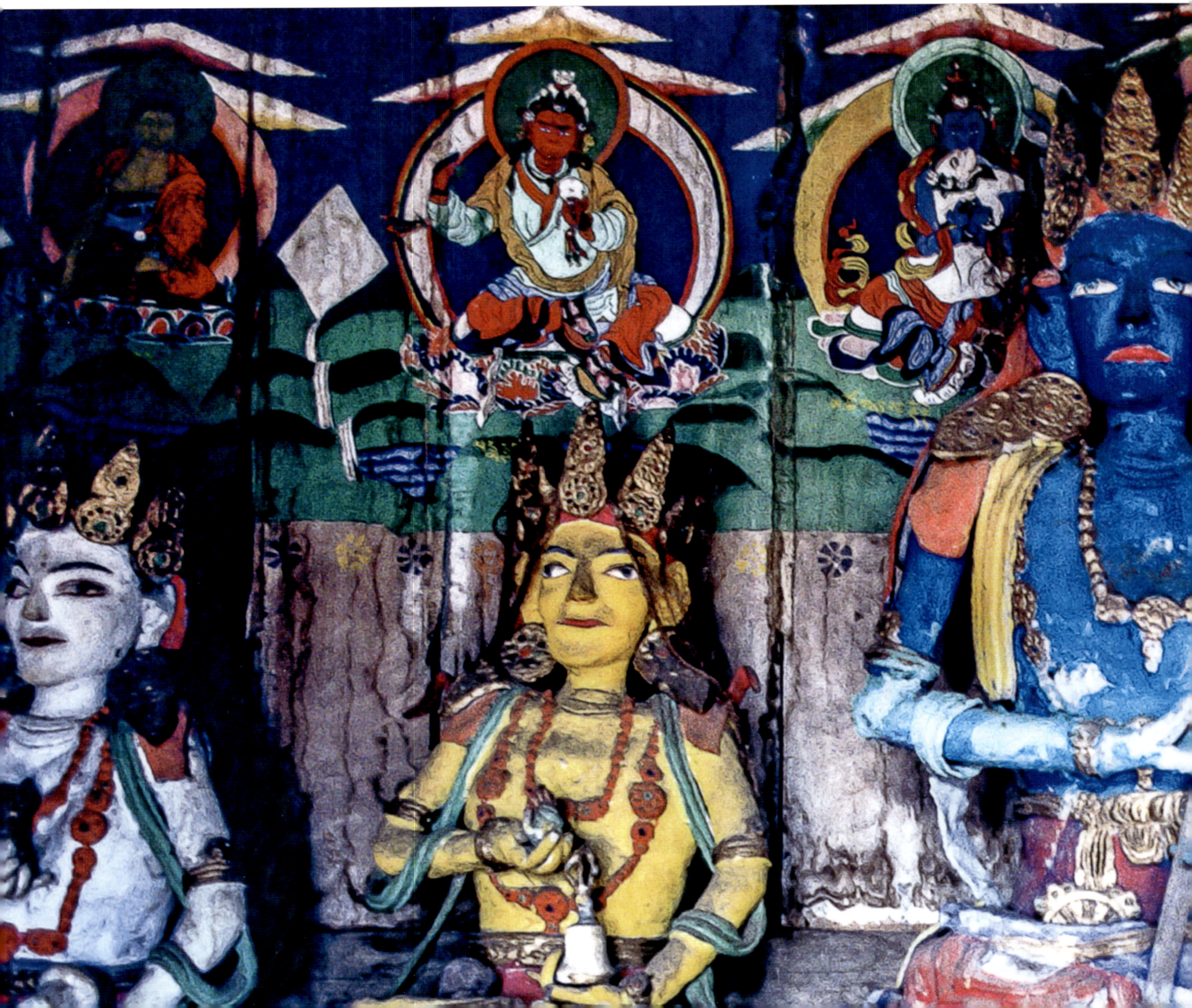

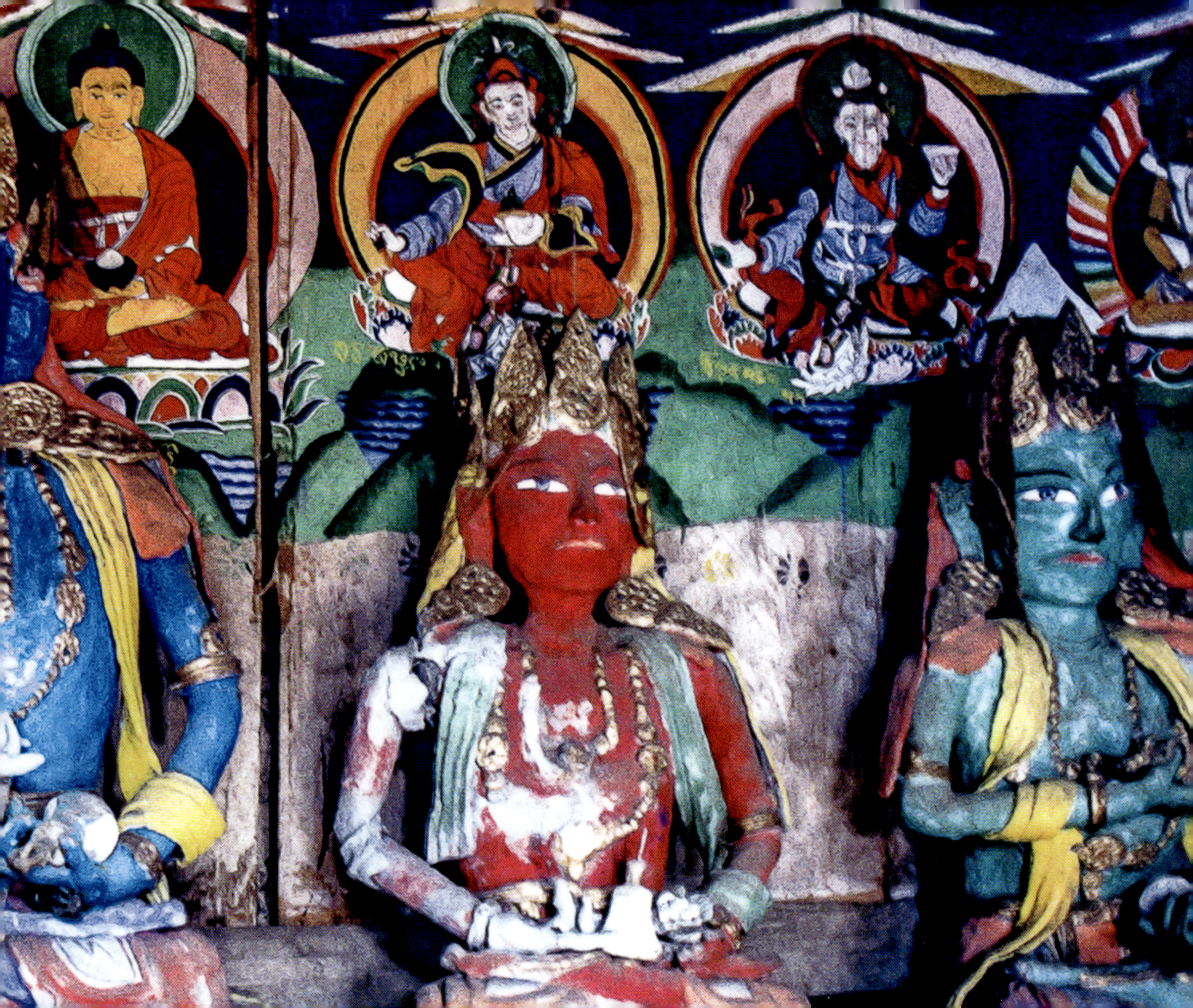

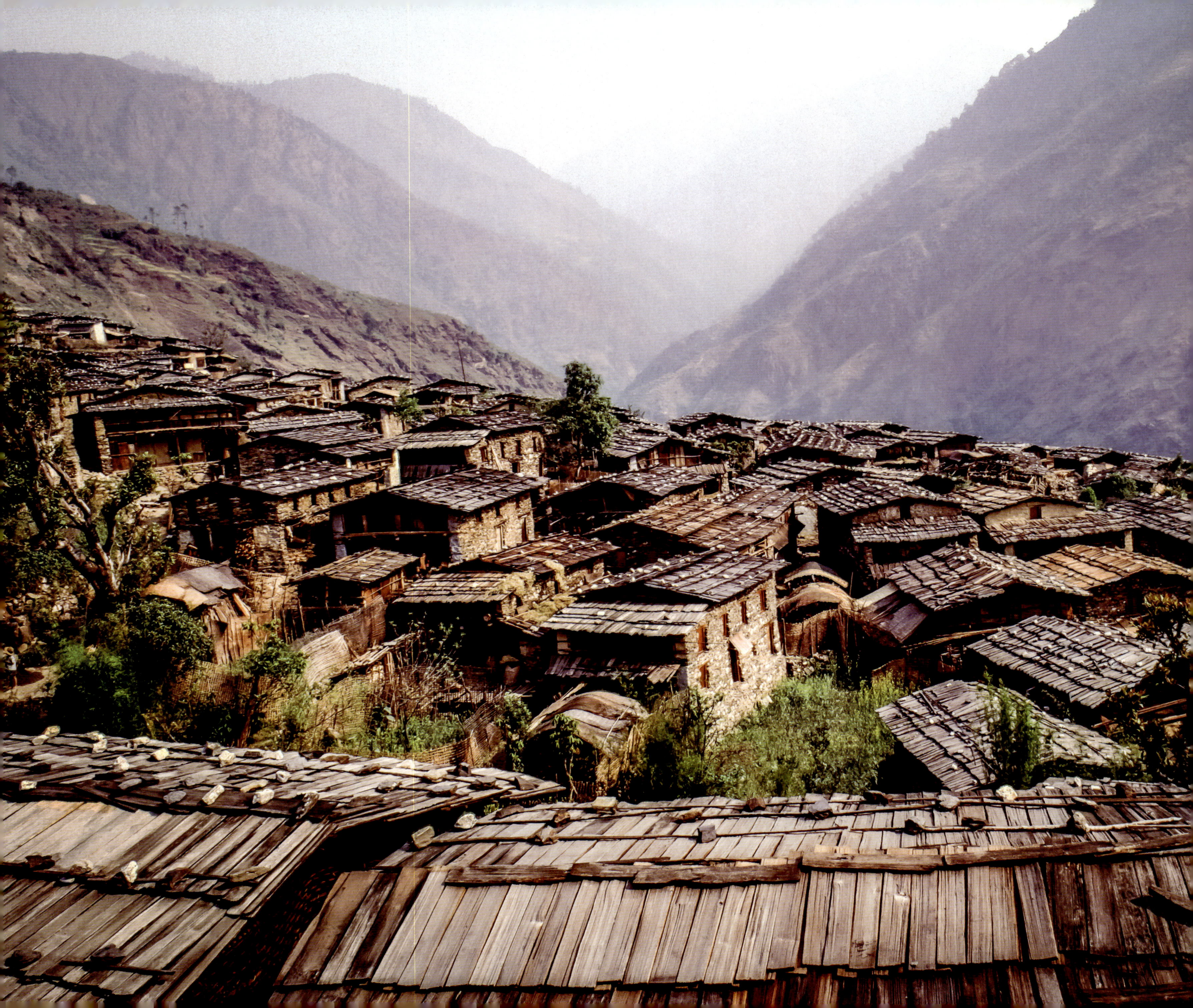

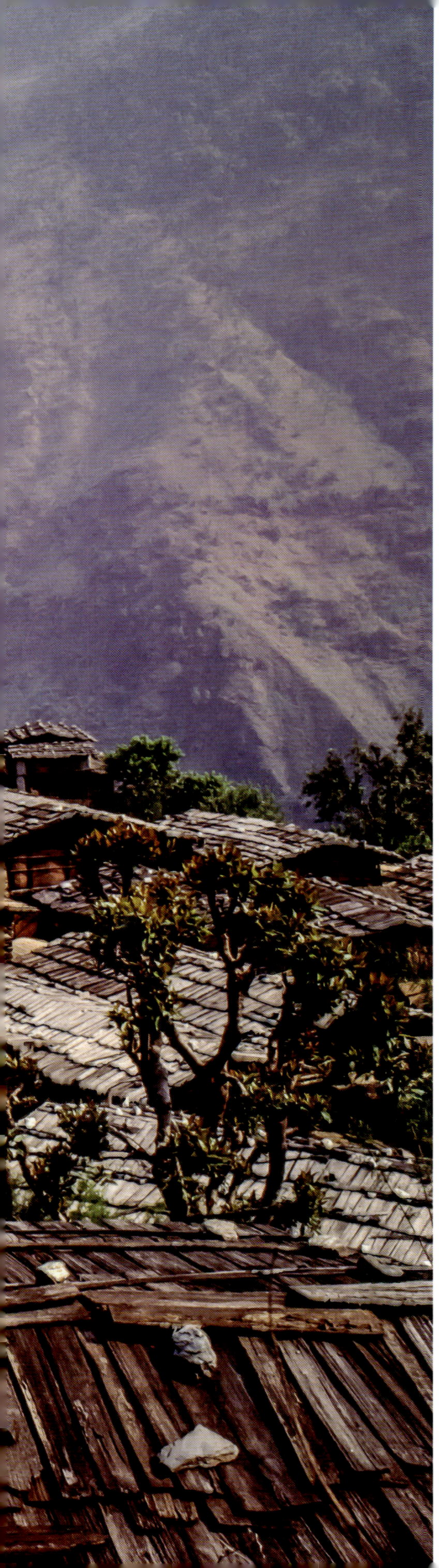

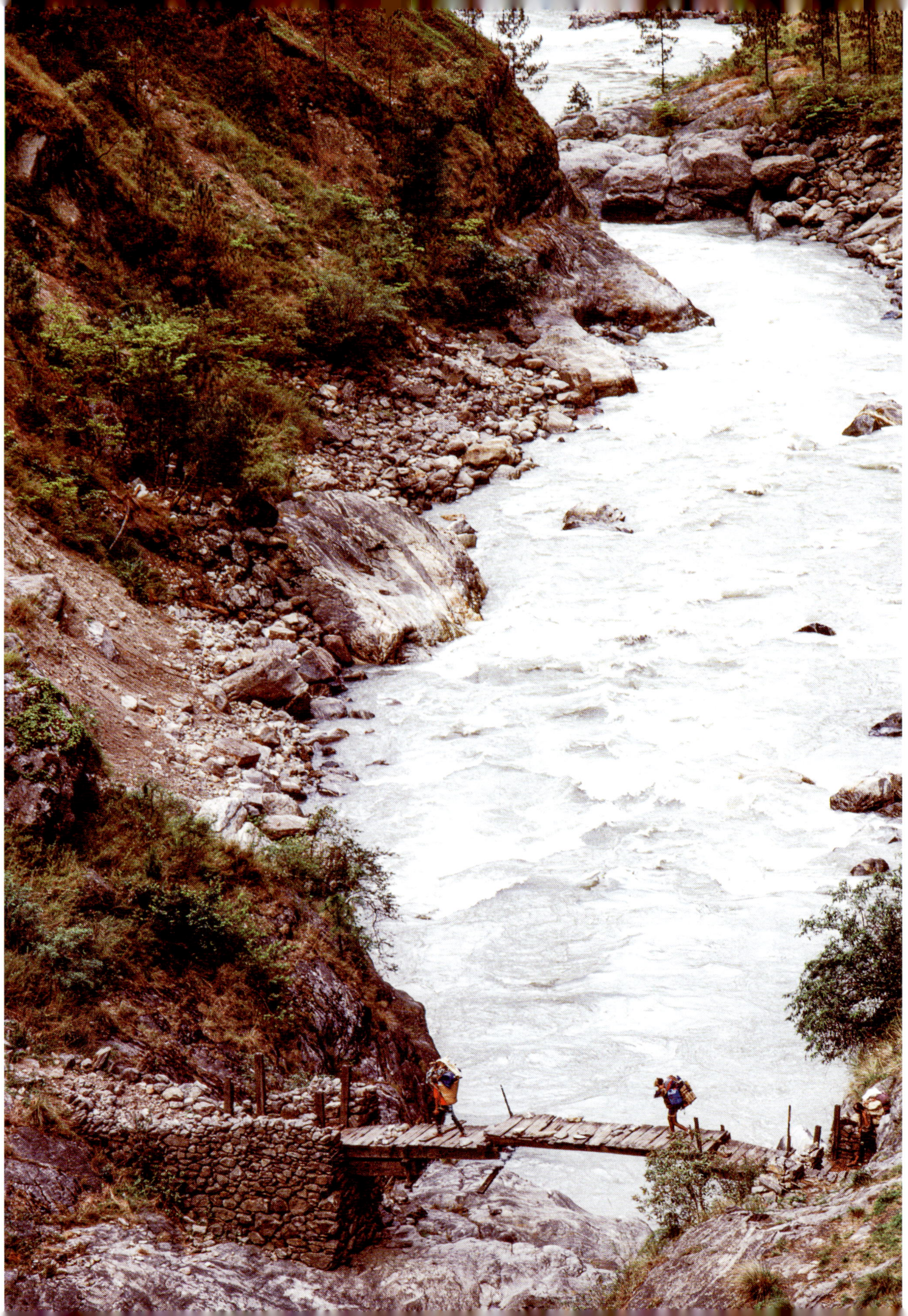

Porters crossing a tributary of the Buri Gandaki River, Nepal, 1991

Opposite: Laprak village, Nepal, 1991

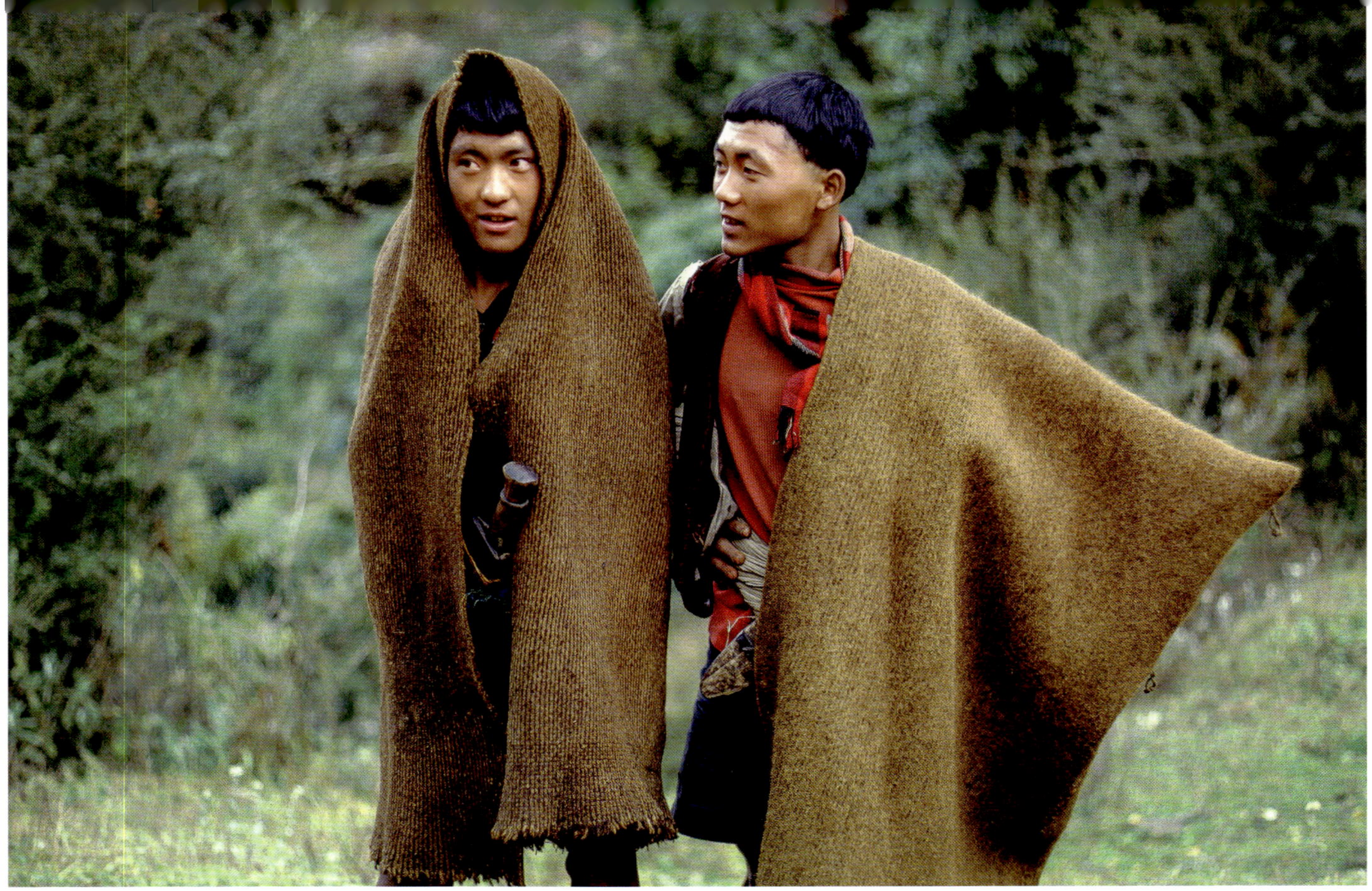

Laprak villagers, Gorkha District, Nepal 1991

Opposite:Nepali shaman, above the
Buri Gandaki River, Nepal, 1991

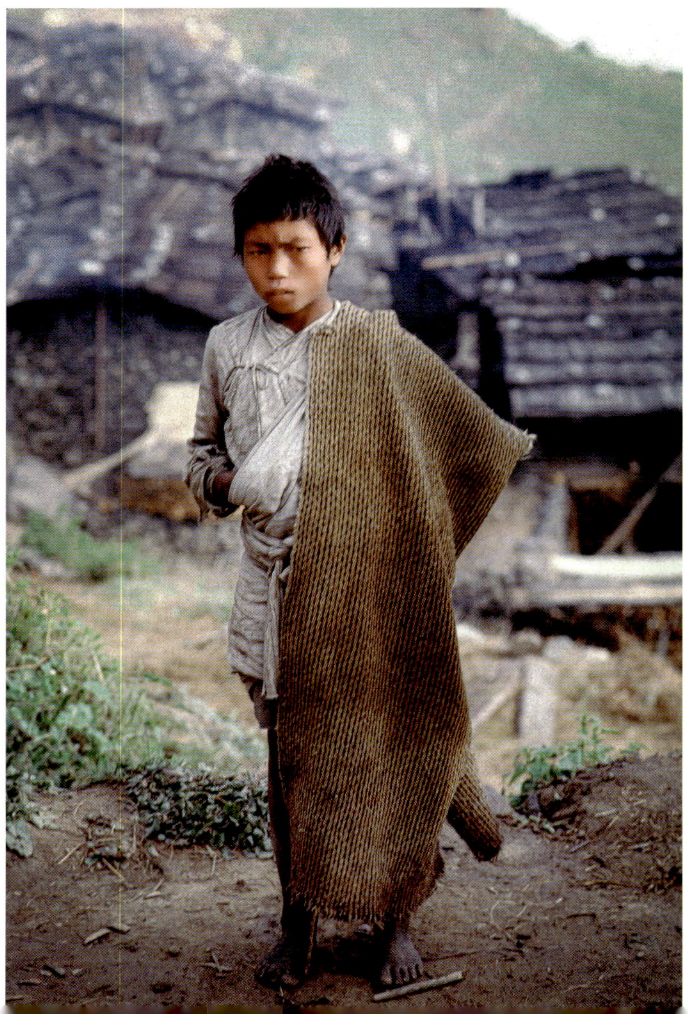

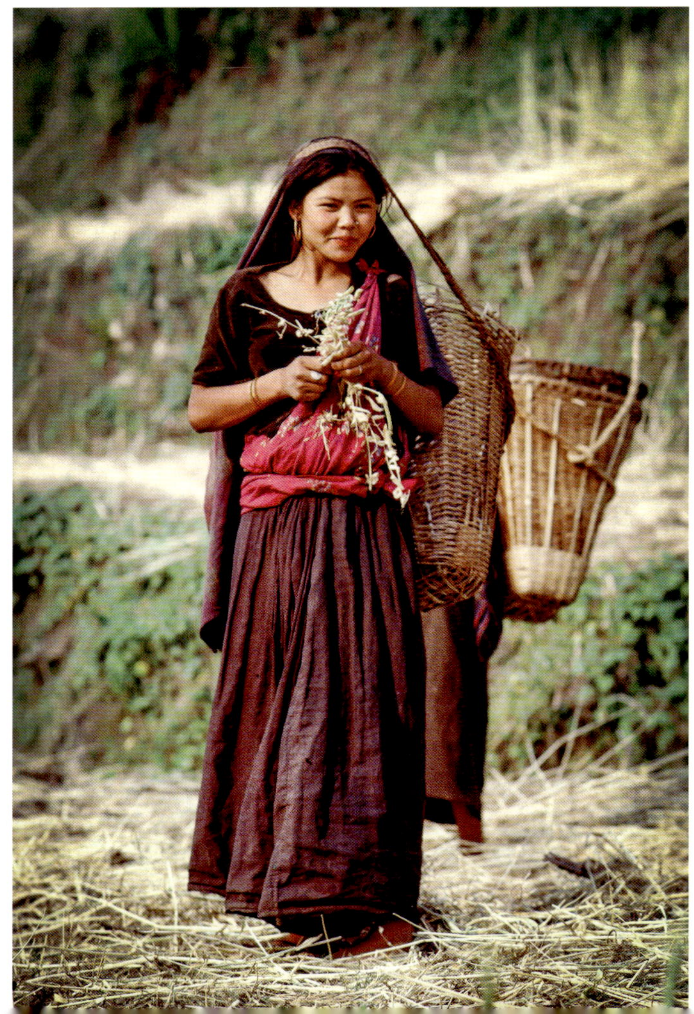

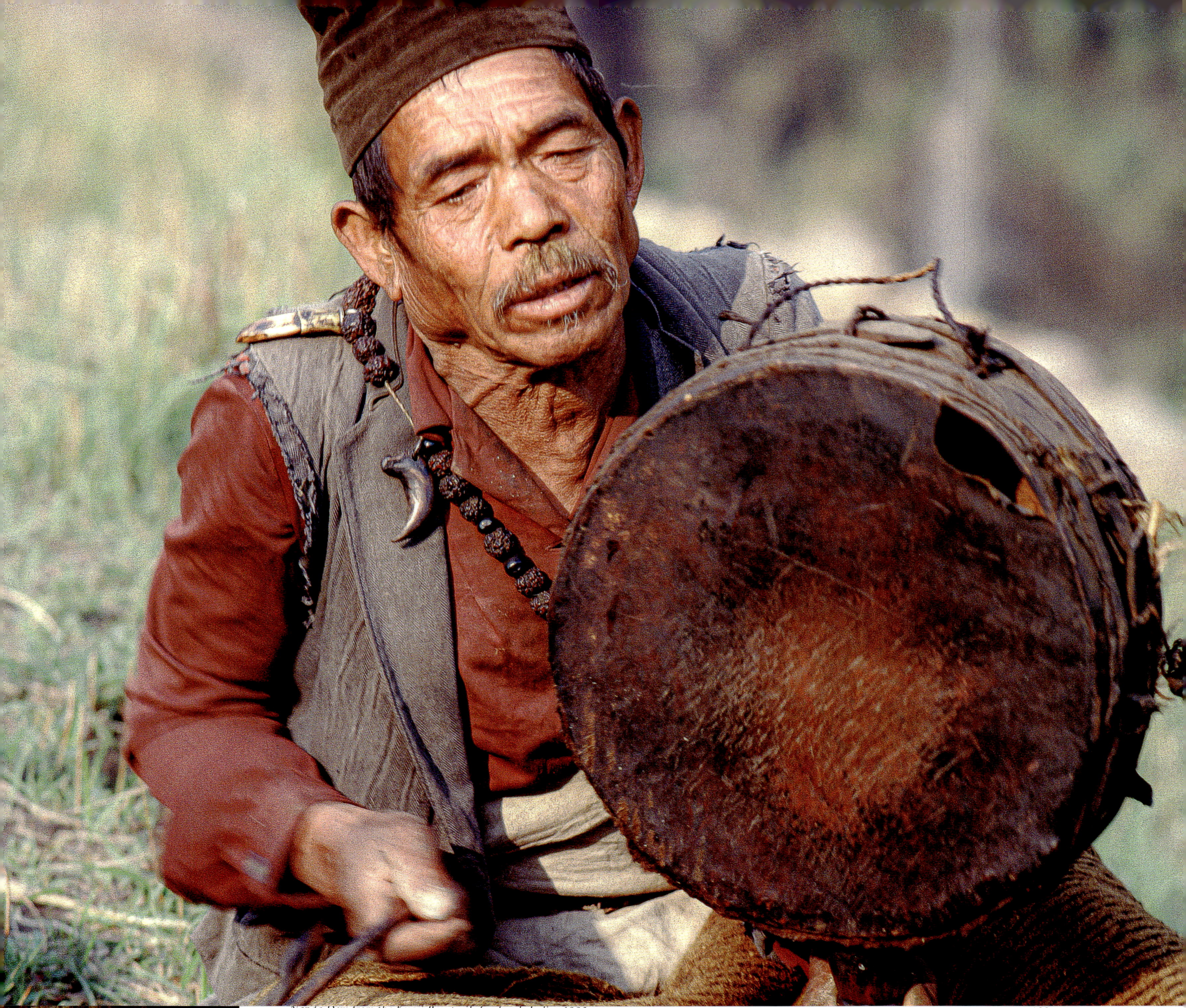

Nanda Devi from the Kuari Pass trail, Garhwal, India, 1993

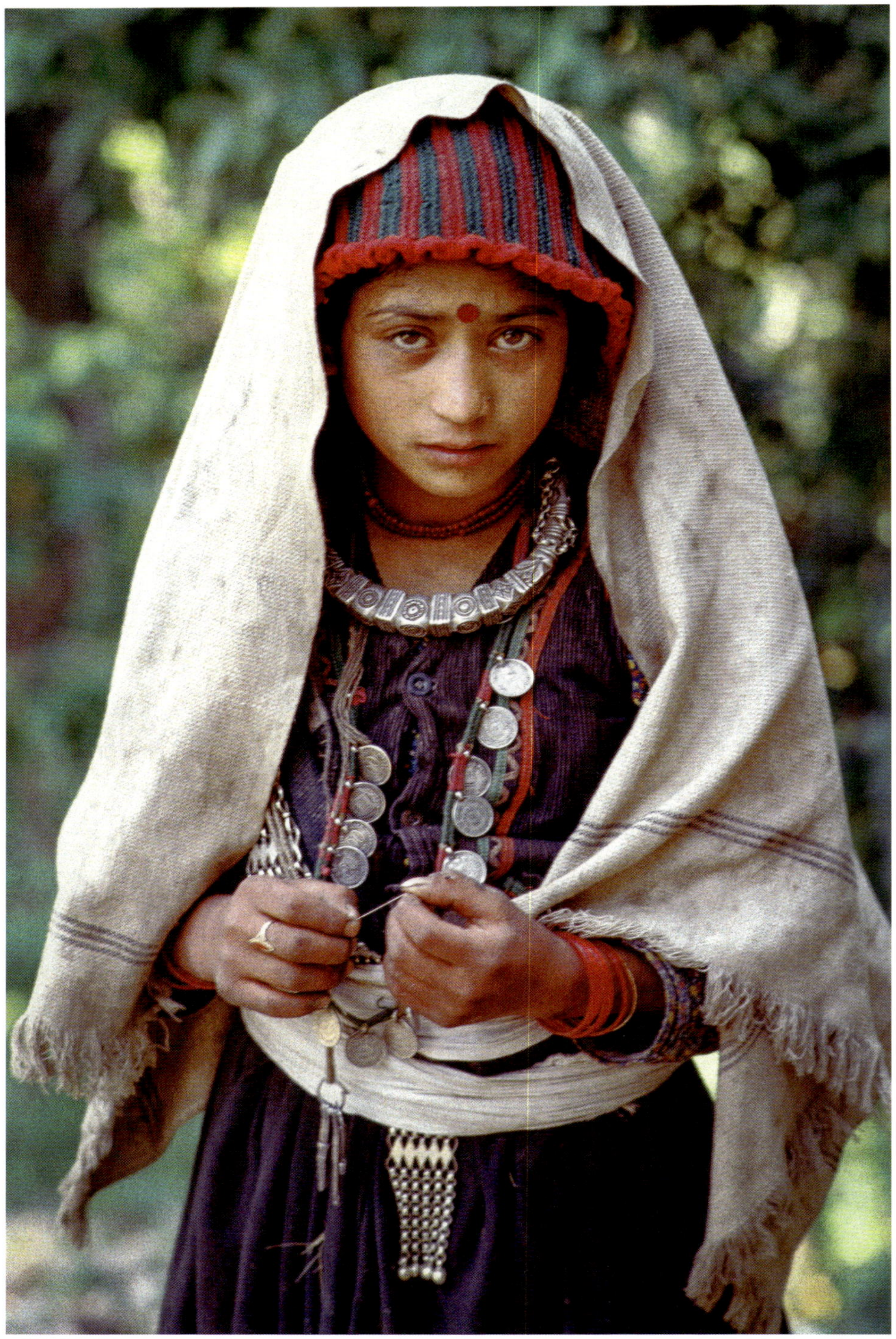

A villager from Wan, Garhwal, India, 1993

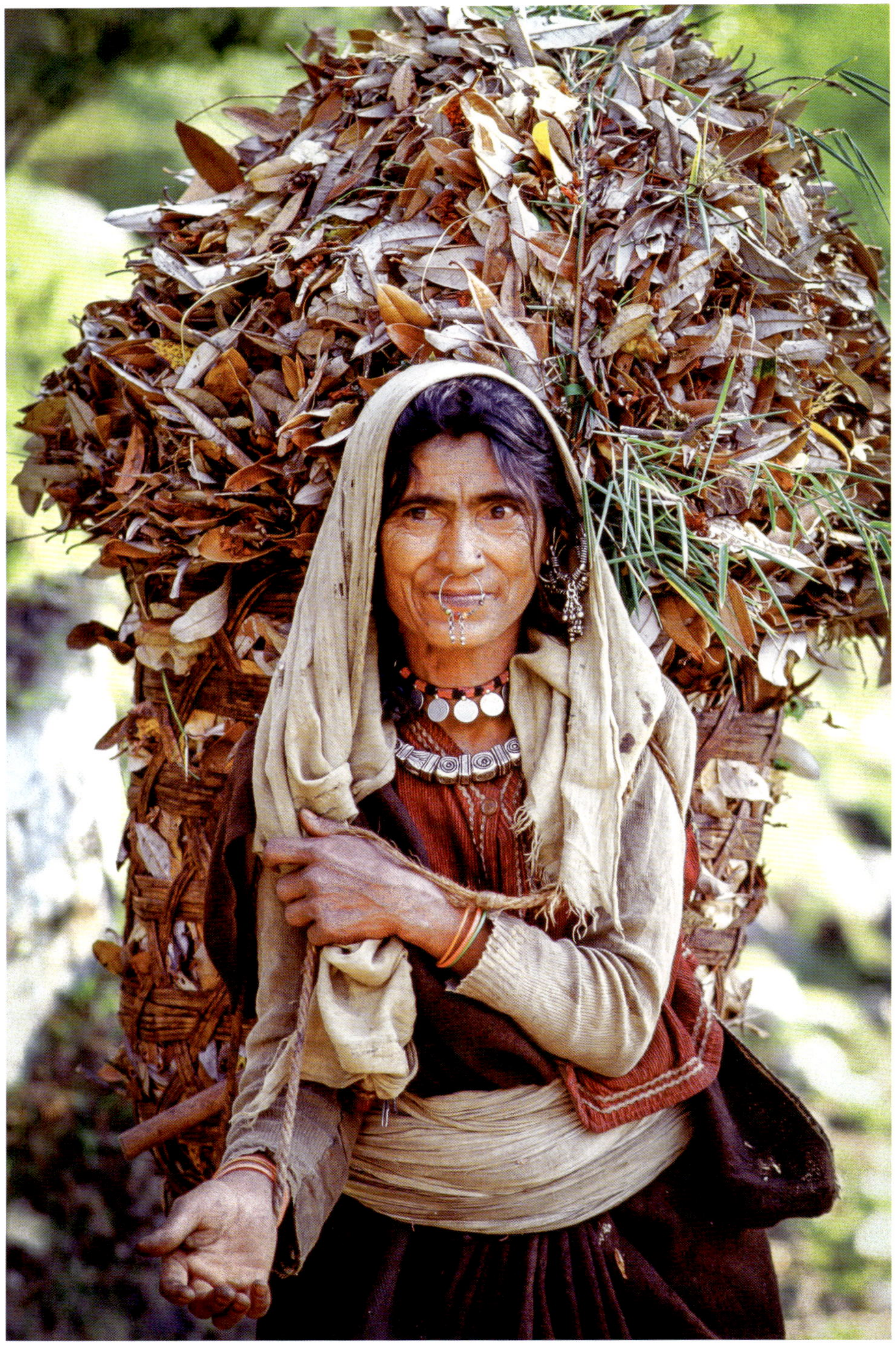

Villager near Sutol, Garhwal, India, 1993

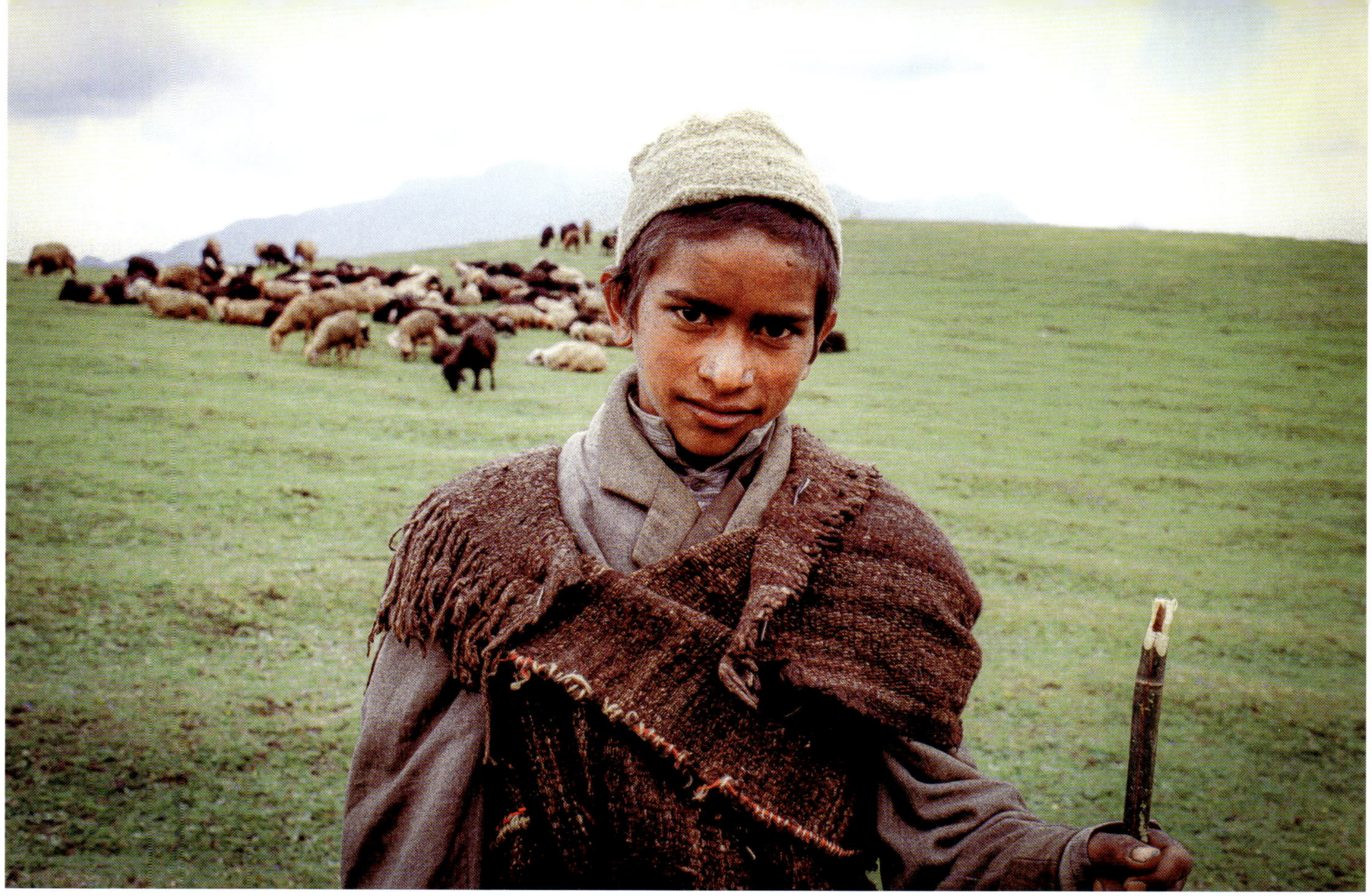

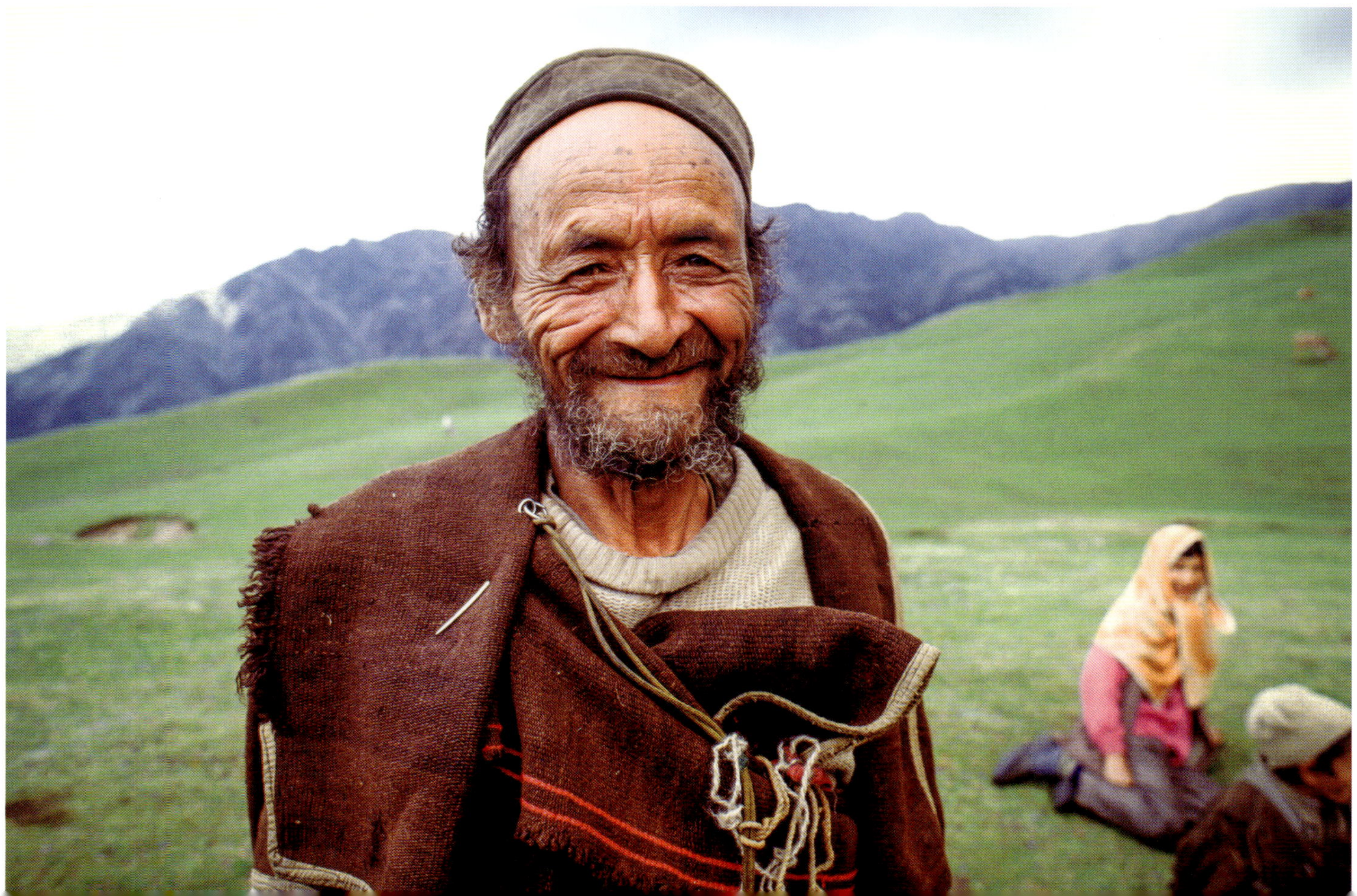

Sheep herders in the Garhwal, India, 1993

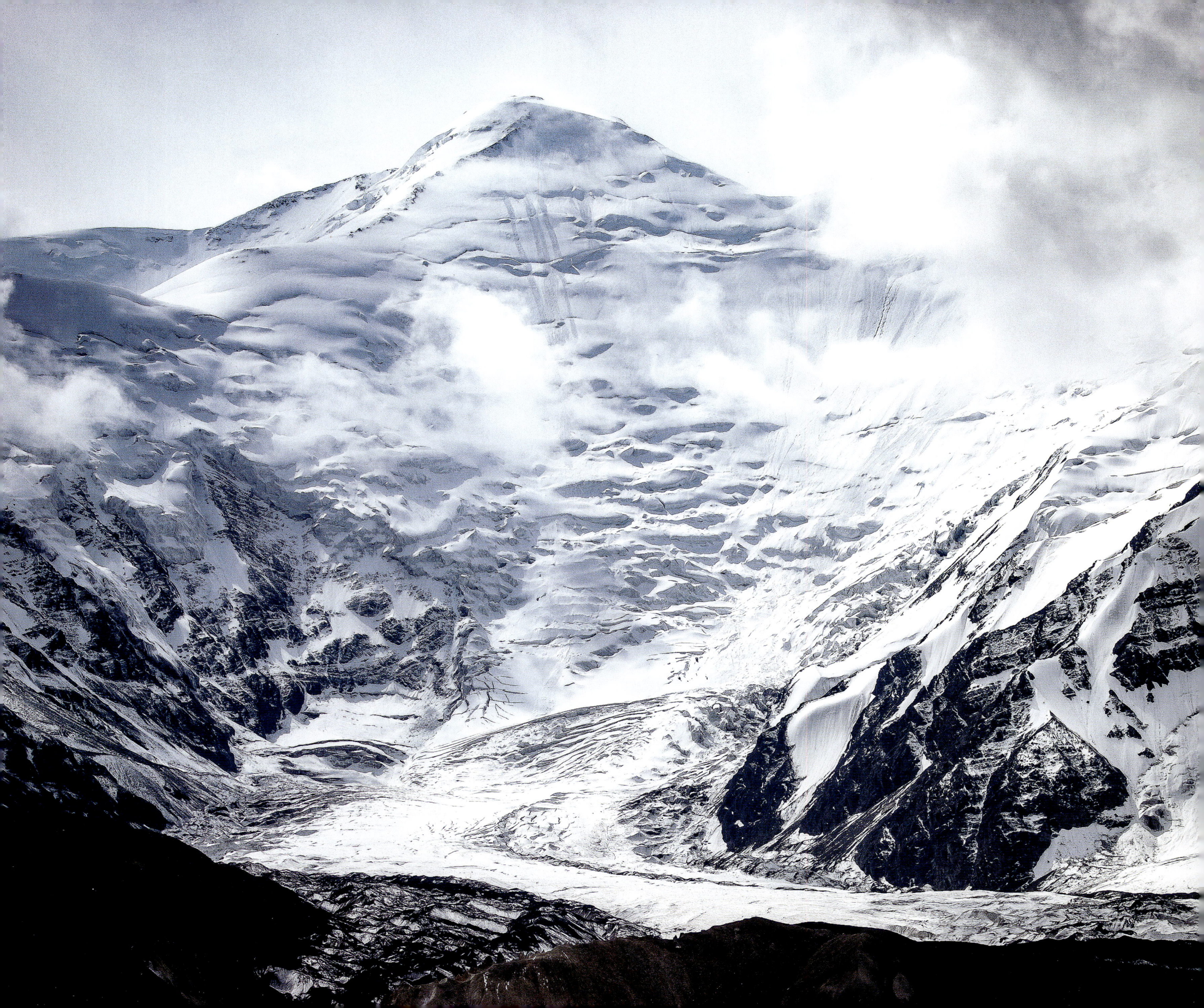

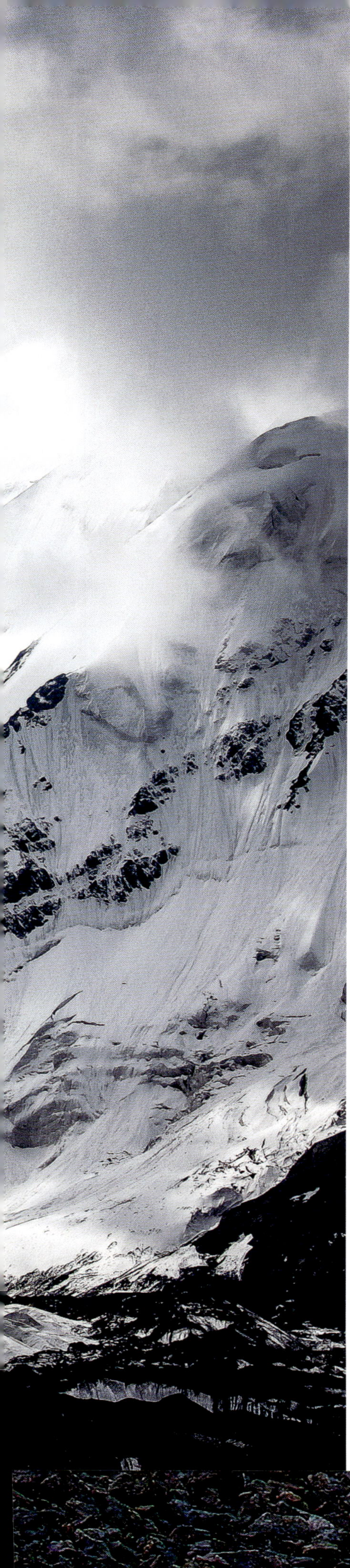

At the Edge of the Pamir, Kyrgyzstan

AUGUST 2008

To reach Peak Lenin (elevation 23,406 feet) in southern Kyrgyzstan, we took a round-about route through a bewildering maze of mountain ranges from the Central Tian Shan Mountains, where we started, to the Pamir. Our trekking adventure began in a little camp called Jeti Oguz just below Issyk Kul (*kul* means lake) in the eastern edge of Kyrgyzstan. Only a day's drive from Bishkek, the capital city, it was another world of wide vistas punctuated with yurts. We lived and worked in the neighboring country of Kazakhstan at the time, since Bill was the director of the USAID program in Central Asia, but this was the closest we had ever been to yurts that were actually occupied. It was easy to see why these dome-shaped tents with collapsible wooden frames covered by canvas and wool felt are still ubiquitous in Kyrgyzstan and the rest of Central Asia. They are strong and portable enough to be disassembled and easily carried to the next pasture. Inside it's always cozy, with a small fire pit for cooking and piles of carpets and quilts to sit and sleep on.

For a week we walked through fir forests and up high mountain passes to the upper reaches of the Karakol Gorge, often passing small collections of yurts and tents where nomad families were busy chopping wood, rounding up their animals, and cooking. At one camp an old train boxcar was rigged up as a sauna, and, after we paid a young boy to use

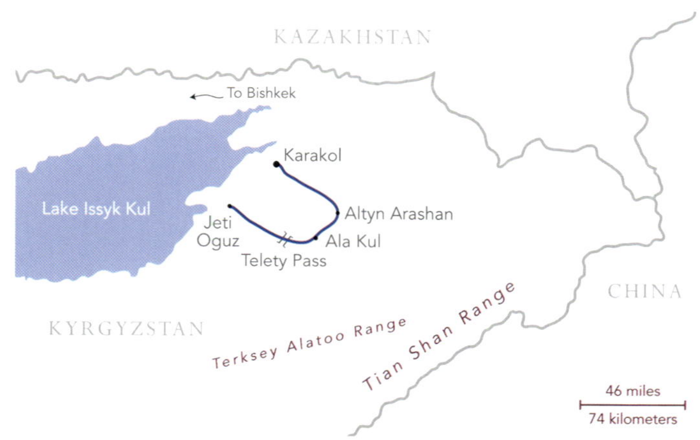

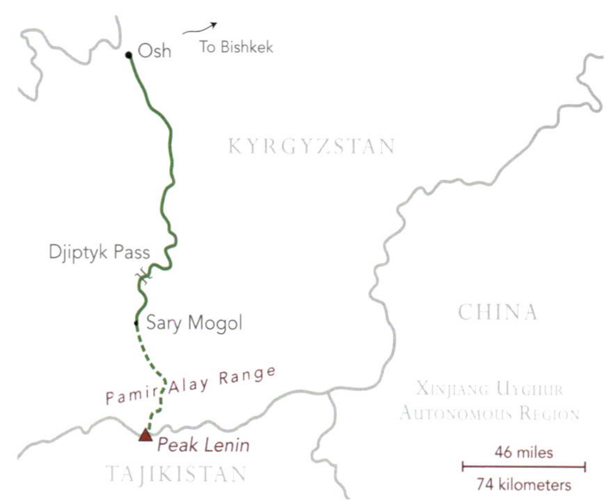

Peak Lenin, Kyrgyzstan, 2008

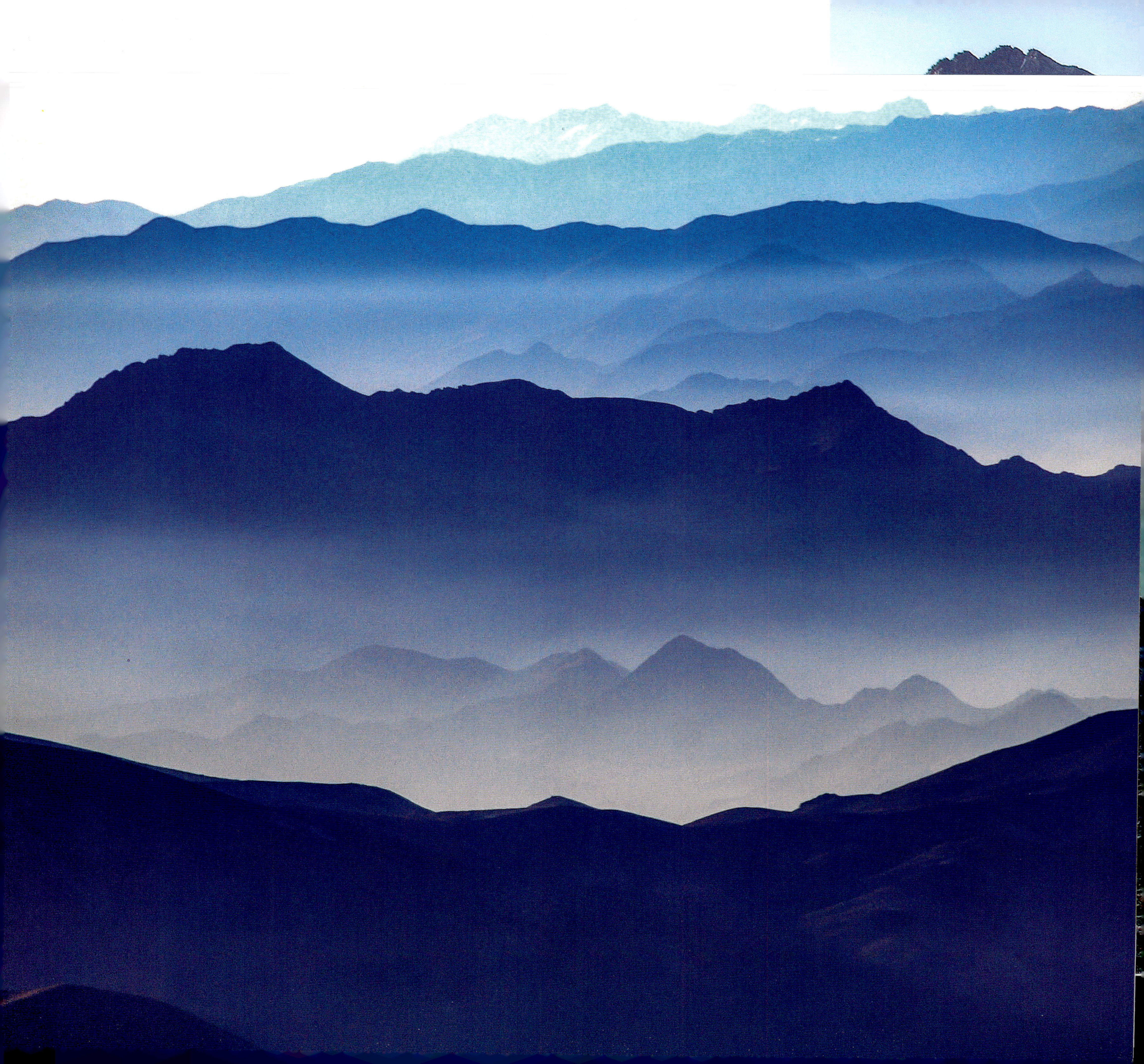

Around and Above the Hindu Kush, Afghanistan

2009–2010

By the time we got to Afghanistan in early 2009, it was a very different place than it had been during the Hippie Trail days of the 1960s and 1970s. Years of occupation by the Soviet Union and then control by the Taliban had stripped away most of the frivolity that tourists had experienced in those earlier days. The city of Kabul was still exotic but was now a more solemn place whose inhabitants had heart-breaking stories of hardships endured over the previous decades. We arrived not as tourists but as employees at the U.S. Embassy. Bill was the director of USAID Afghanistan, the multi-million-dollar U.S. government assistance program, and I was a cultural affairs specialist for the State Department. We lived within the walls of the U.S. Embassy, but from our small, third-floor apartment we could see the tops of the peaks of the Hindu Kush Range, which stretches for hundreds of miles from the eastern edge of Afghanistan to Pakistan.

We longed to get out and see more of Afghanistan. Bill visited aid projects throughout the country to monitor the establishment of medical facilities, schools, and other facilities designed to help the Afghan people. He often travelled by helicopter and always with his camera and a team of seven Special Ops shooters and one medic. Returning back to our apartment, still in his bulletproof vest and metal helmet after a one- or two-day trip, he would describe the incredible scenery he saw from above or on the ground as well as the dignified

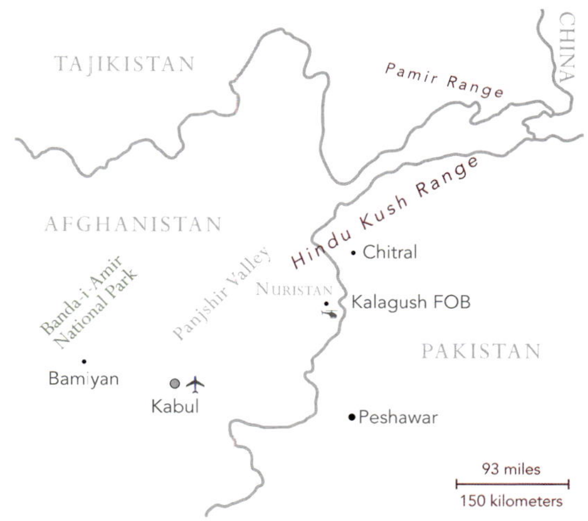

government officials who were doing their best to serve their constituencies in the face of very difficult political obstacles.

While living in Kabul, we learned that Sir Aurel Stein (1862–1943), the famed archaeologist and explorer of Central Asia, had spent a lifetime hoping to visit Afghanistan. He finally made it in 1943, but, soon after his arrival, he died and was buried in the British cemetery there. Stretching the truth a little, we told the Embassy that he was our great-great uncle, so we were able to visit his grave and pay homage to this intrepid explorer.

On an official visit to the Banda-i-Amir for the opening of Afghanistan's first national park, a contingent of U.S. Embassy staff and Afghan government officials converged on

The Hindu Kush, north of Kabul, Afghanistan, 2009

the shores of a bright turquoise lake reminiscent of those we had seen on the Tibetan Plateau. After the speeches we wandered along the lake, took rides on an incongruous, duck-shaped paddleboat, and mingled with members of the Hazara ethnic group who lived in the area.

We also went with a small group to Bamiyan, the site of the colossal carved-stone Buddhas dating from the sixth to seventh centuries CE. The Taliban had notoriously destroyed the two Buddhas in 2001, but the niches they once stood inside were still intact. We admired them from a distance and then got a closer look by climbing into a system of tunnels and rooms throughout the sandstone cliffs in which the Buddhas were carved. Nearby was Shahr-i-Gholghola, also known as the "City of Screams." This former fortress-city was destroyed by Ghengis Khan (ca. 1158–1227) in 1222 CE. Its hillsides had recently been cleared after the Taliban planted mines there, so we carefully stayed on the trail and tiptoed through the ruins and piles of pottery shards along the trail.

One Sunday afternoon a group of embassy staff jumped into helicopters and flew a short distance to a large open field outside Kabul to watch a *buzkashi* match. This ferocious sport, played throughout Central Asia, has two teams of players on horseback who compete to win possession of a headless goat carcass. The action was chaotic, and it was clear that few rules applied. Men and horses crashed together in a tangled pile. From where we sat the view was mostly of horses with nostrils flaring and the faces of wild-eyed men urging them on. The players wore heavily padded jackets and fur-rimmed hats that resembled those we had seen in Kashgar and in old Soviet tank-driver helmets. The whole scene had a dystopian feel. I sat next to the sports announcer for the game, who kindly taught me a few words of Dari to shout over the PA system.

Later, I volunteered to visit eastern Afghanistan to inspect sites for an American-sponsored student study center in Nuristan Province, which is enclosed by the Hindu Kush mountains in the northeast of Afghanistan. The area was previously called Kafiristan, "Land of the Unbelievers," because the inhabitants followed an animistic religion rather than Islam. In 1895, the king of Afghanistan used his military might to convert the population, and today Nuristan, which means "Land of Light," is known for its fervent Islamism. Due to its strategic location, Nuristan is a Taliban stronghold and base for movements back and forth across the border with Pakistan.

As I got off the military helicopter at the Kalagush Forward Operating Base in Nuristan, I looked around and suddenly realized why no one else from my office had wanted to go there: The small military base that houses less than 100 State Department and U.S. military personnel sat in the middle of a narrow valley surrounded by mountains, the perfect target for Taliban snipers. I was scheduled to spend seven days there since the military helicopters only came in from Kabul once a week. Although a well-armed group of us left the base several times to look at buildings that might fit the needs of the study center, I spent most of the time within the confines of the base, either in the dining hall or my plywood-framed room.

It drove me crazy knowing what was just outside the boundaries of the tiny base. I had seen fantastic wood carvings from Nuristan at the National Museum in Kabul and had read about them in Eric Newby's classic travel book, *A Short Walk in the Hindu Kush* (1958), as well as in Rudyard Kipling's *The Man Who Would be King* (1888), so I was anxious to see more. The day before I left the base was attacked by shooters hiding in the hills above, so we civilians spent the day sitting in an underground bunker. Several years later this base and others in Nuristan were determined by the U.S. military to be indefensible and were abandoned.

Band-i-Amir National Park, Bamiyan Province, Afghanistan, 2009

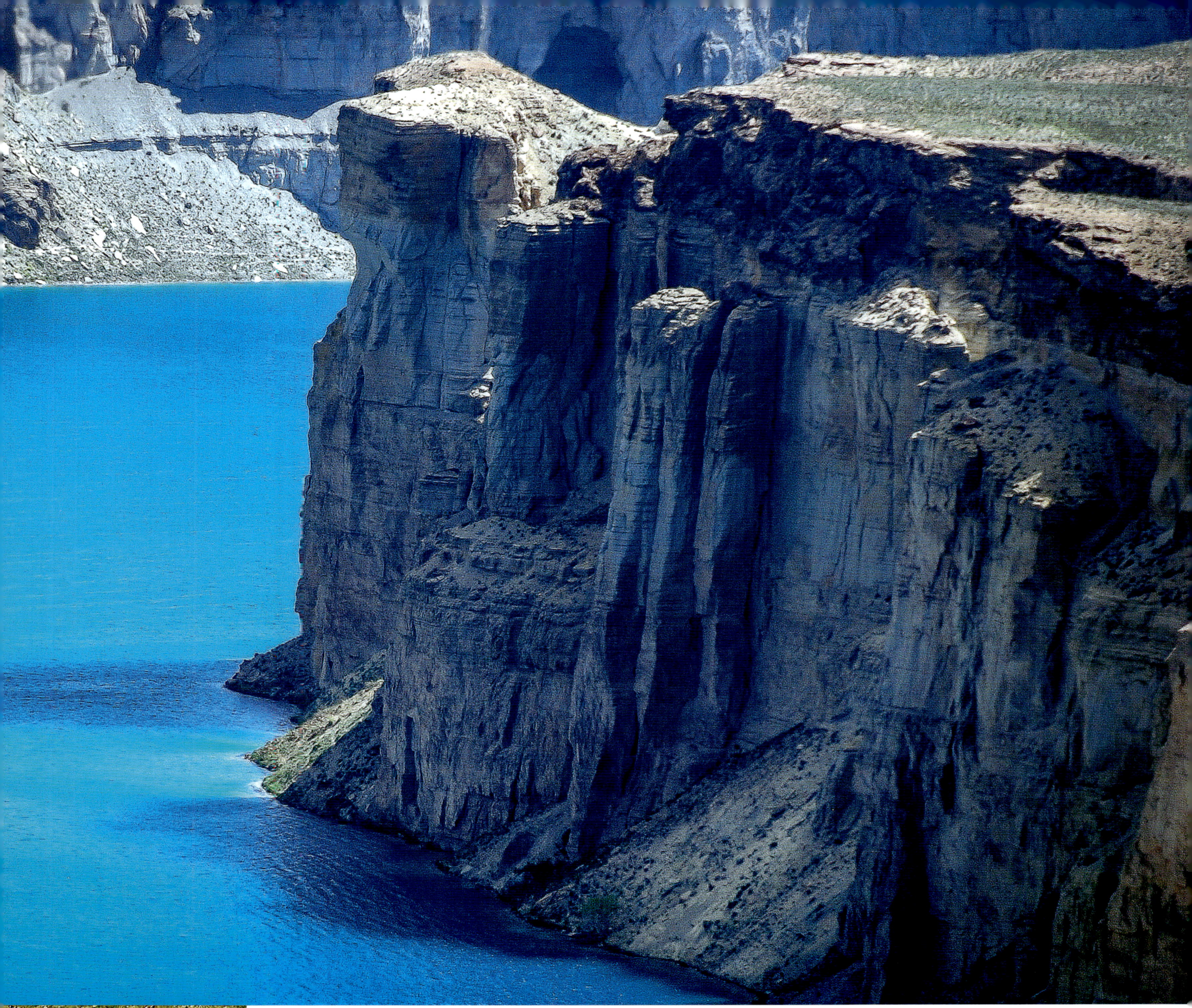

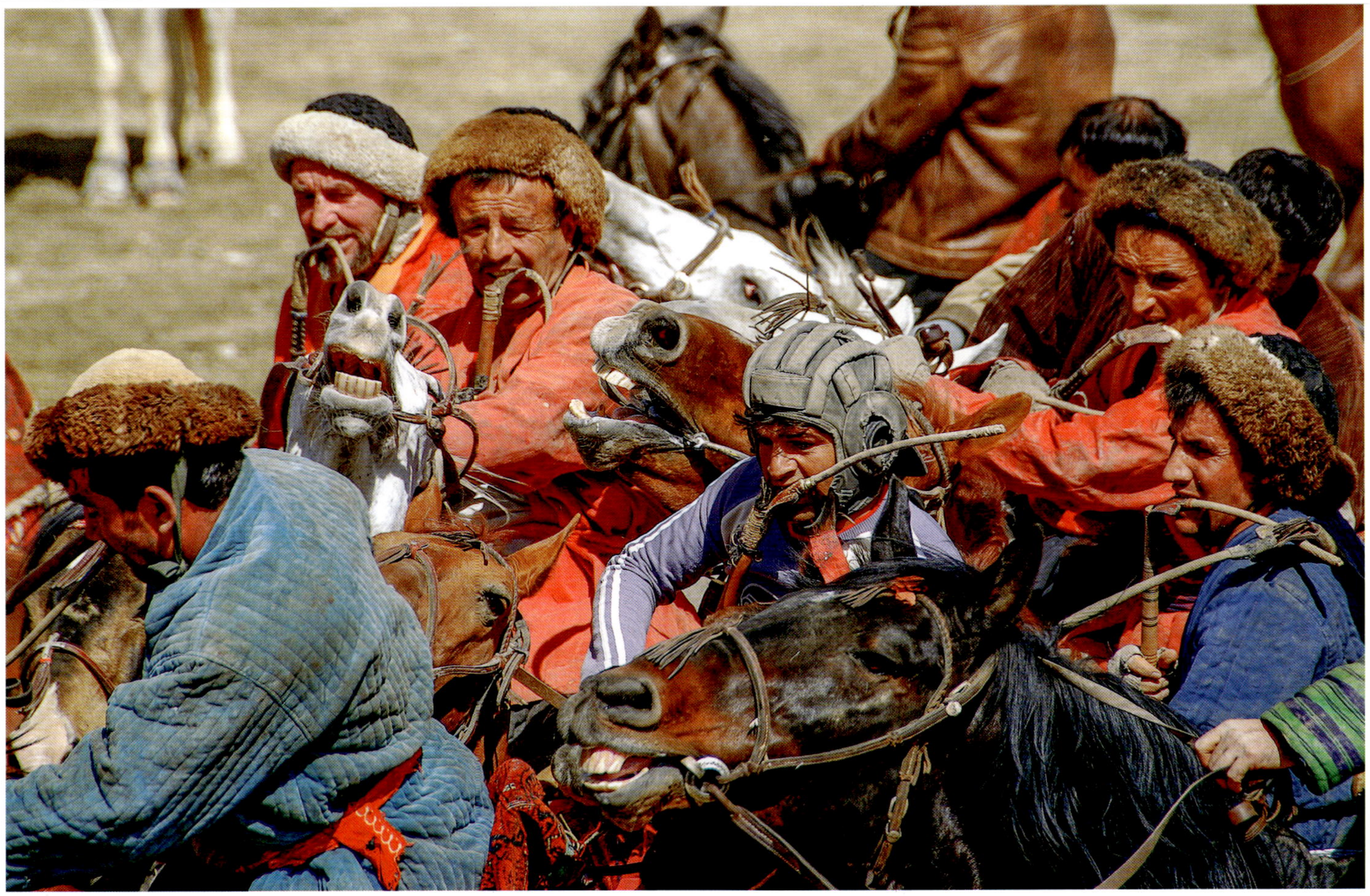

Buzkashi, northern Afghanistan, 2010

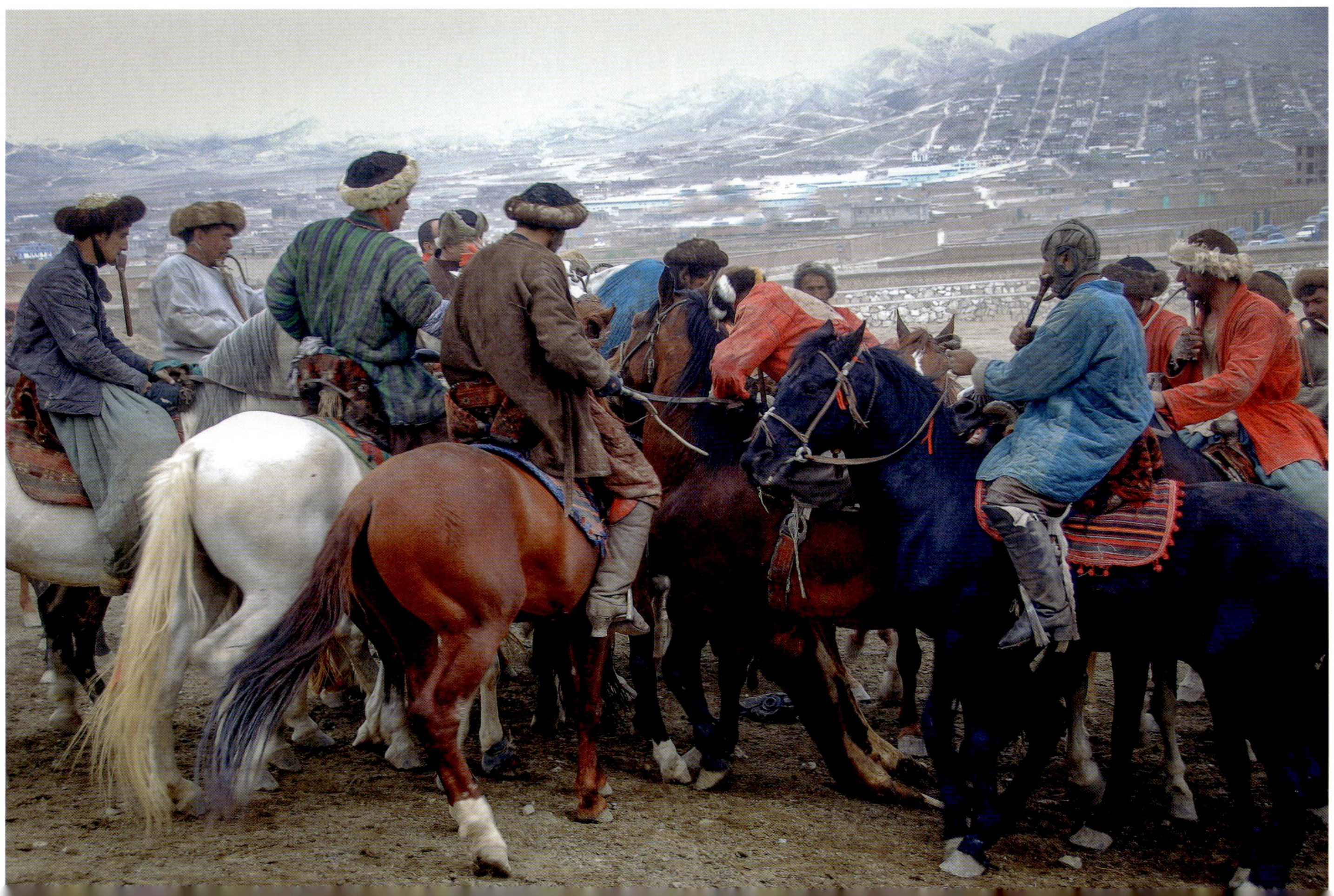

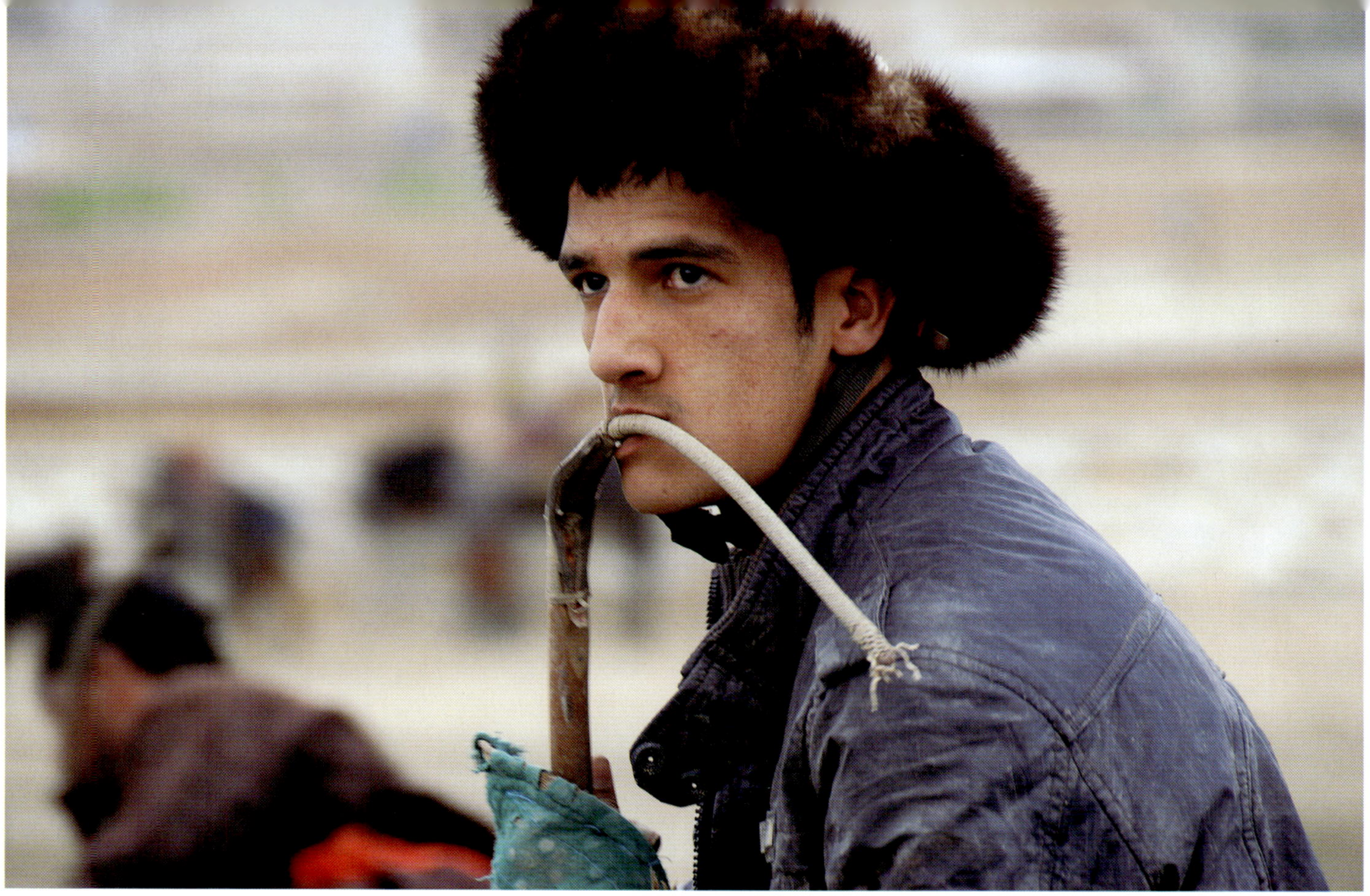

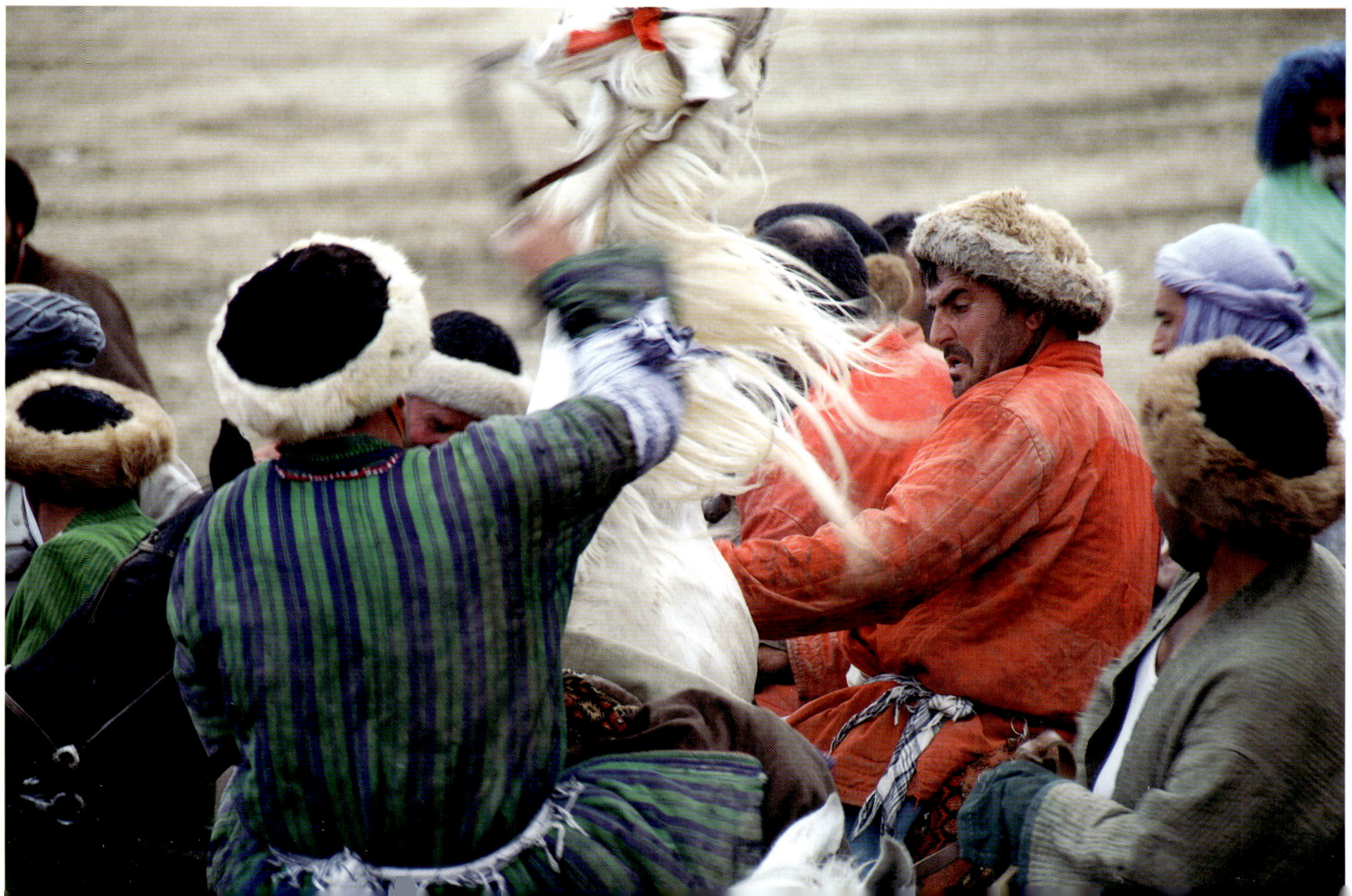

Buzkashi, northern Afghanistan, 2010

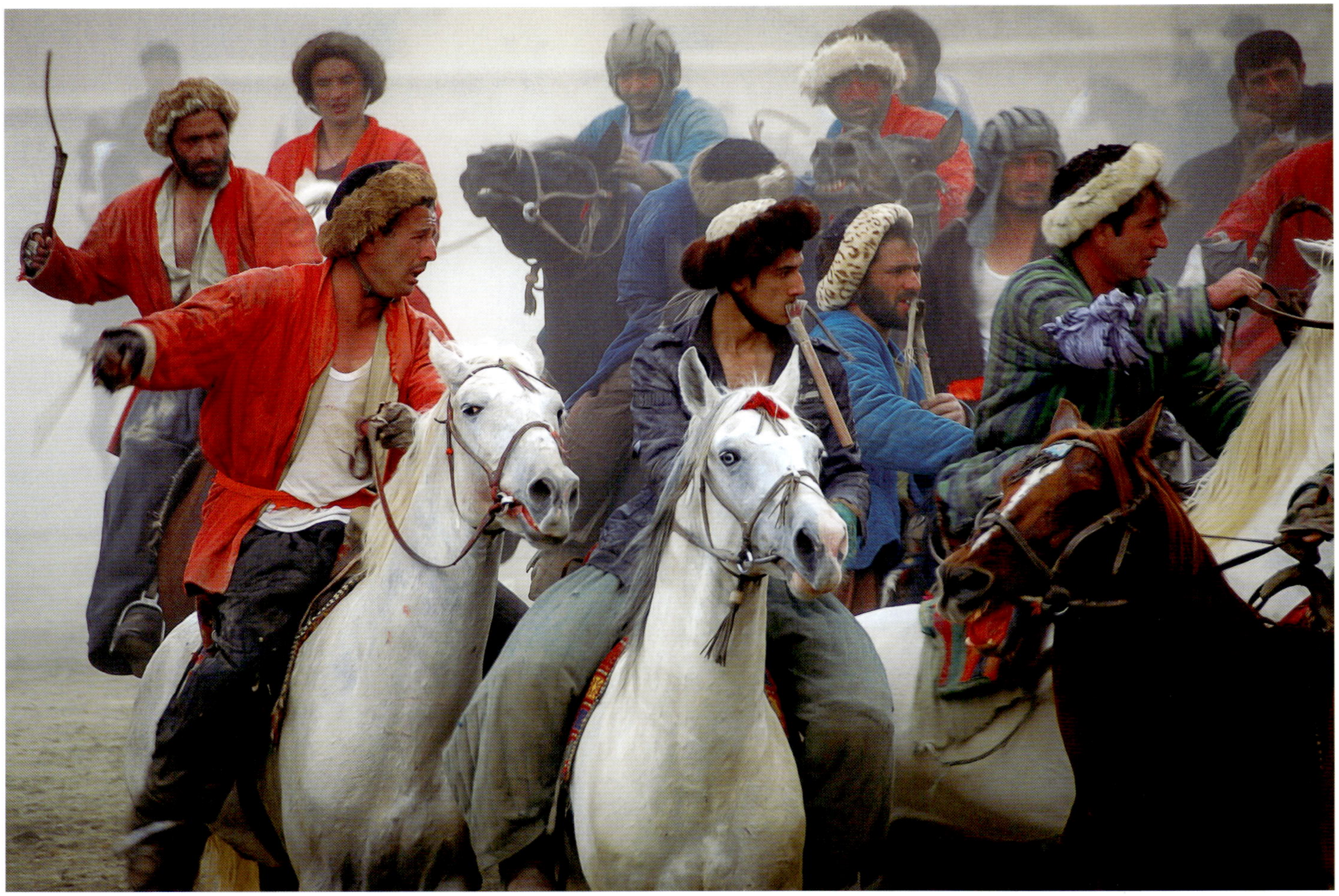

Buzkashi player, Northern Afghanistan, 2010

The site of the Buddhas of Bamiyan after the Taliban, 2010

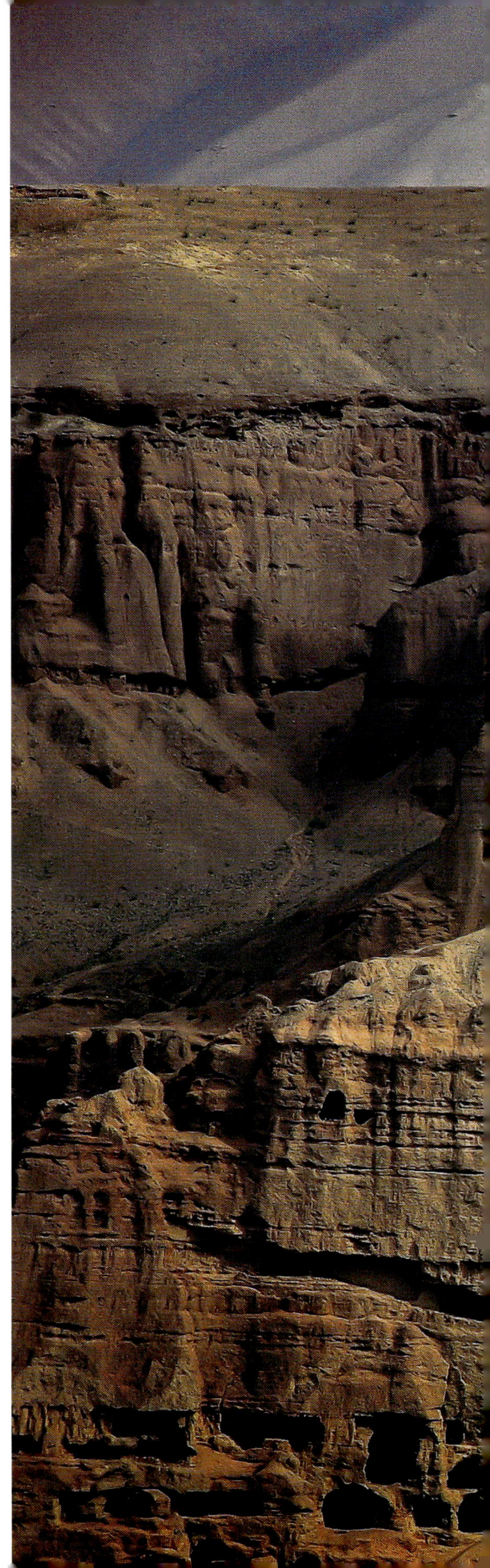

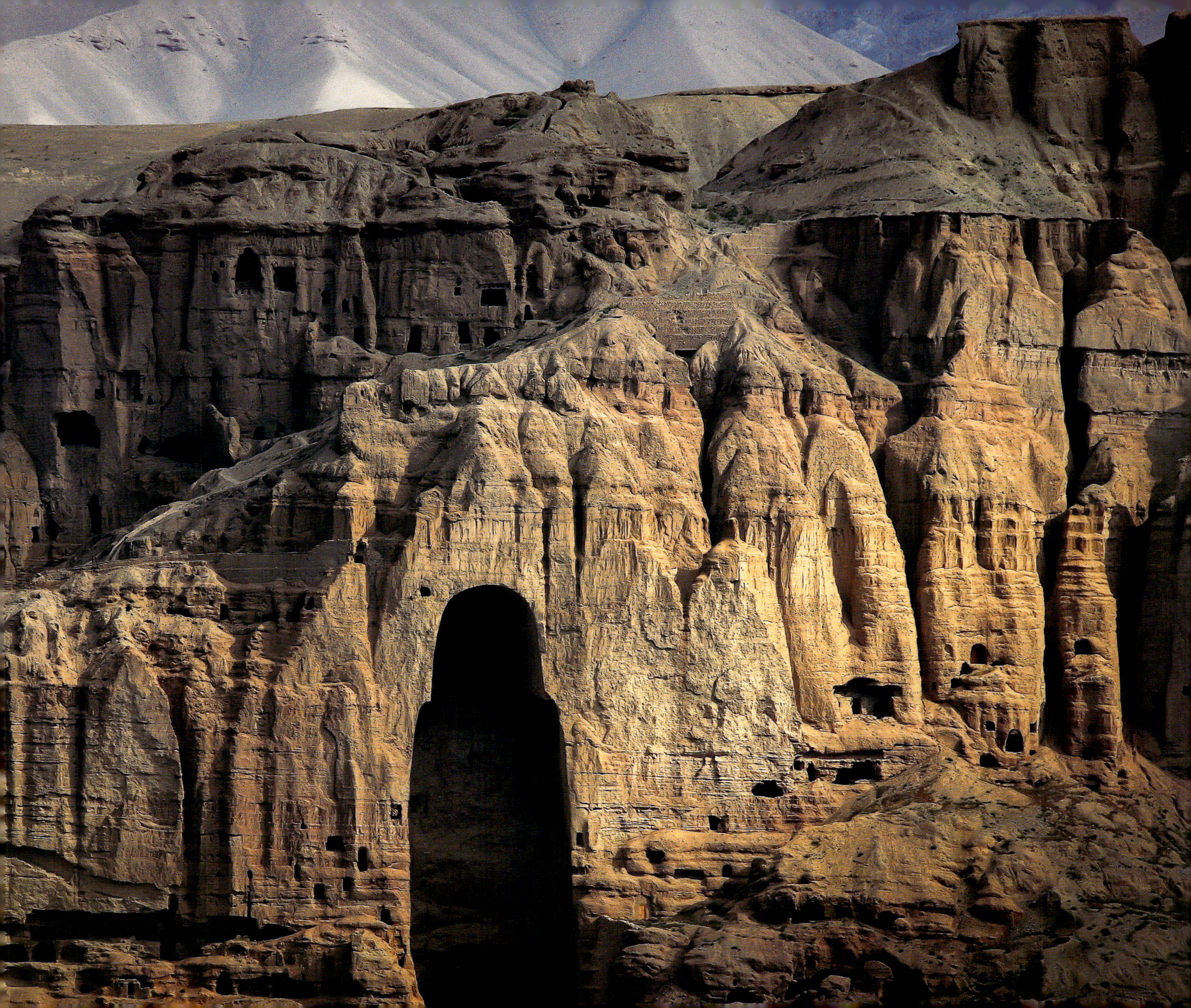

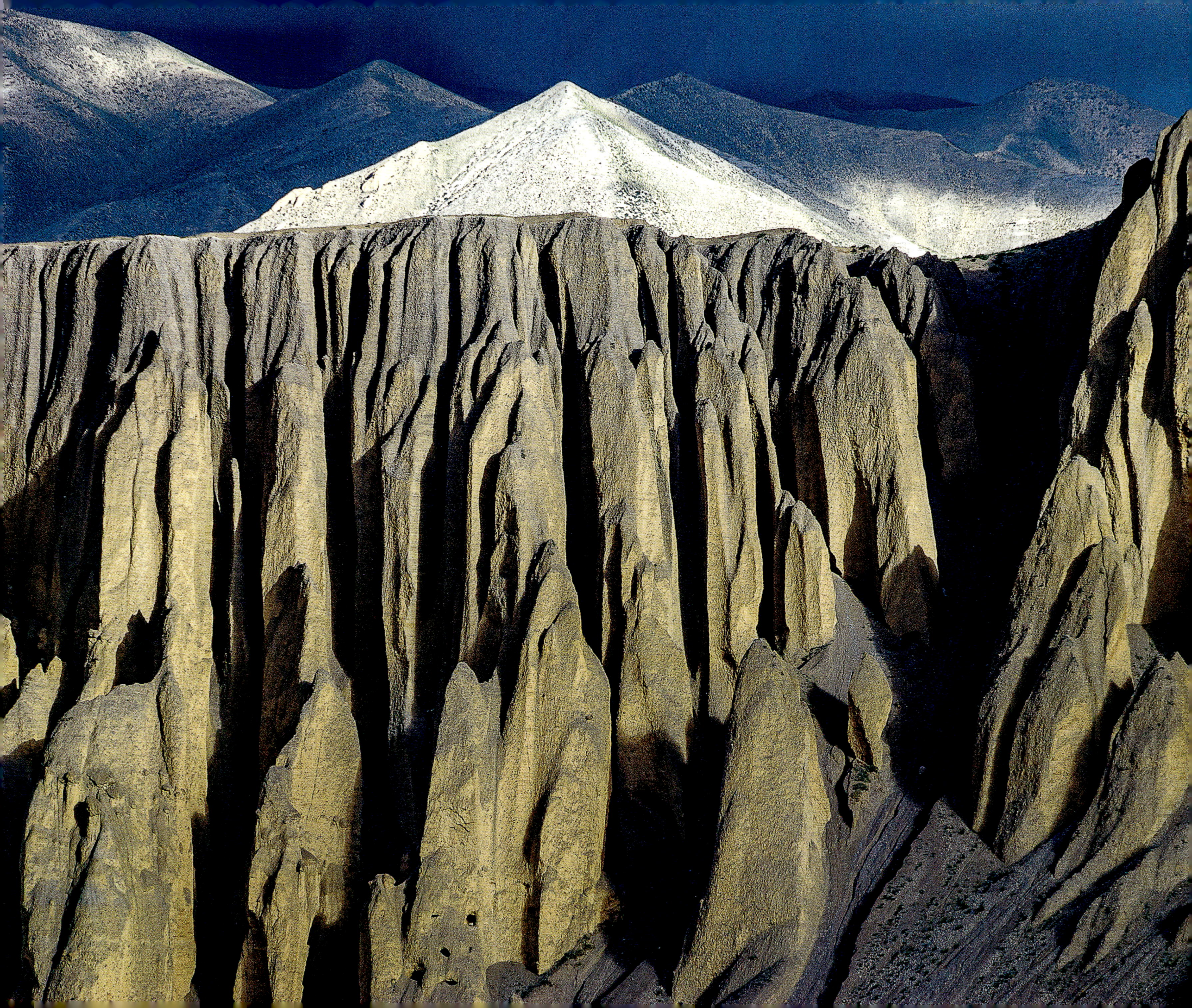

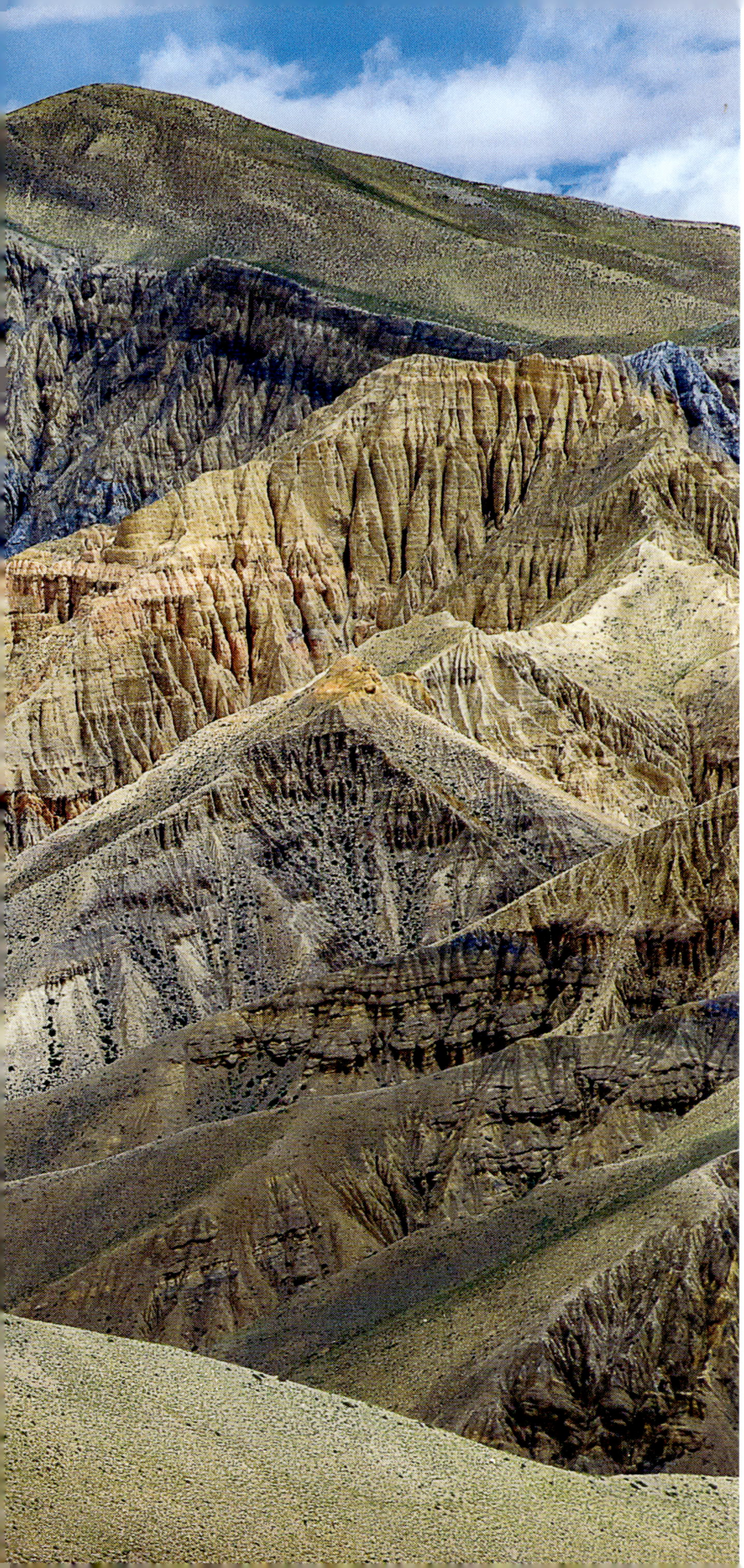

Porter on the trail through the Hidden Valley to Drakmar, Upper Mustang, Nepal, 2013

Following spread: (left) Tsarang Monastery, Upper Mustang, Nepal, 2013
(right) The ancient city of Lo Manthang, Upper Mustang, Nepal, 2013

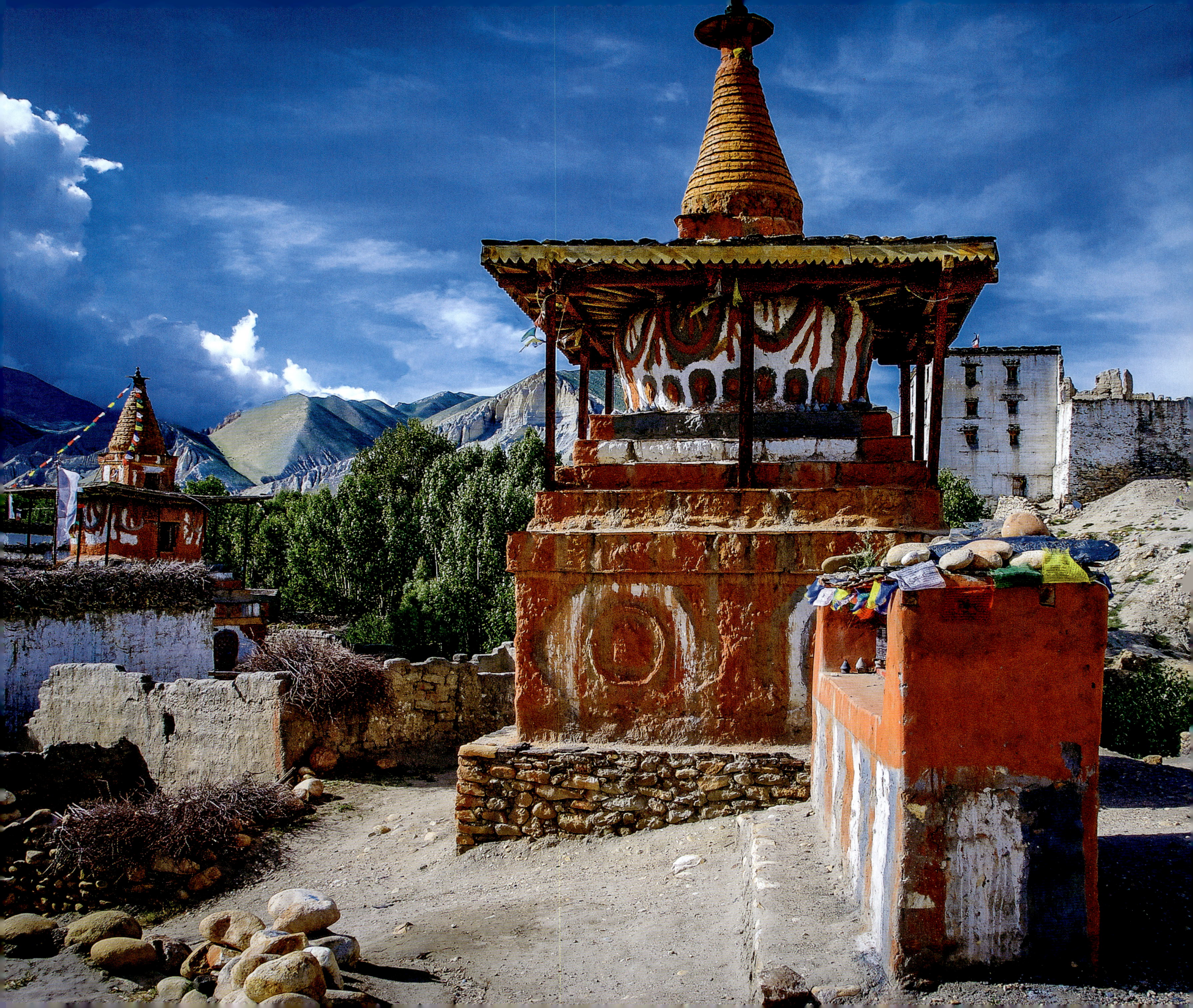

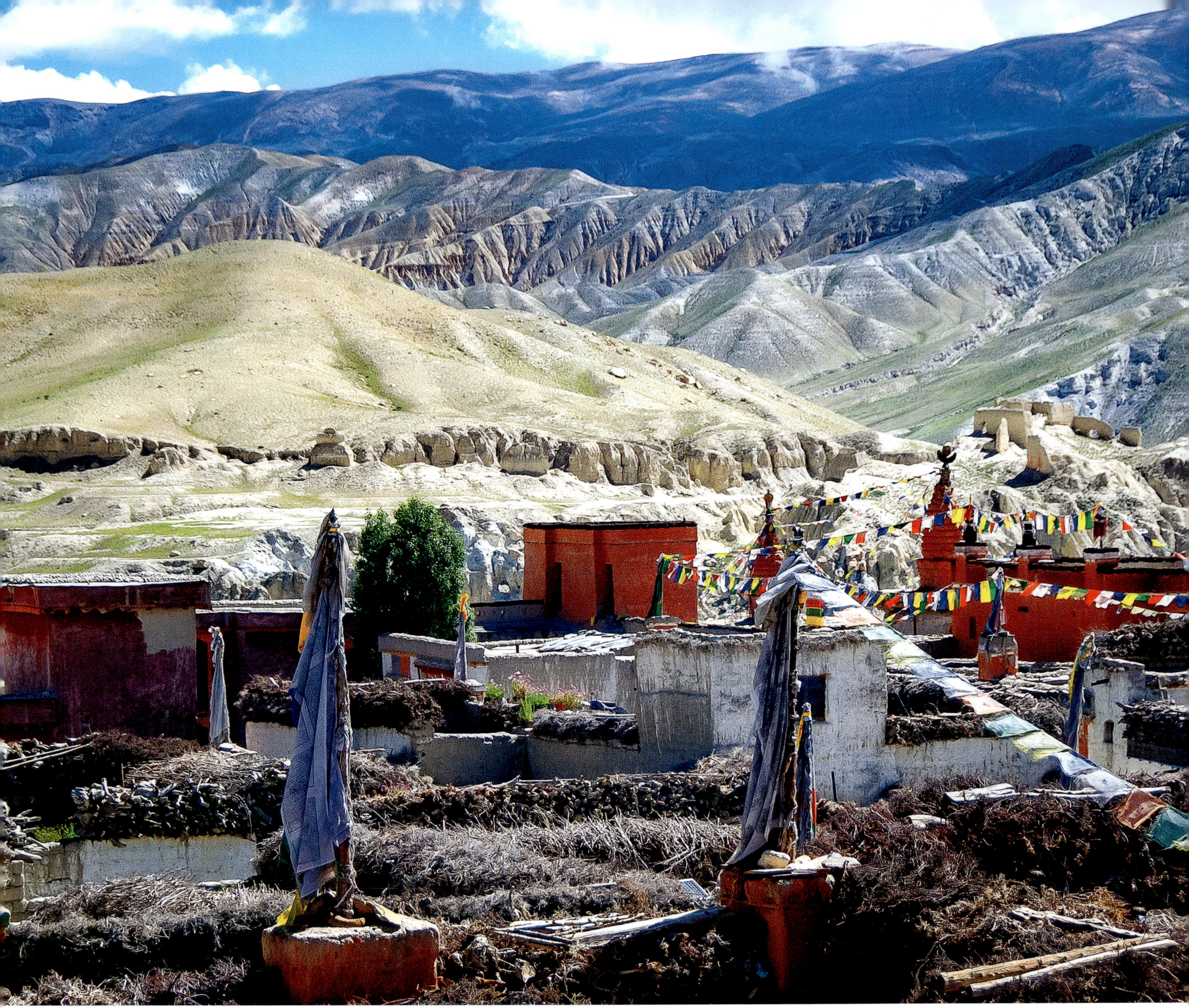

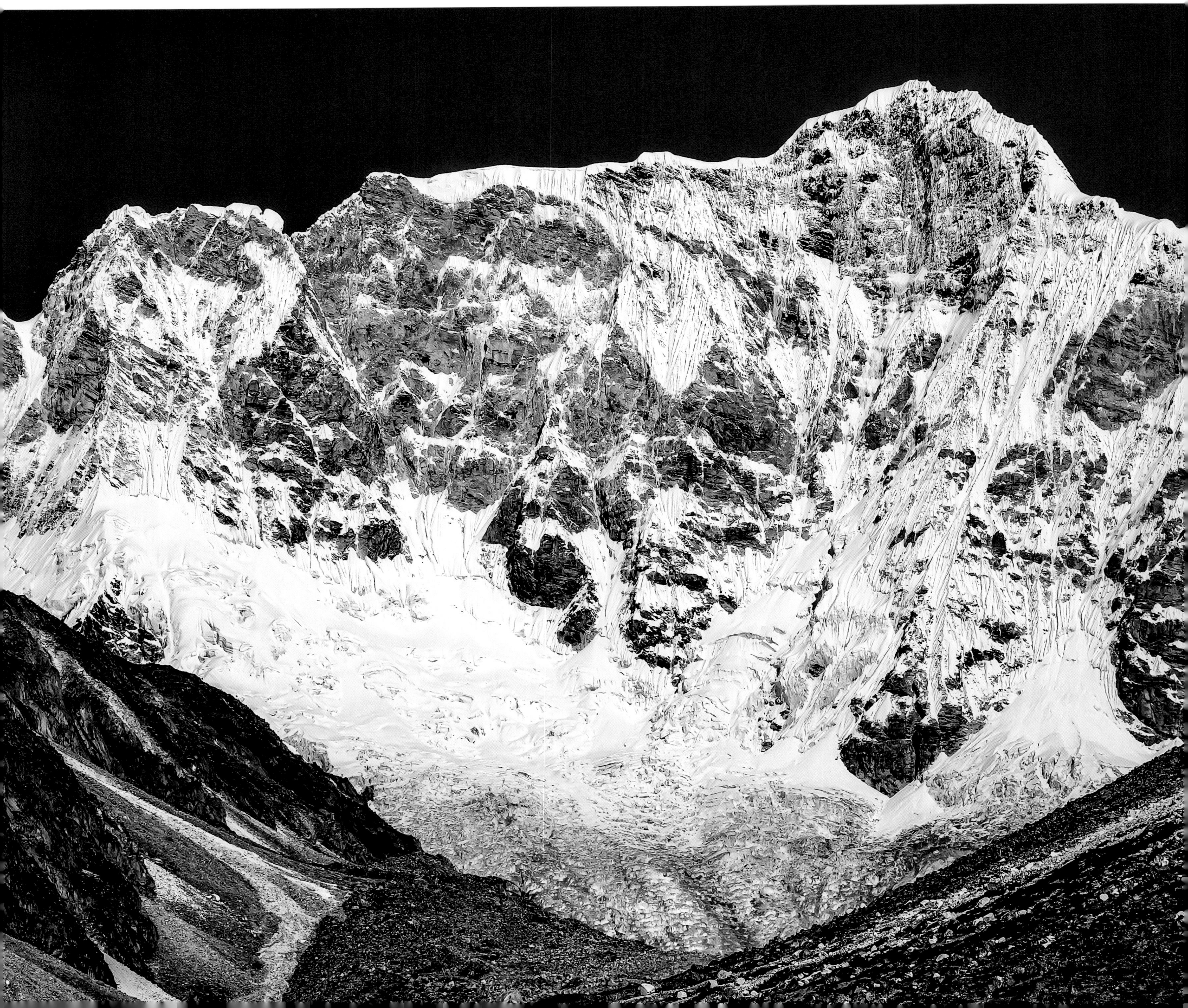

To the Sacred Mountain, Bhutan

OCTOBER 2015

It is not easy to get to Gangkar Puensum, Bhutan's sacred mountain. There are few villages along the way, and the trail is mostly just a rocky path that animals follow on their way up to summer pastures. The name Gangkar Puensum, which translates roughly to "three brothers," refers to the three peaks connected by steep ridges that make up this huge snow-covered massif. At 24,836 feet, it is Earth's tallest mountain that has never been summited.

There were several unsuccessful attempts to climb the mountain during the mid-1980s. When an American expedition tried, the only available maps were sketchy, and their guides took them up the wrong valley, perhaps purposely, so they never made it. Several years later, the Bhutanese government declared all mountains there off-limits to all climbers out of respect for local religious beliefs. A Japanese team tried to get around this restriction in 1998 by approaching Gangkar Puensum from the Chinese side, but they were forced to turn back when their permit was canceled at the urging of the Bhutanese.

Our trek to the base of Gangkar Puensum began in Dur Zam, a drive of several hours from the town of Bumthang. Pouring rain created an inauspicious start made even more miserable by the food poisoning that overtook our group. Two nights earlier, a member of our group stood up from the dining room table at our hotel and immediately collapsed on the floor. Others continued to fall ill one by one in the coming days.

Rain continued off and on for the most of our thirteen-day trek. The rocky trails were muddy and slippery, and we were happy we had brought gaiters, usually used for snow, to keep the mud off our boots. The first day we hiked for eight

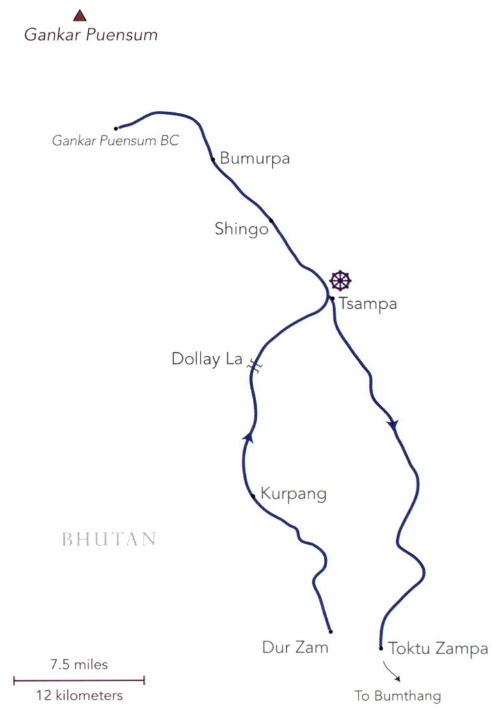

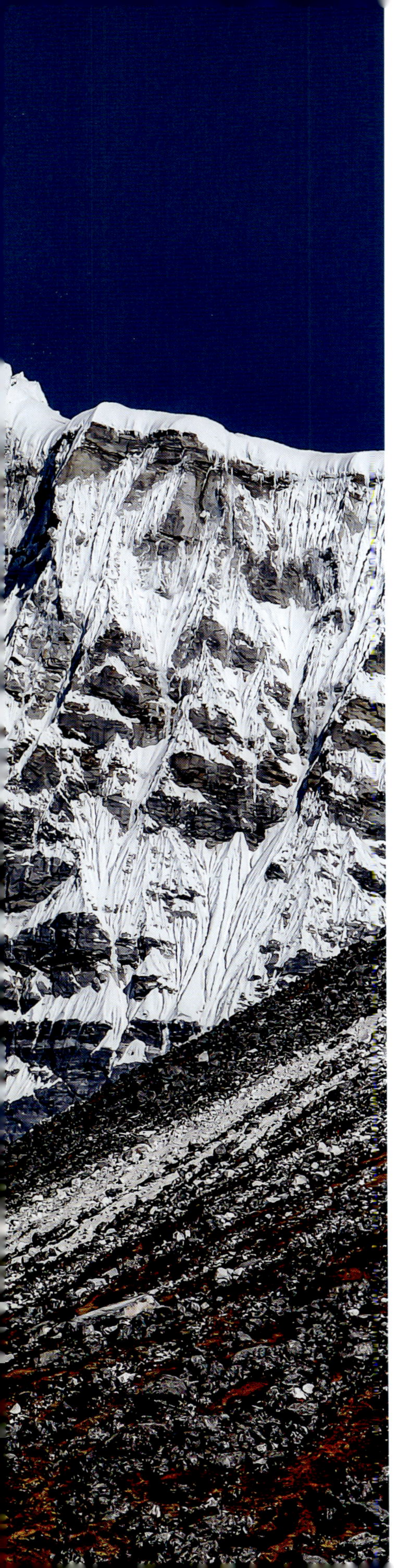

Gangkar Puensum, Bhutan's sacred and unclimbed peak, 2015

hours through forests of bamboo and birch. With an elevation gain of nearly 2,000 feet we ended up in an ecosystem of juniper and fir forests. On the third day we reached the tree line and continued through valleys with little vegetation lined by high rock cliffs. On day four, with eight miles of increasing elevation ahead of us, Bill was hit with the same intestinal bug that had knocked out others in the days before. It was a struggle keeping up with the group, who were hours ahead of us.

In late afternoon we started the long ascent up to Dollay La Pass (elevation 15,700 feet). When we finally reached the top at dusk, we could barely see our camp far in the distance. Our group leader had thoughtfully retraced his steps back to check on us earlier, but now we were alone and wondering how we could navigate the marshy landscape between us and the camp. We headed toward a dry-looking section and were relieved to see Tashi, our Bhutanese guide, who led us through the maze of spongy grass, mud, and water.

All of our trials were rewarded the next day with our first glimpses of the massive, snow-covered expanse of Gangkar Puensum in the distance. The next day we arrived at Bumurpa, an alpine tundra meadow just below the South Face at the eastern edge of the massif. We put up our tent facing the mountain and marveled at the site. Bill and I have seen a lot of majestic mountain peaks, but Gangkar Puensum was something special because it was so hard to get to.

On the way back we stopped at Tsampa Monastery, where a small celebration was taking place. The monks invited us in and told us that it was a blessing for animals, especially yaks, which play an important role in this remote region as pack animals and food sources.

After our trek ended we flew to Paro and made a pilgrimage to the Taktsang Monastery. Situated high on a steep cliffside outside Paro, it is popularly known as the Tiger's Nest. It is said that reincarnated Guru Padmasambhava, who introduced Buddhism to Tibet, meditated at this location during the eighth century CE, arriving on the back of a tigress. The original structure was built in 1692, but a fire destroyed the main building in 1998. Thankfully, it has been beautifully rebuilt. It took us several hours to walk all the way up, and we were happy that we were in good shape after fourteen days of hard hiking.

Isolated canyon on the trail from Gangkar Puensum, 2015

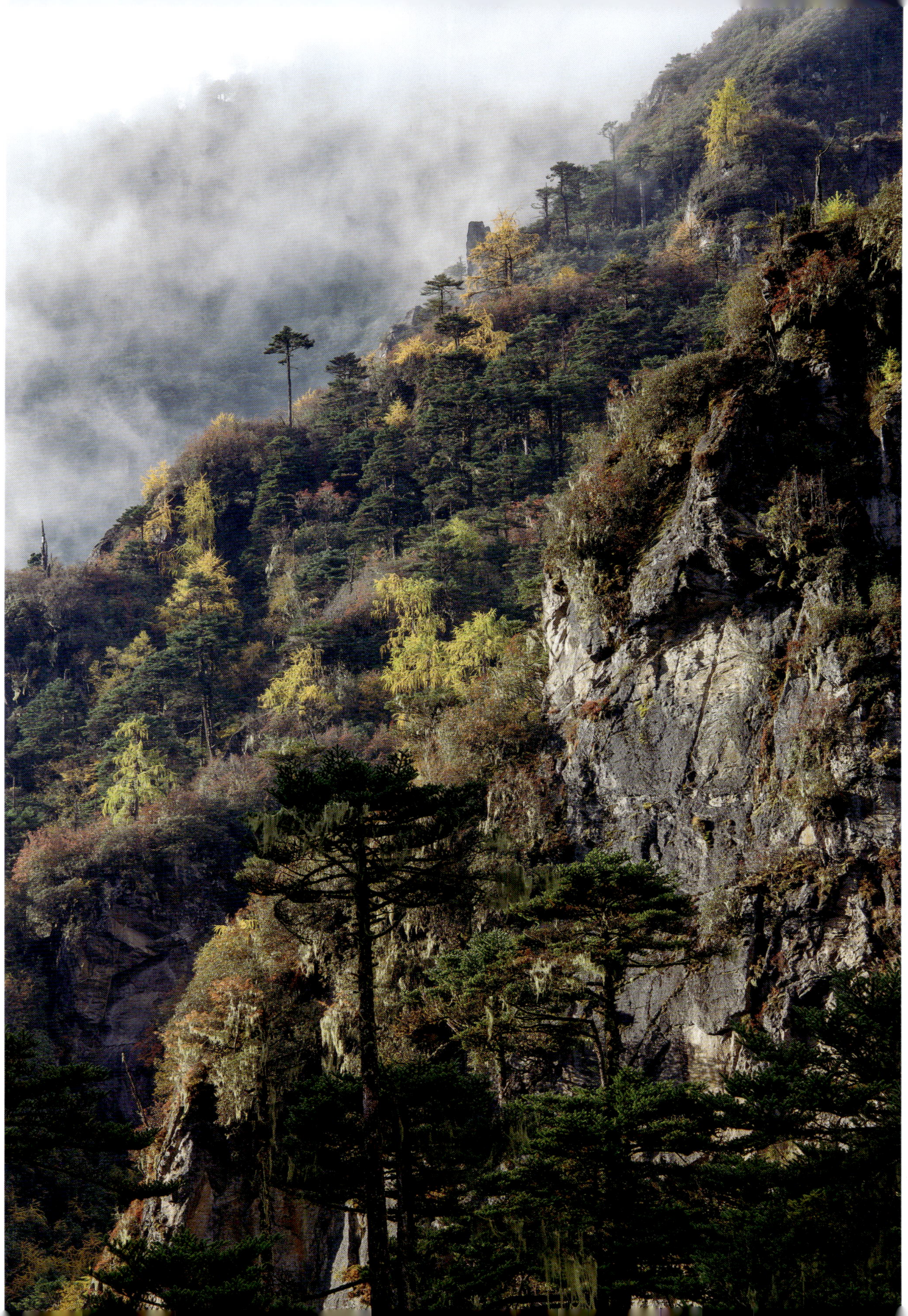

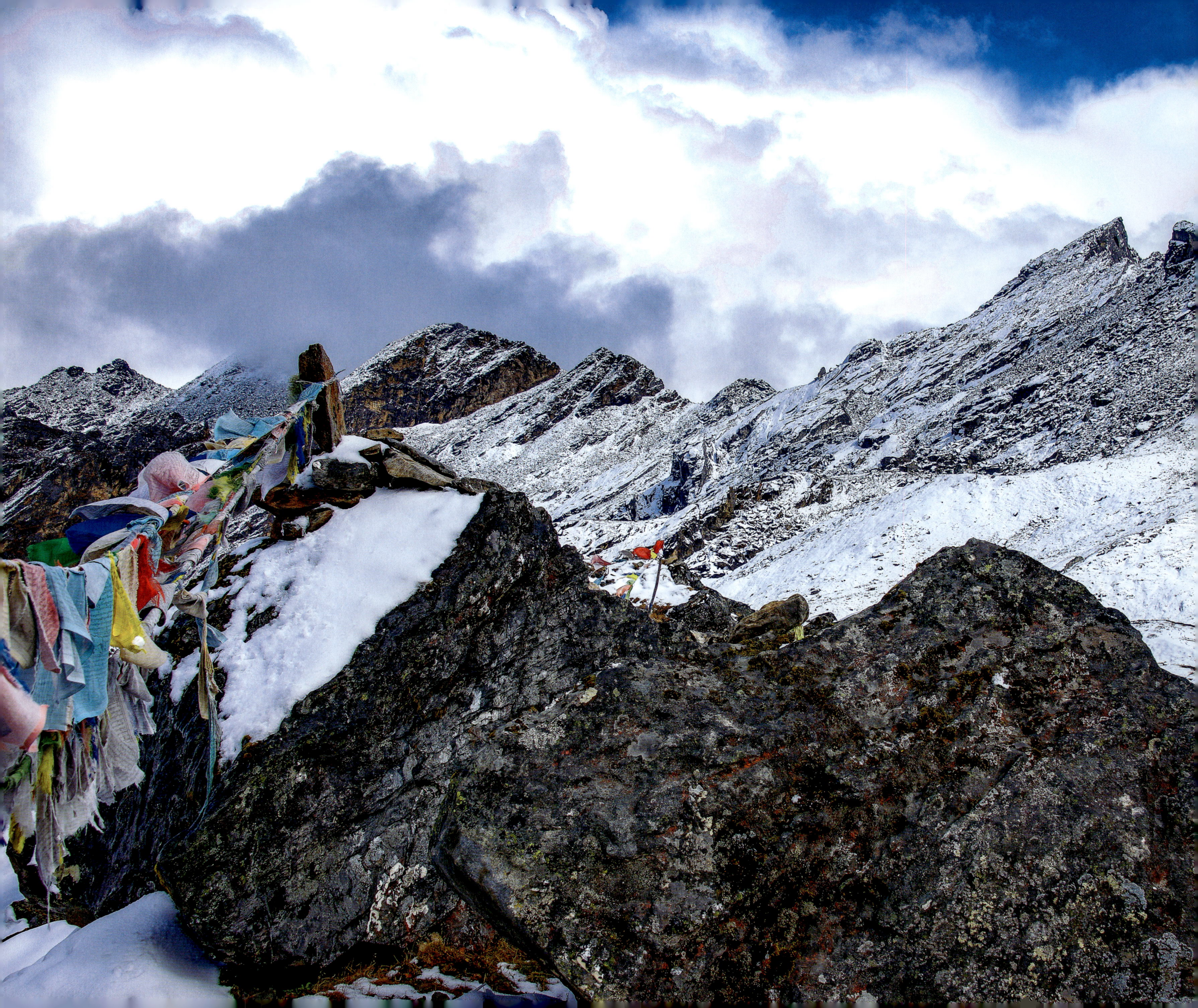

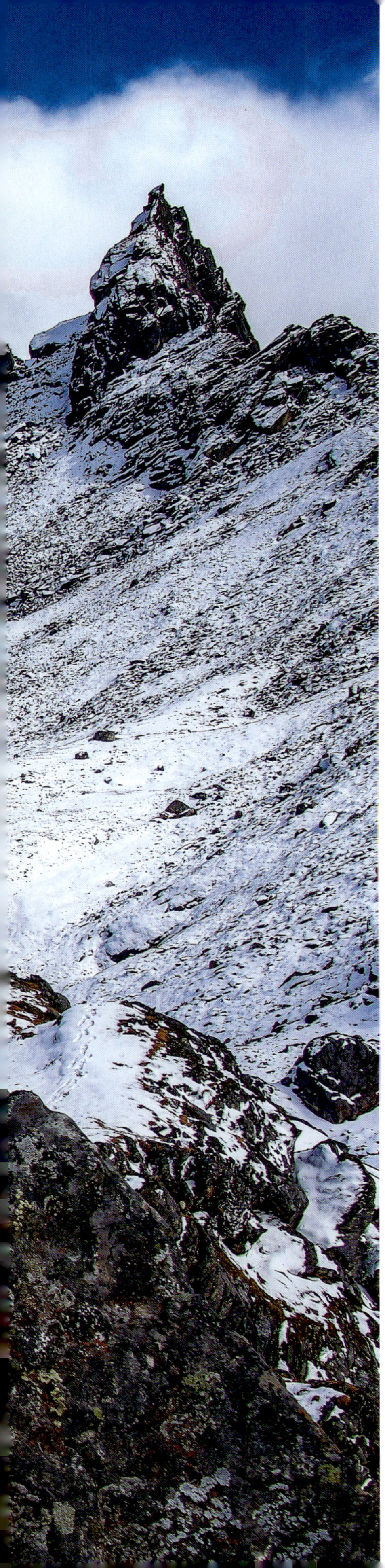

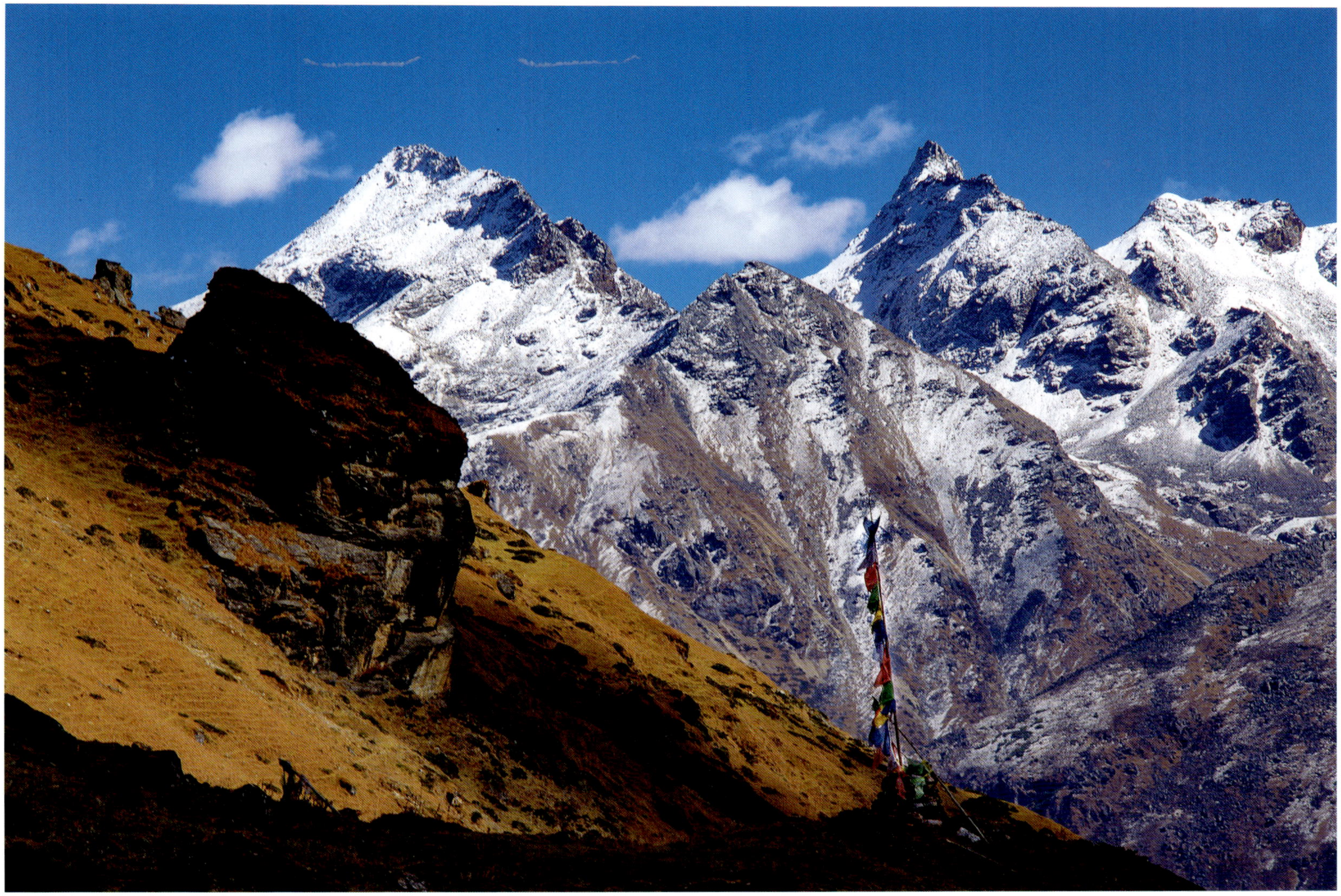

On the trail to Gangkar Puensum, 2015

Crossing Dollay La, Bhutan, 2015

Following spread: Prayer wheel (left) and prayer flags (right), Bumthang, Bhutan, 2015

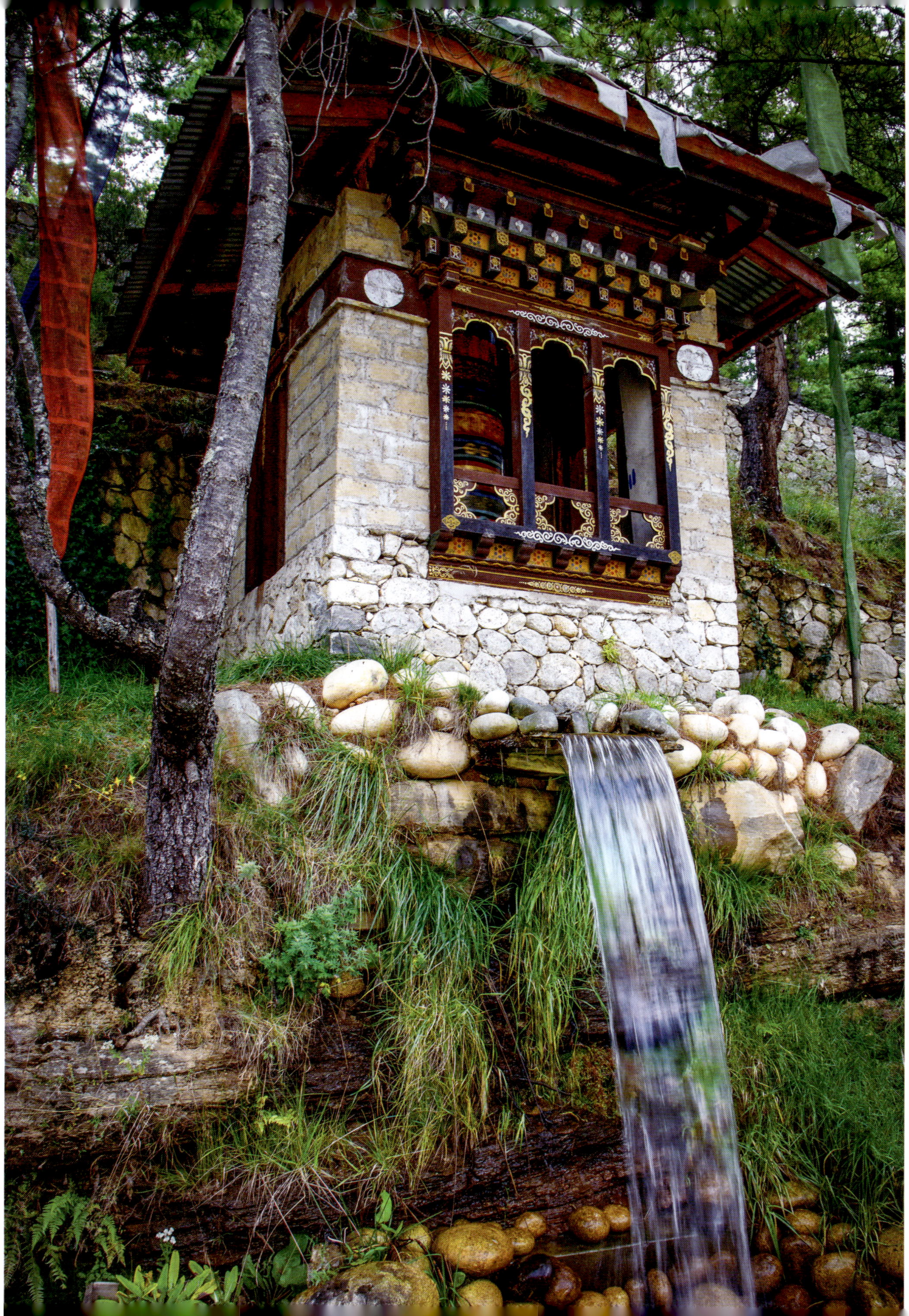
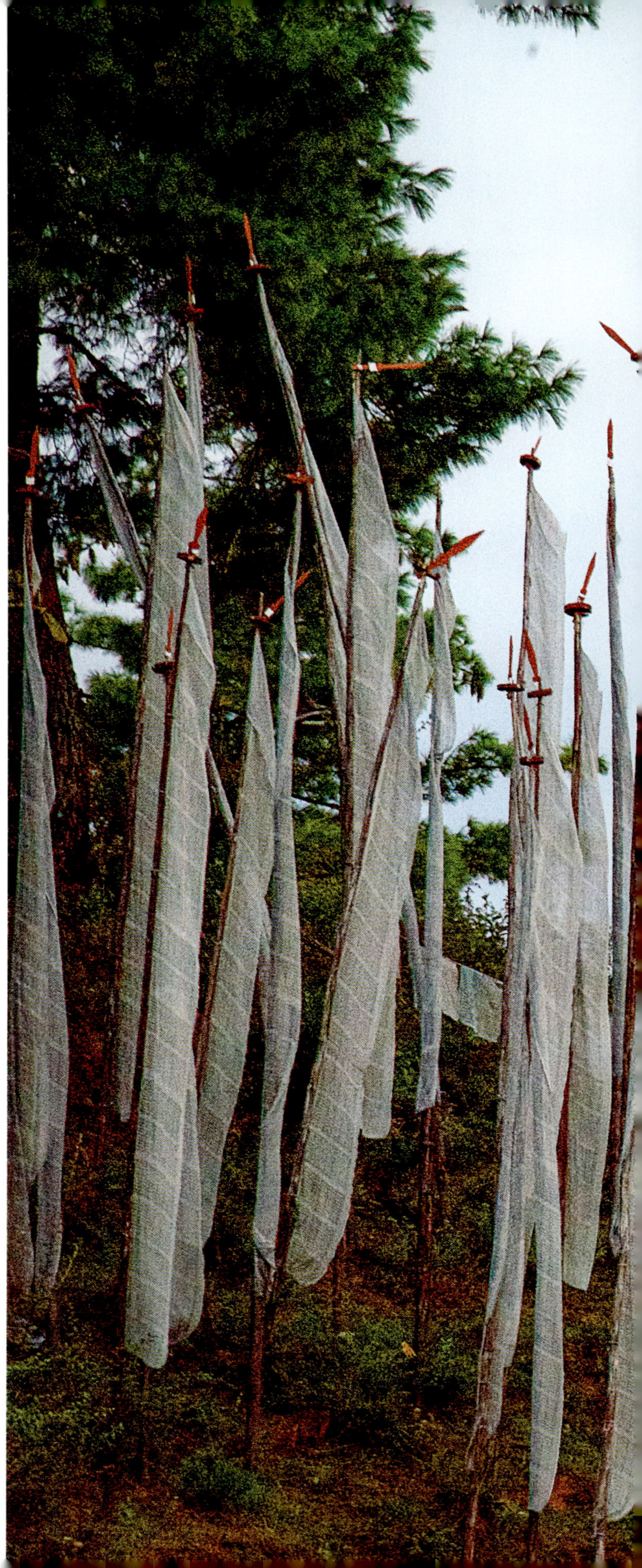

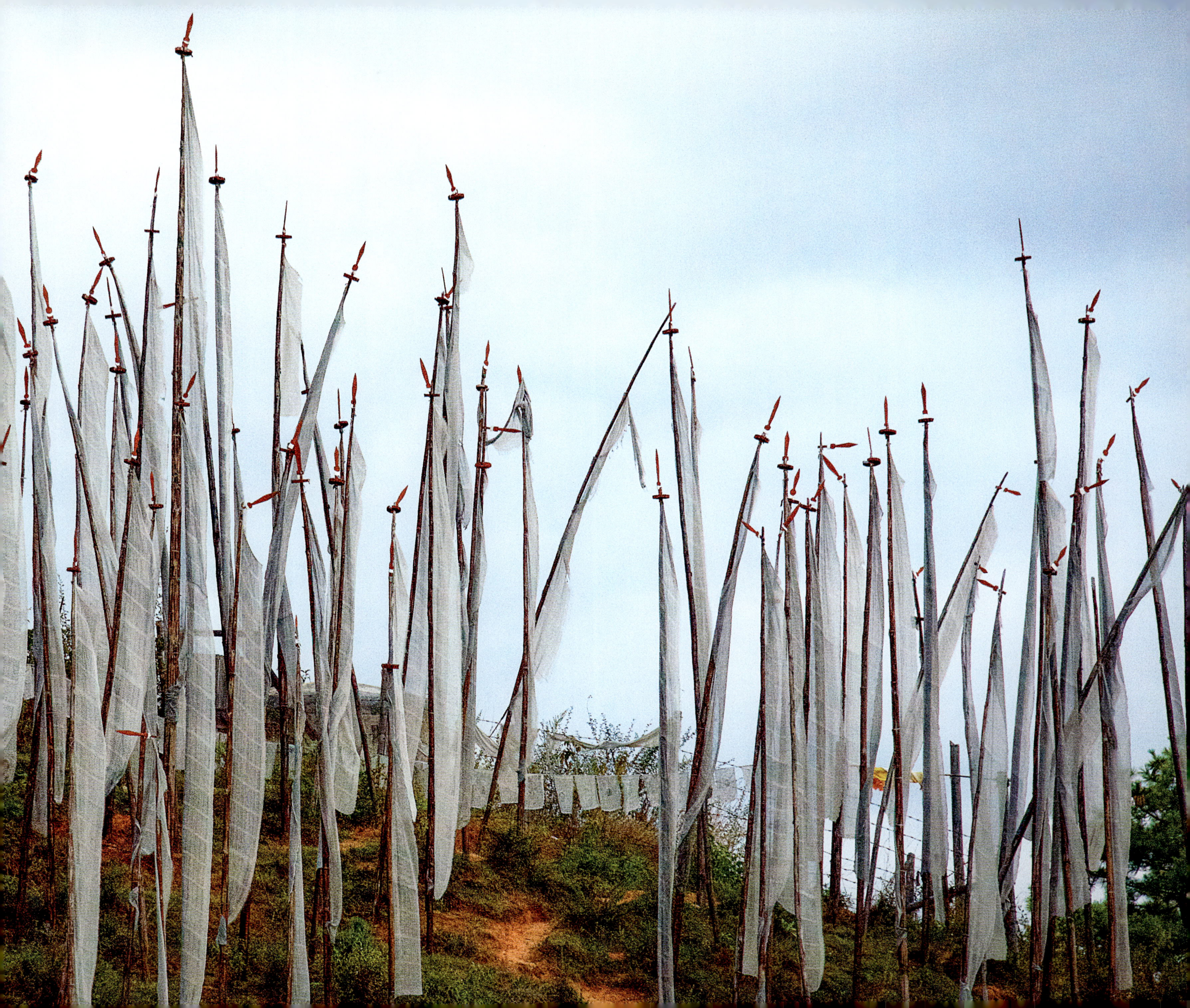

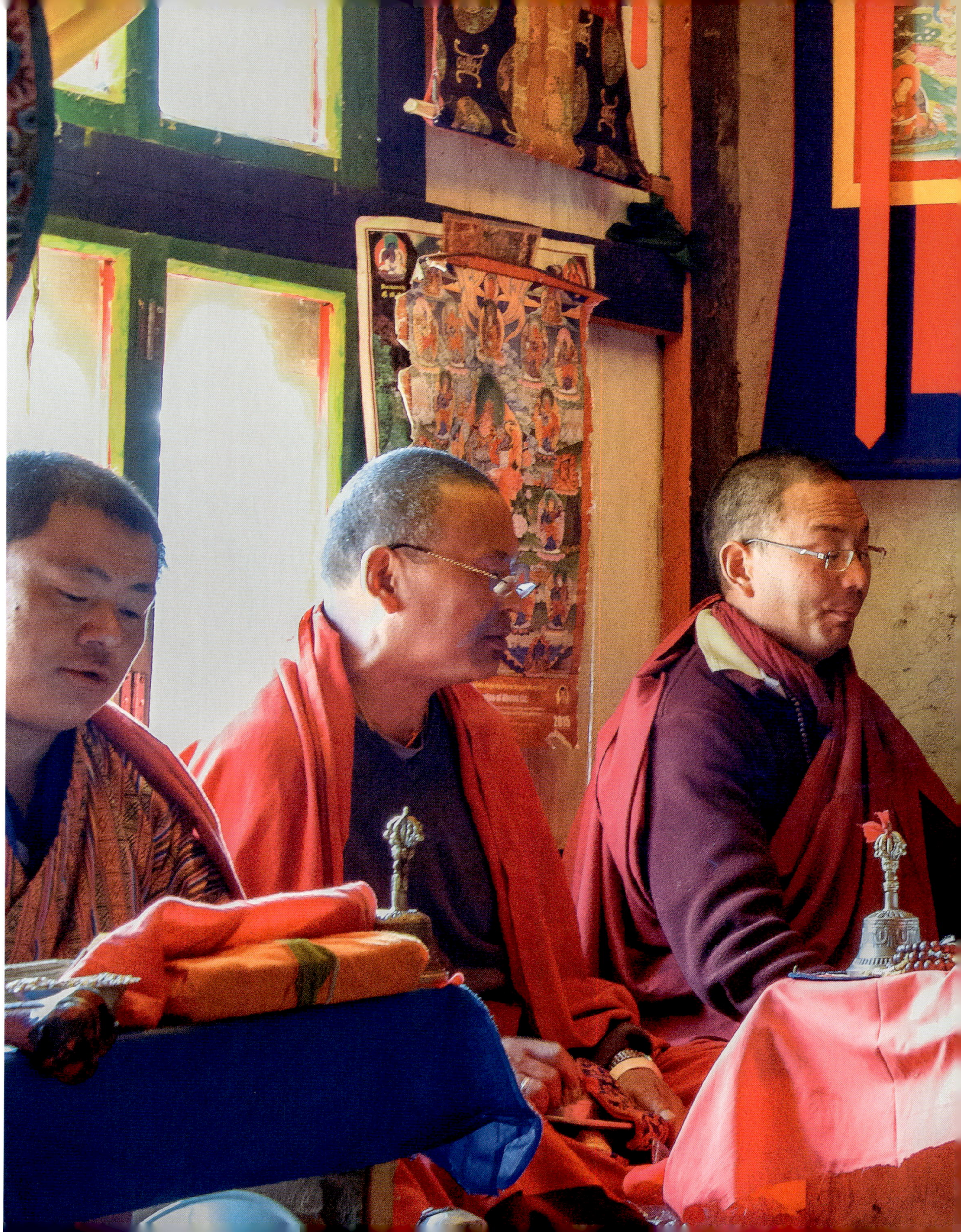

Tsampa Monastery, below Gangkar Puensum, Bhutan, 2015

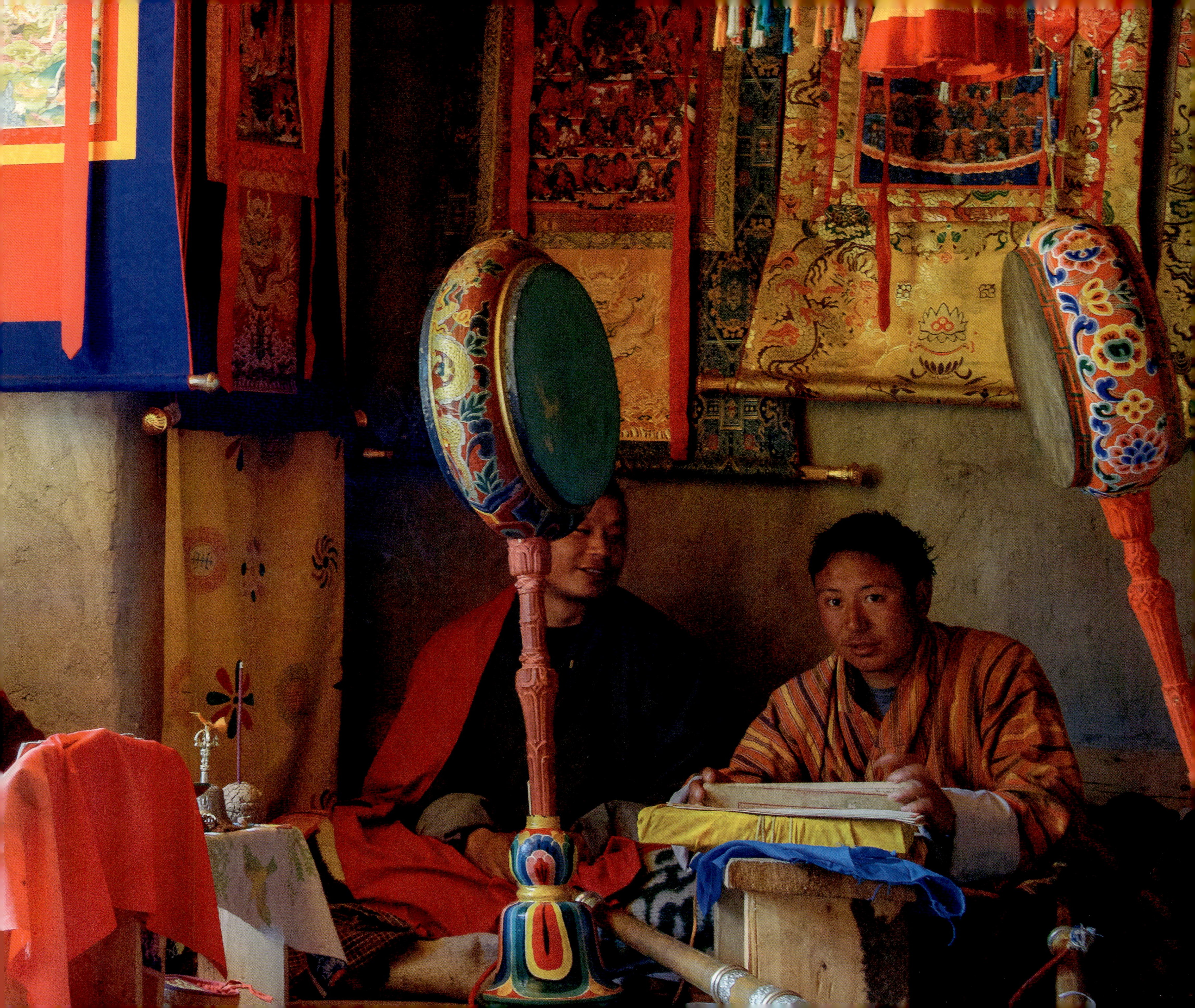

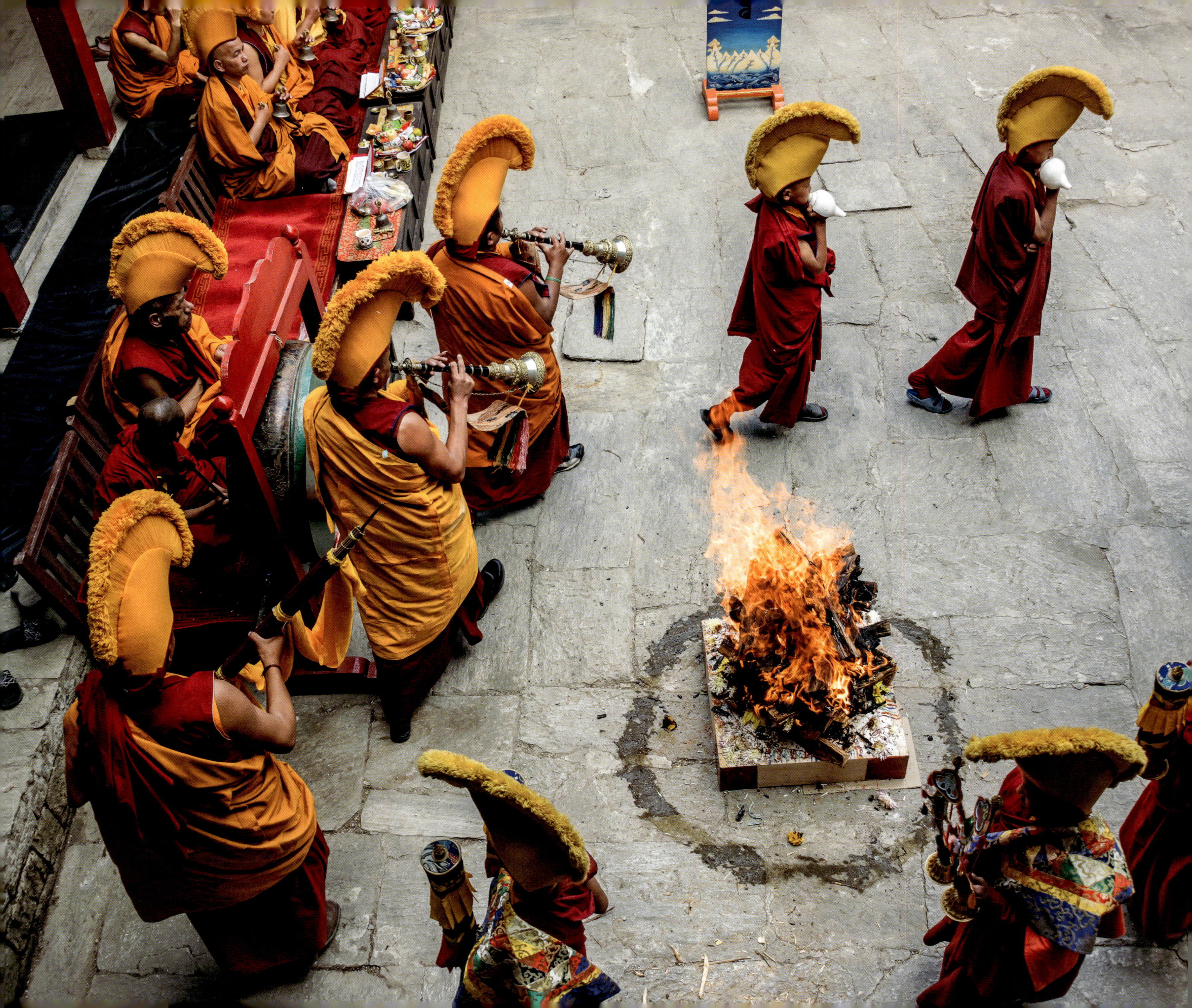

Mani Rimdu at Chiwong Monastery, Nepal

NOVEMBER 2017

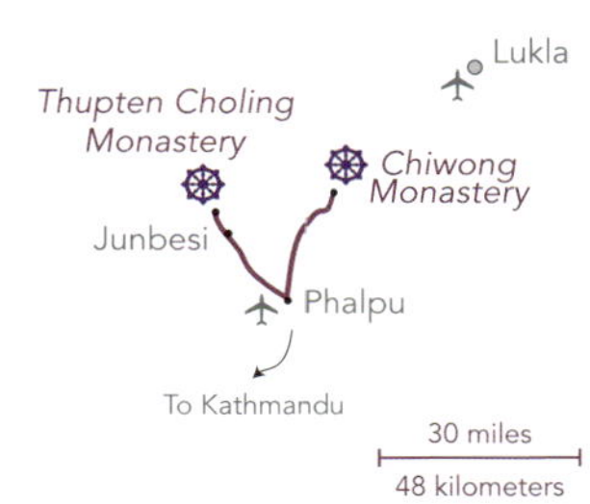

FROM MY PERCH on the second-story wooden mezzanine of Chiwong Monastery, I looked into the stone courtyard below: Monks had lit a fire as a *puja* (offering) to the gods. It was the last day of Mani Rimdu ceremonies, and soon the crowd—a few tourists like us, local people, and dozens of monks—would disperse, ending the festivities for another year.

The sacred Mani Rimdu ceremonies traditionally last for nineteen days; the last three, which we attended, are the most accessible to the public. Four days earlier we had driven ten hours from Kathmandu with Ram and Chandra to the village of Phalpu in the Solo Khumbu region south of Mount Everest. It is less than a one-day hike from Phalpu to Chiwong Monastery, where a more hallowed world awaited us. Perched on a steep hillside overlooking the valley, the two-story, white stucco-covered structure with carved wooden lattice windows looked like many of the other Tibetan Buddhist monasteries we had visited in Tibet, northern India, Bhutan, and Nepal. The monastery was founded in 1924, so it is relatively new but still an important place, due to its association with important spiritual teachers.

There was no more room for guests in the monastery dormitory, so we pitched our tent in a small flat spot just below the entrance to the monastery. At night, the muffled sound of monks chanting in the monastery blending with loud rock music from a makeshift hostel in the field below seemed symbolic of the times we live in.

Like other Tibetan Buddhist events we had attended throughout the Himalaya, the atmosphere was welcoming. Often we were asked to join the monks for meals, and on the second day we were invited to an audience with Sang Sang Rinpoche, the religious leader of the monastery, a gregarious man with a big smile who blessed us and tied red-string prayer necklaces around our necks. During the ceremonies, he presided from an elaborately carved wooden seat and altar covered with embroidered silk, prayer cloths, and strands of marigolds. Over three days we watched a succession of masked dances said to re-create legendary events, including the establishment of Buddhism in Tibet by the great saint, Guru Padmasambhava. We remembered what we had discovered at the Kalachakra Initiation given by the Dalai Lama in Bodhgaya in 1985, where we joined thousands of Tibetan Buddhist pilgrims. Many of them, like us, were oblivious to the intricacies of this complicated teaching, but we all felt the significance of traveling to this sacred place, paying respects, and collectively receiving blessings.

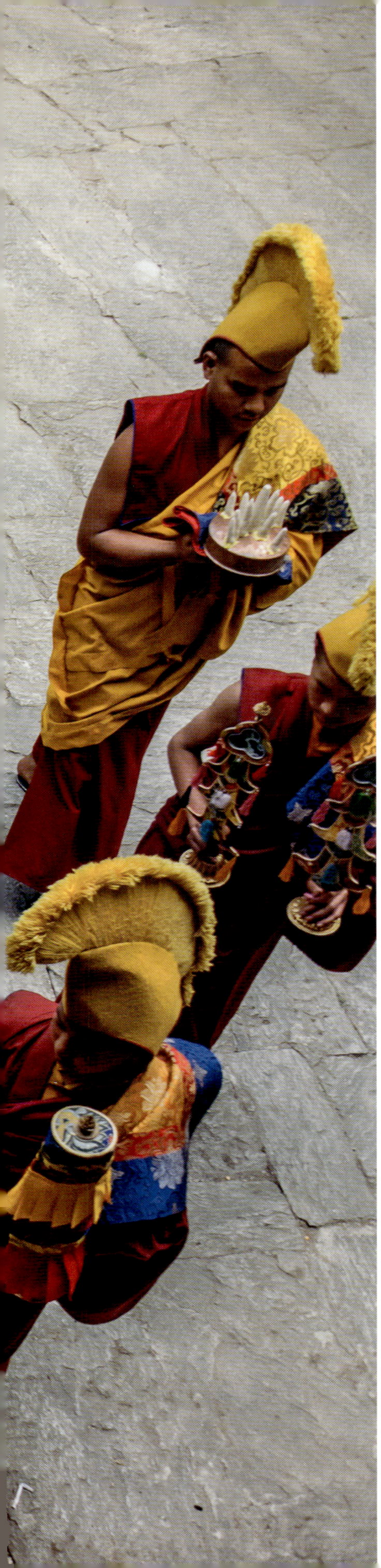

Fire puja, Mani Rimdu ceremony, Chiwong Monastery, Nepal, 2017

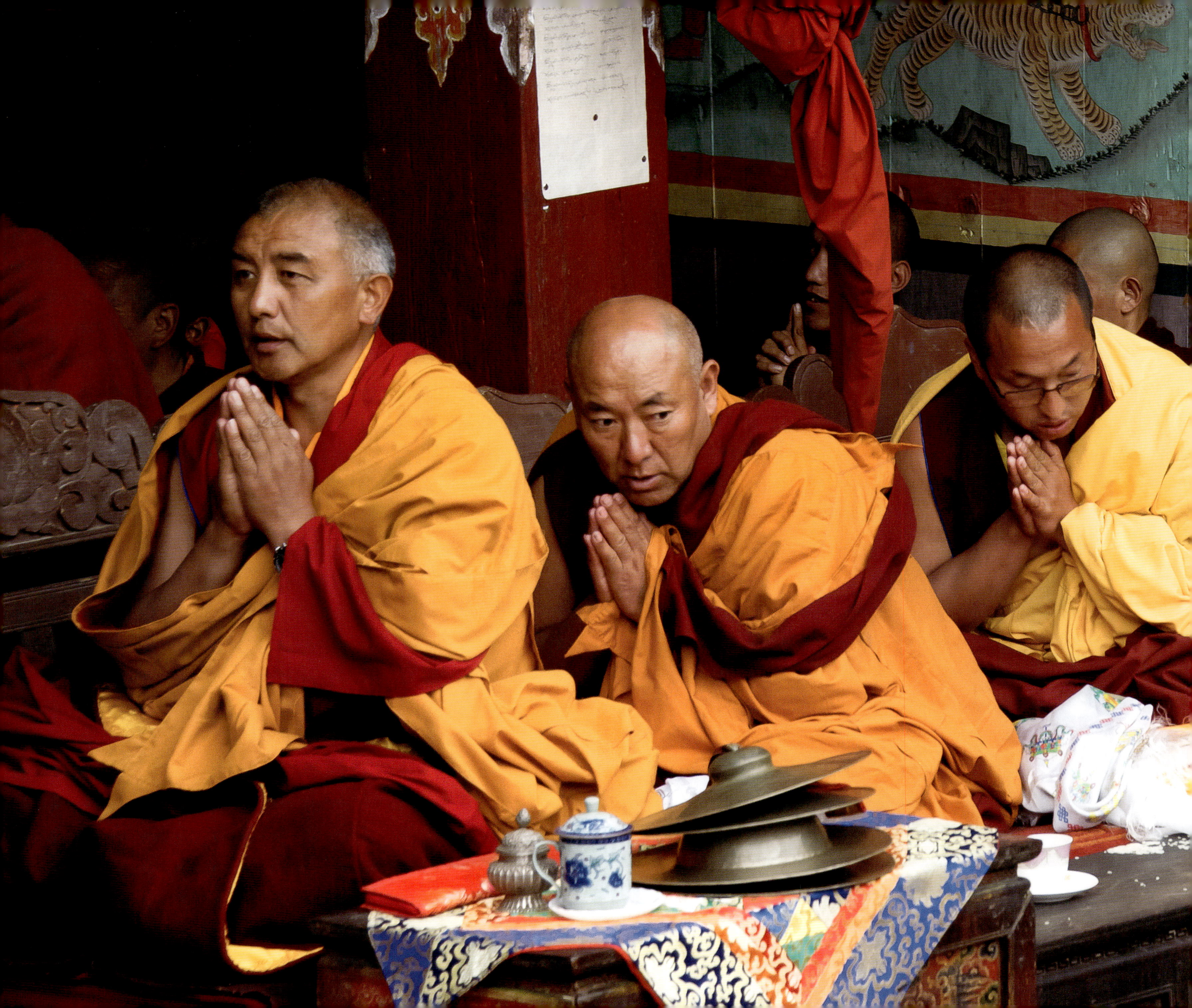

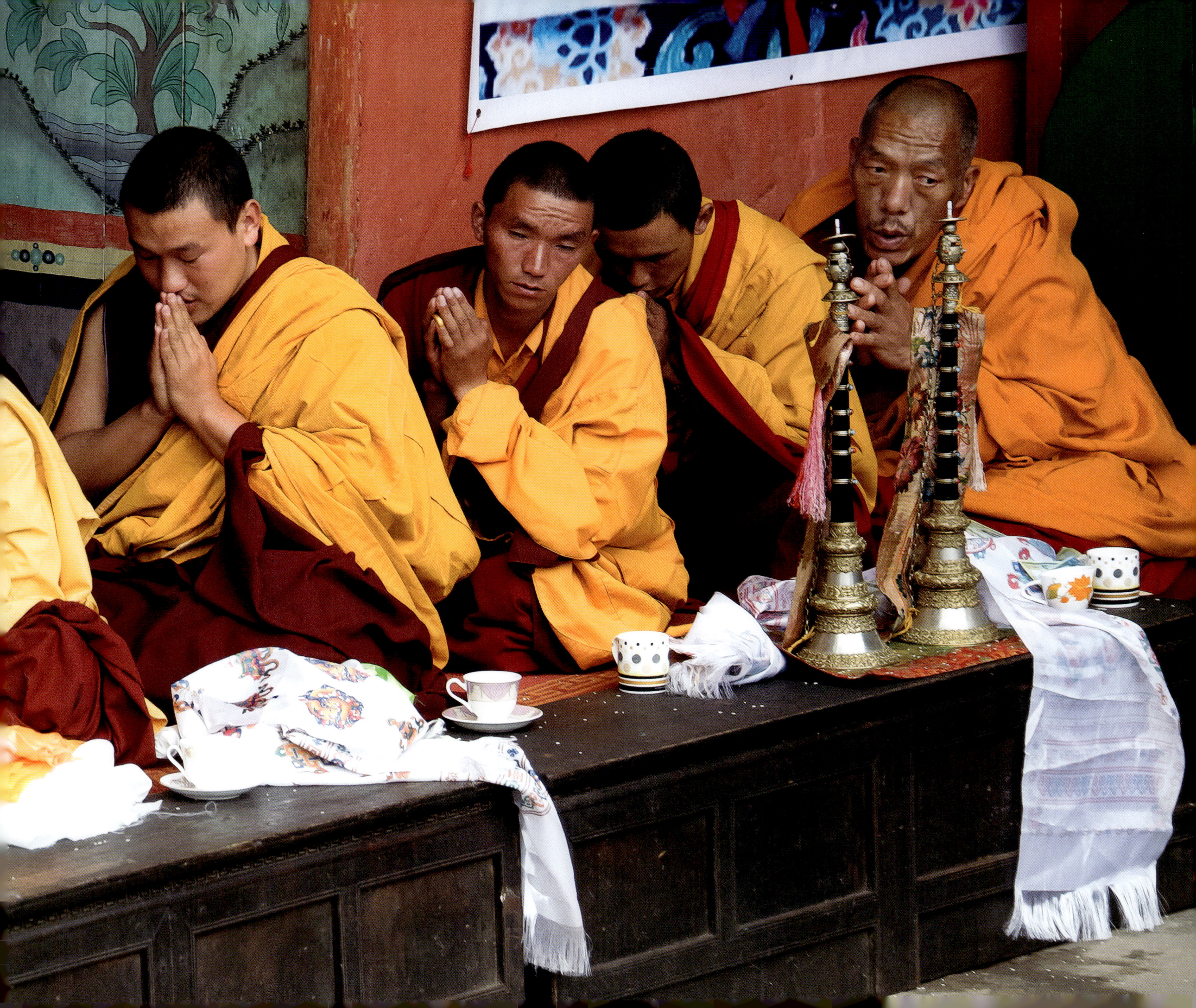

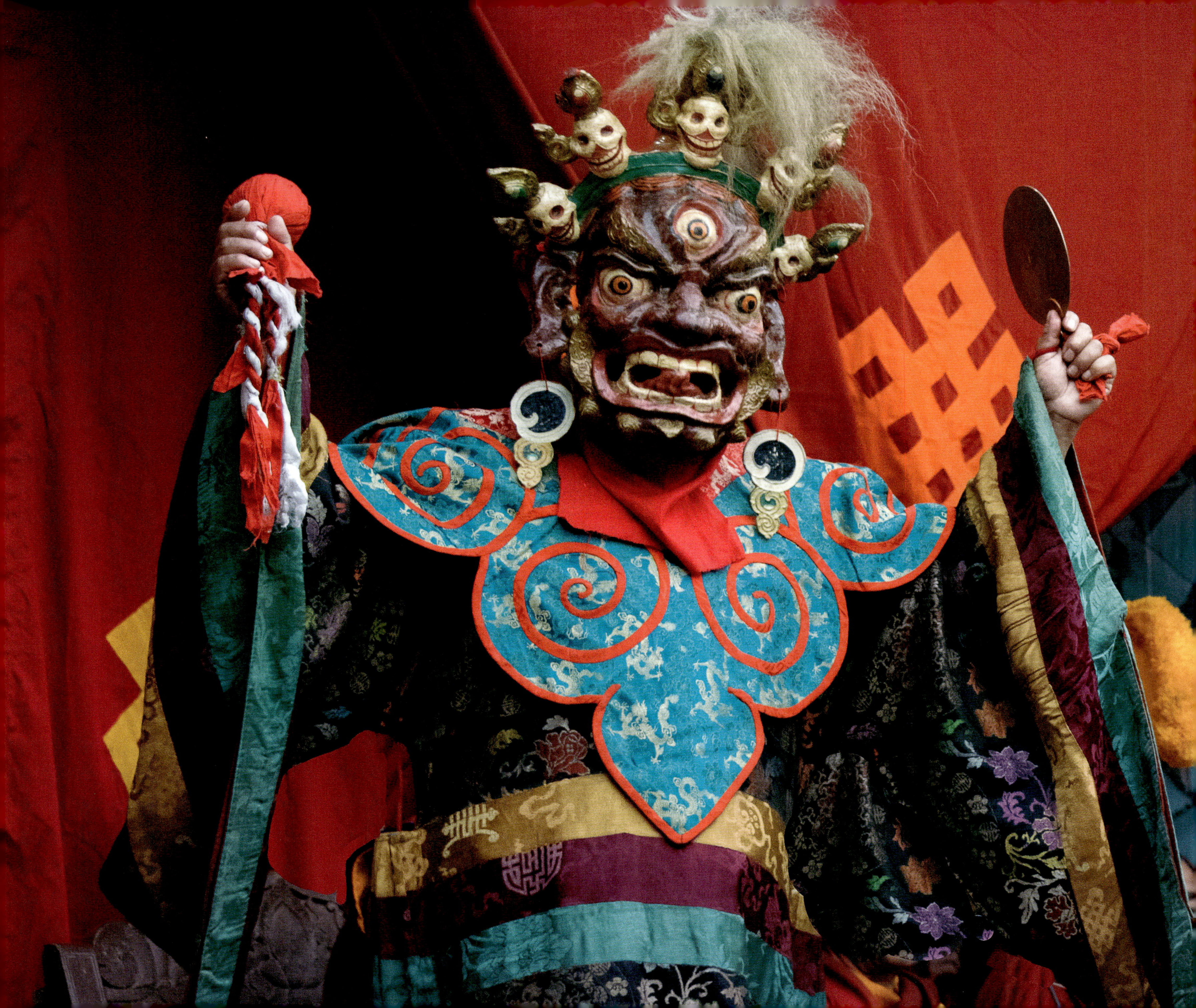

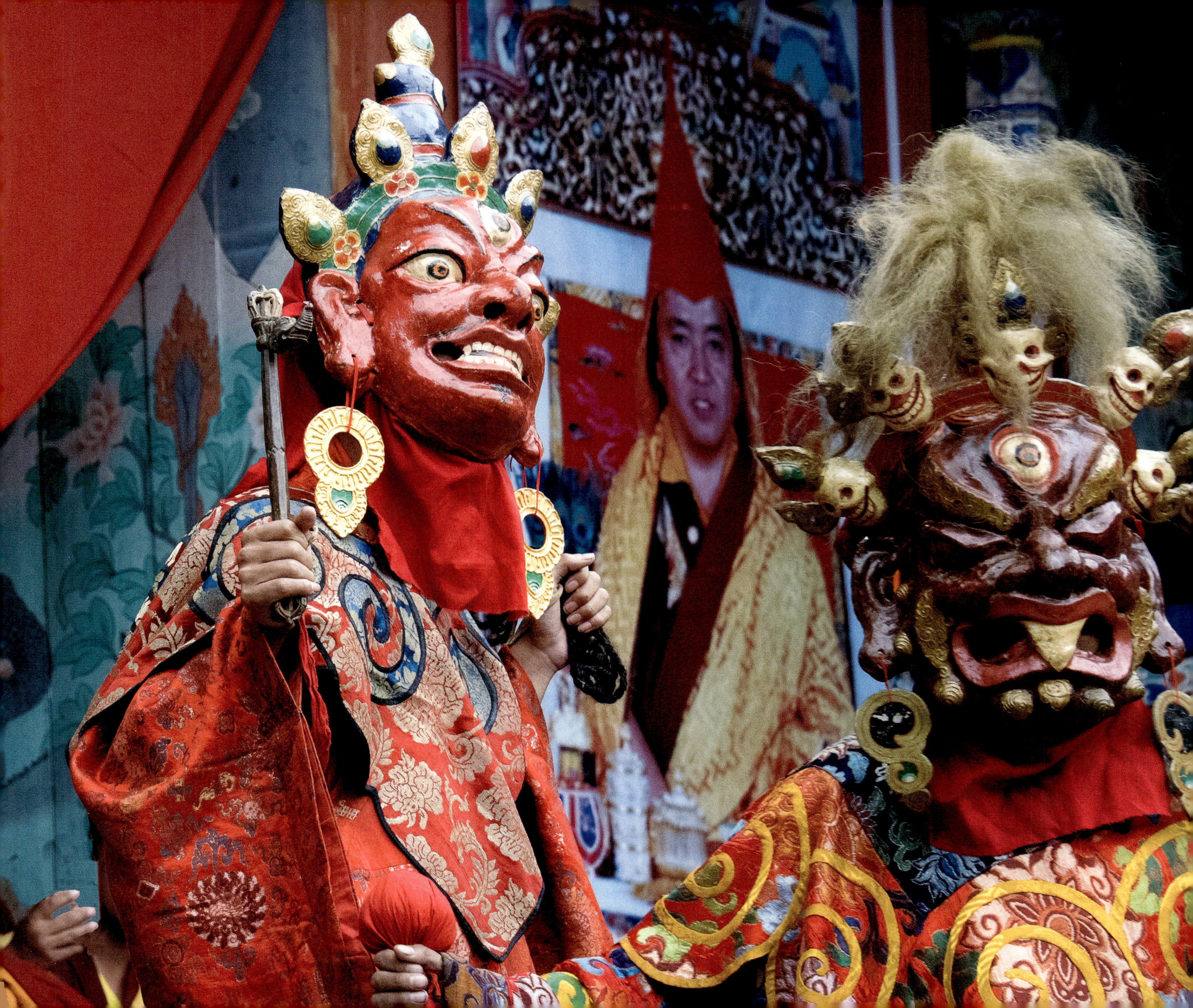

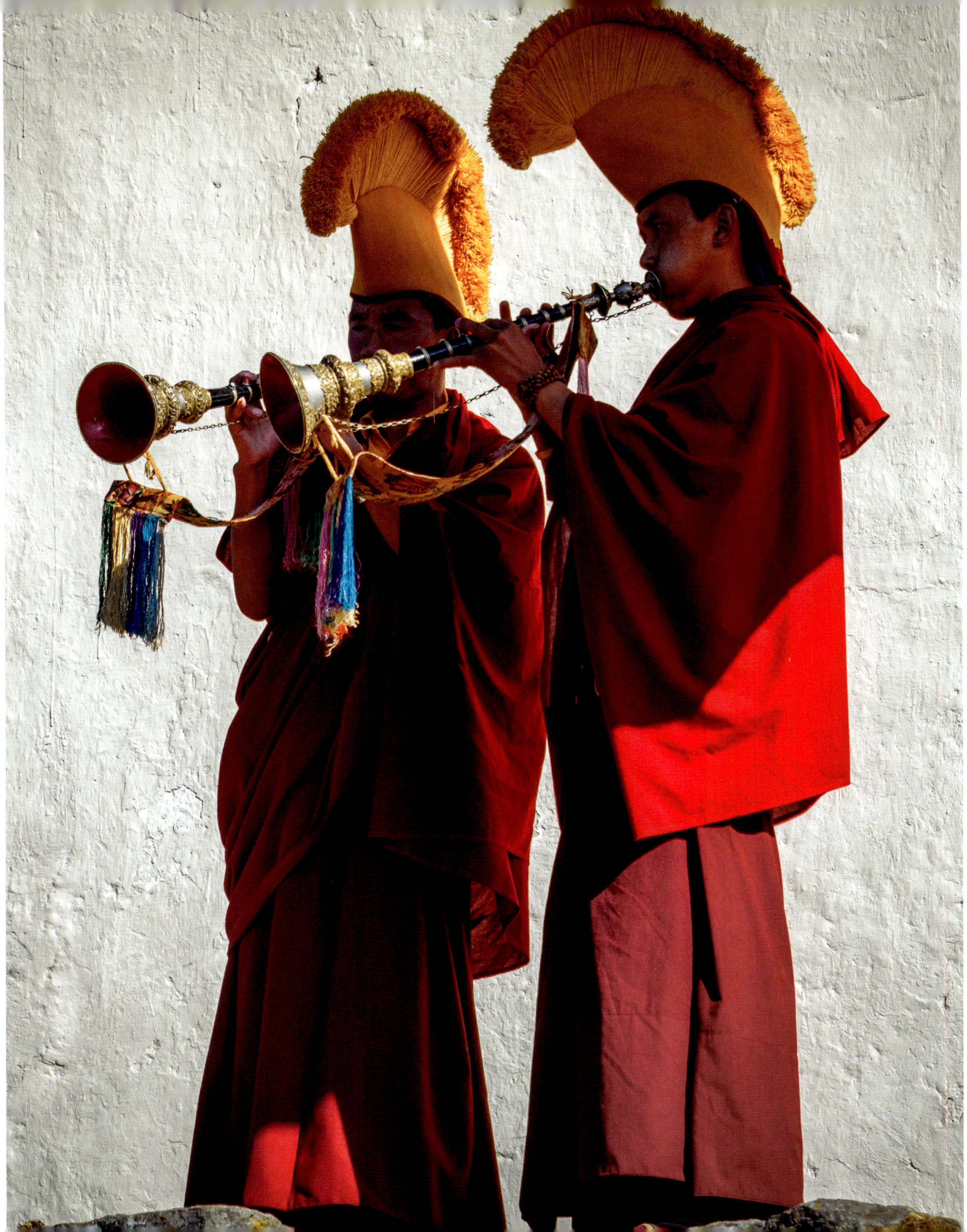

Opening of the Mani Rimdu ceremony, Chiwong Monastery, Nepal, 2017

Opposite: Mani Rimdu ceremony, Chiwong Monastery, Nepal, 2017

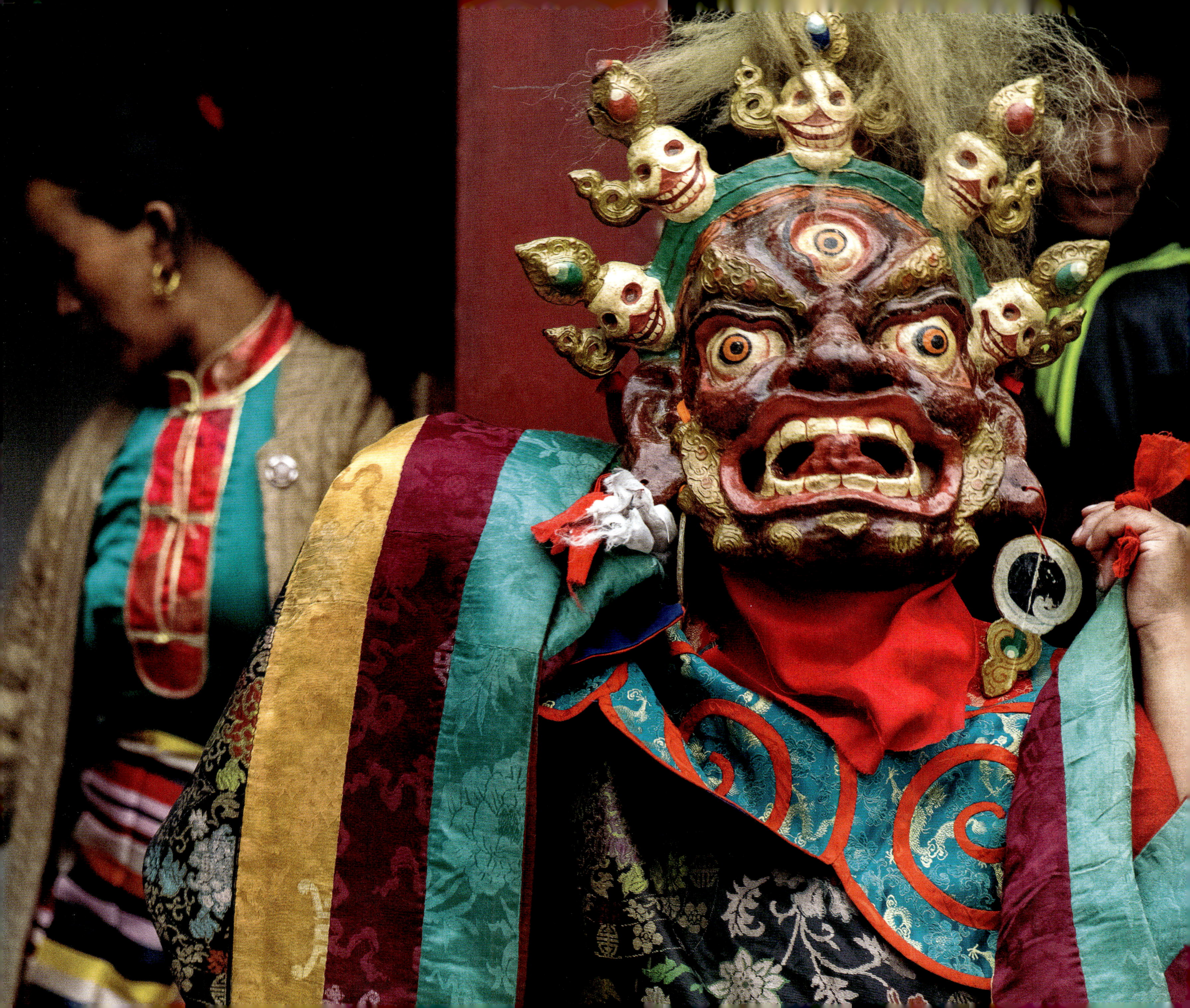

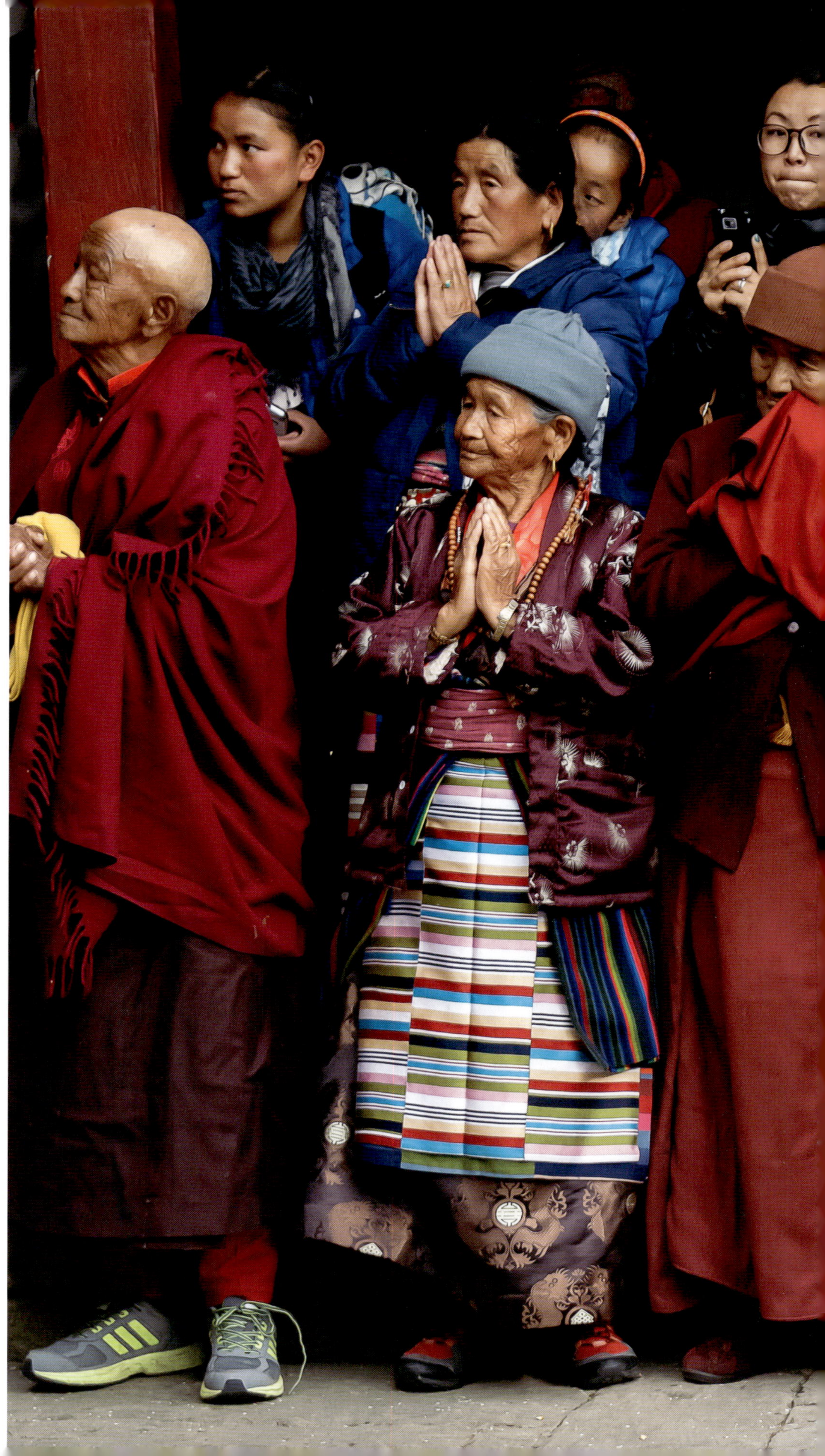

Mani Rimdu ceremony, Chiwong Monastery, Nepal, 2017

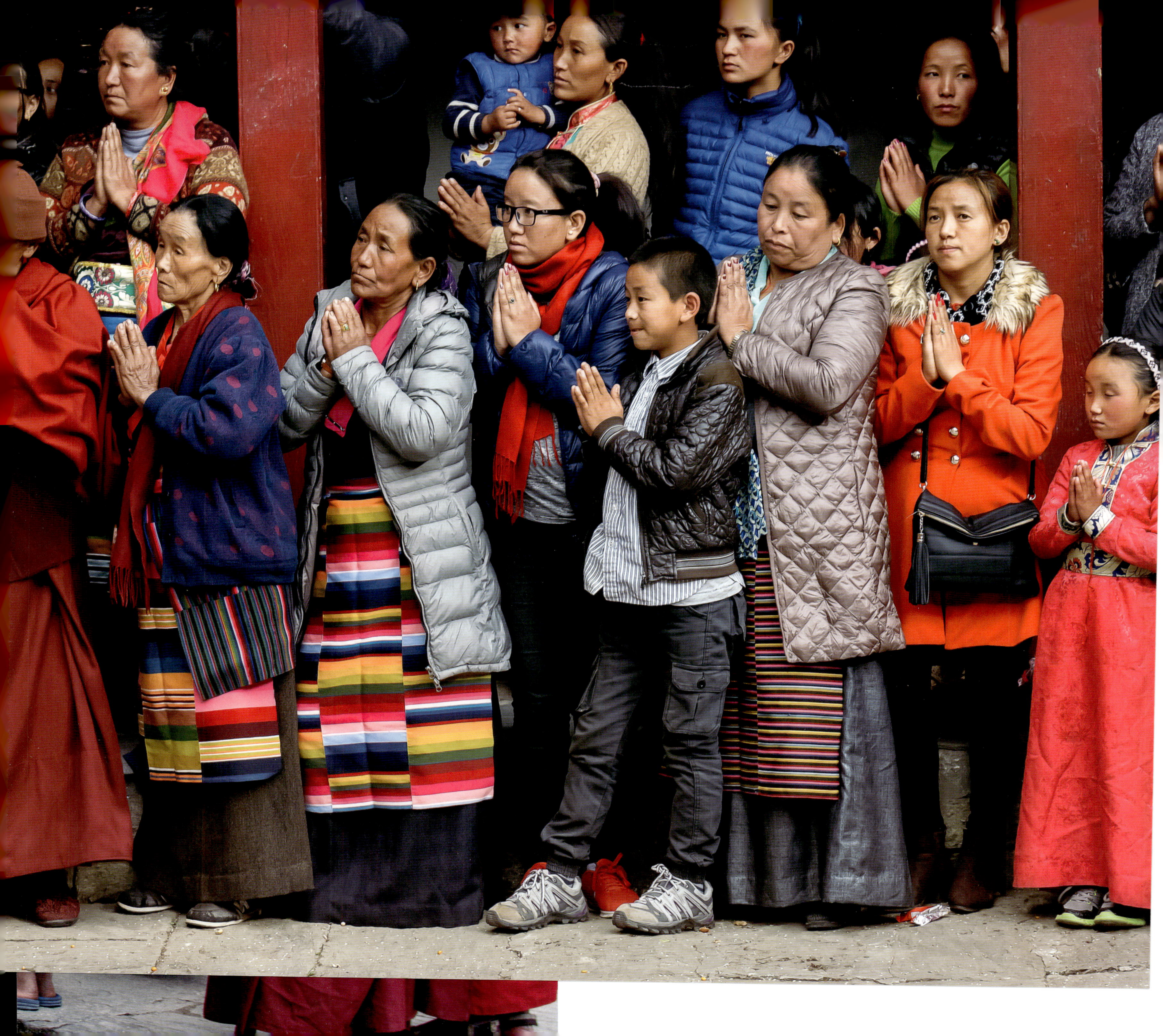

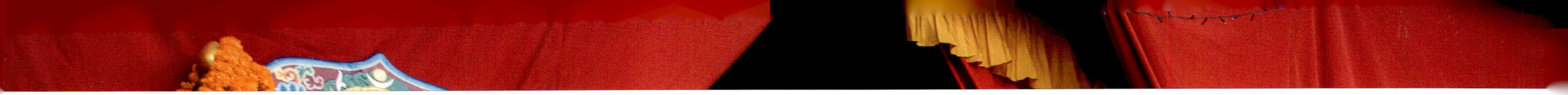

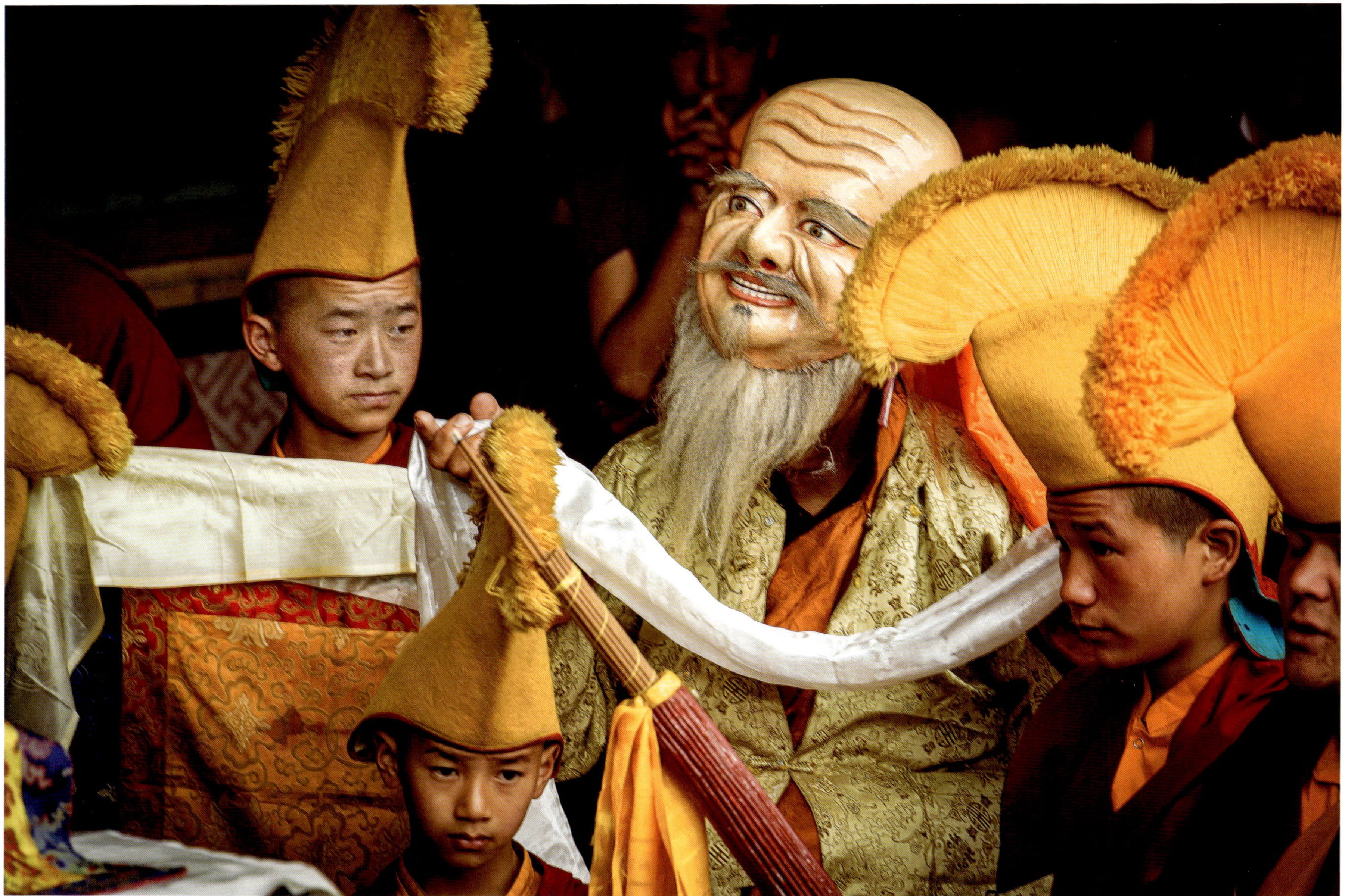

Mani Rimdu ceremony, Chiwong Monastery, Nepal, 2017

The closing of the Mani Rimdu ceremony, Chiwong Monastery, Nepal, 2017

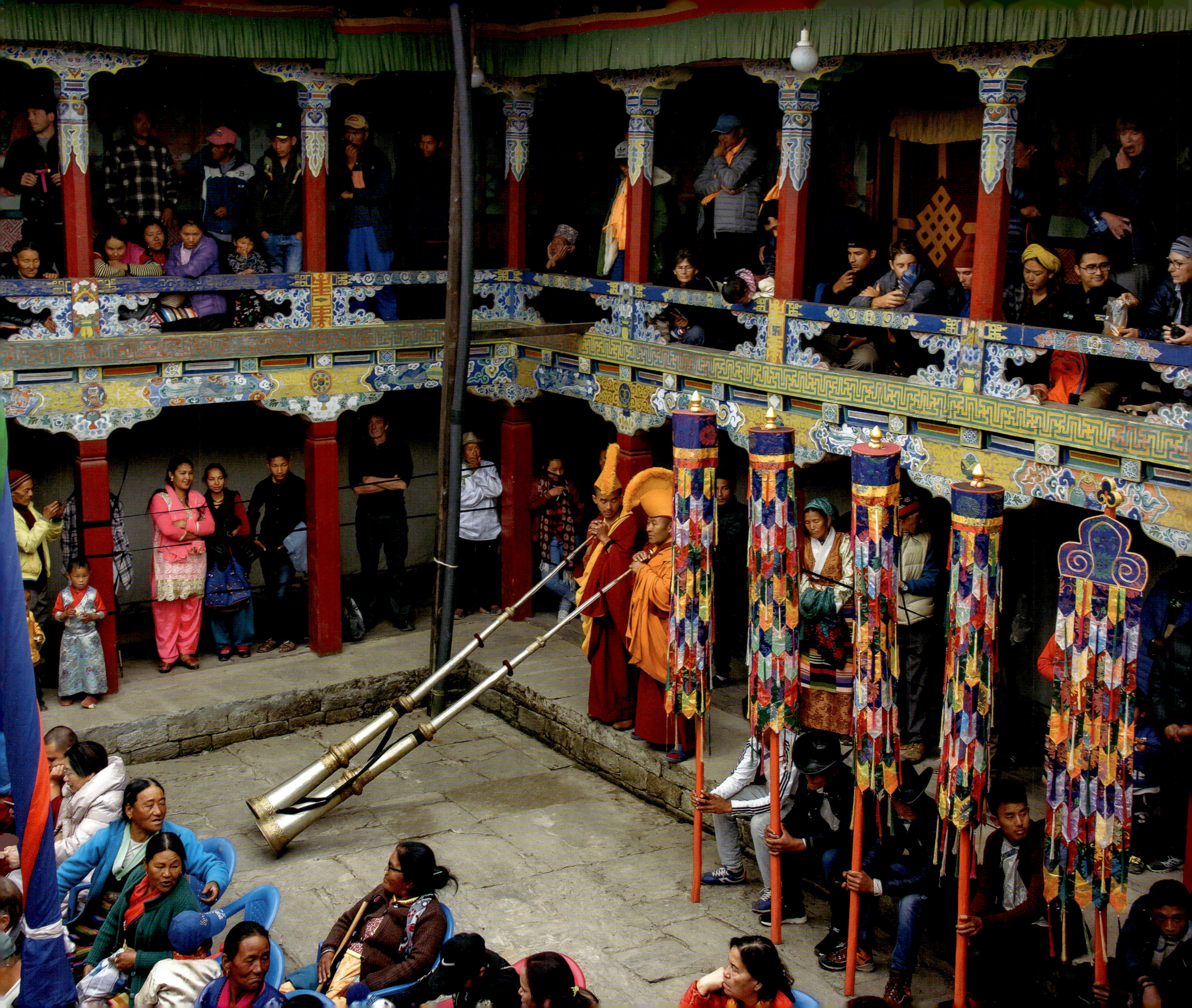

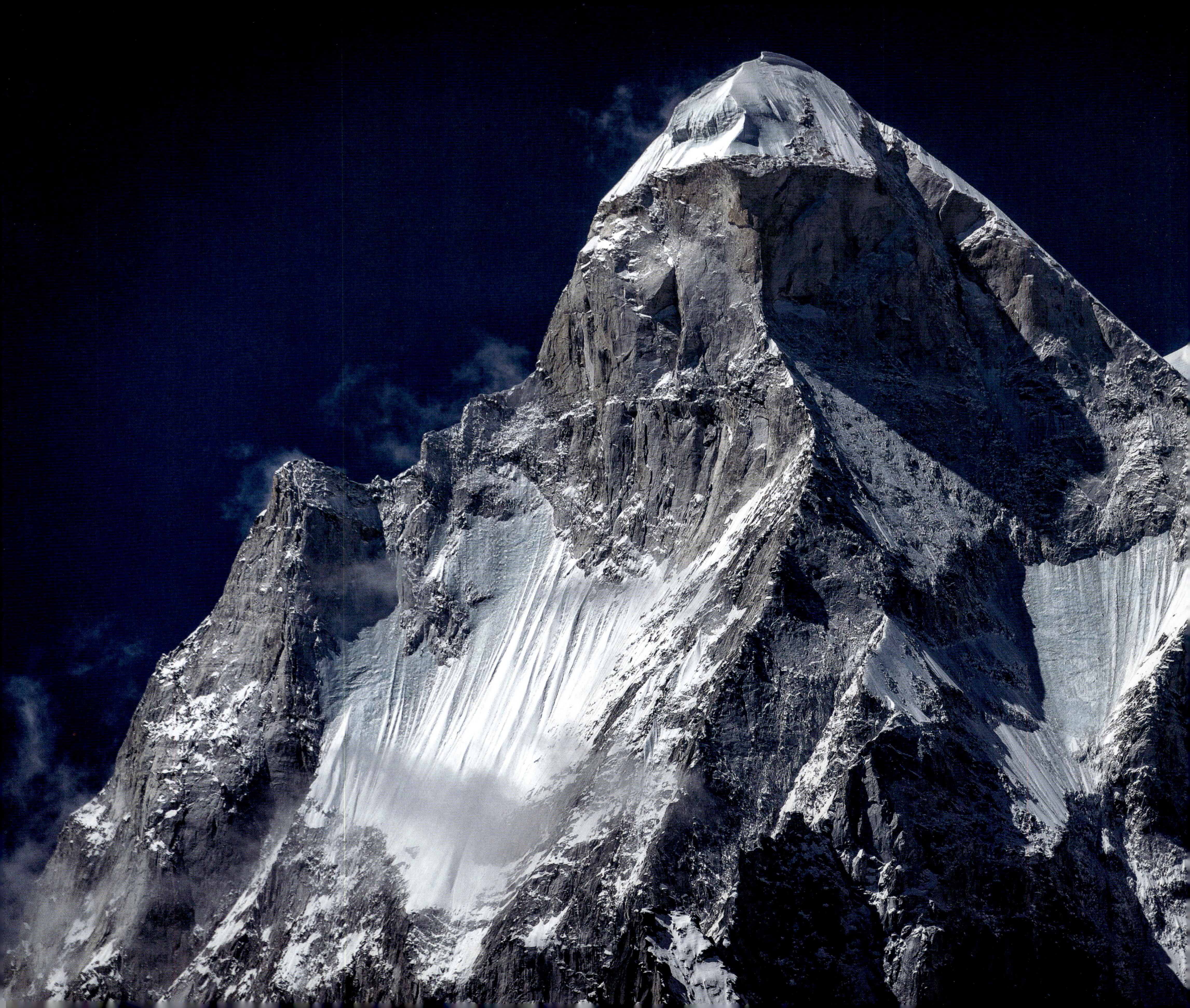

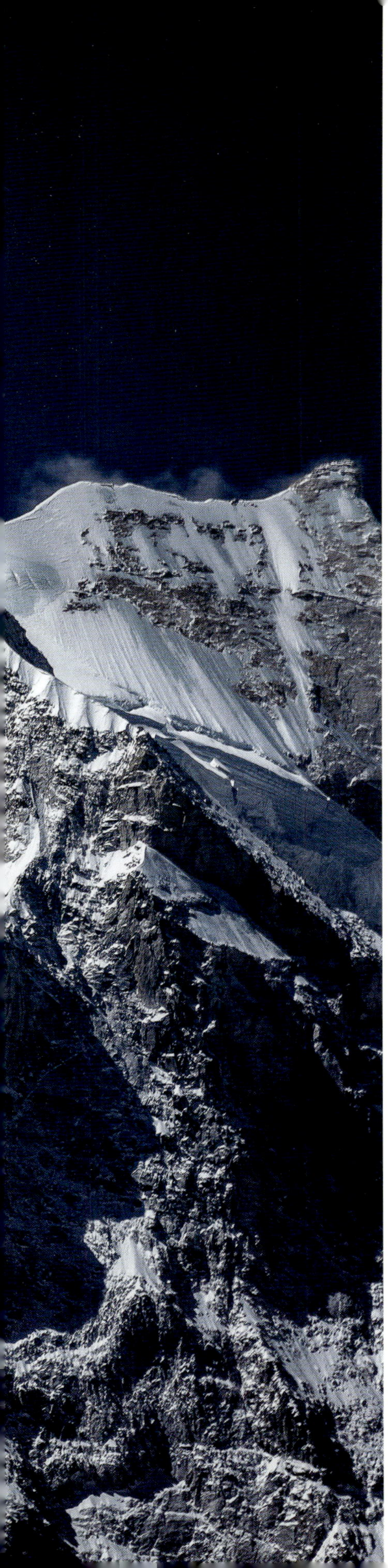

To the Source of the Ganges, India

JUNE 2018

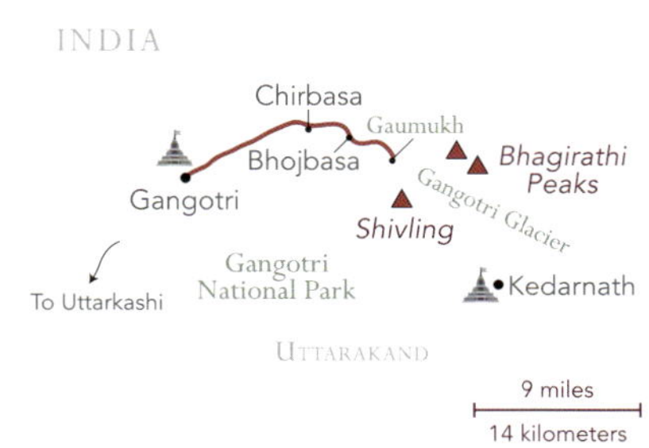

The sacred Ganges River winds its way throughout India, past the holy cities of Varanasi and Rishikesh and through low-lying lands that flood every year. The source of this iconic and symbolic river is Gaumukh, the snout of the Gangotri Glacier in northern India. There, snowmelt from the Himalaya pours out of an opening at the bottom of the glacier into the Bhagirathi River, a principal tributary of the Ganges. Gaumukh means "mouth of the cow," which the snout of the glacier was believed to resemble until several years before our visit, when huge chunks of it fell away. In spite of this, many *sadhus* (Hindu holy men) as well as everyday believers travel to the town of Gangotri, eleven miles below the glacier, to visit the temple there and bathe in the river, believing that the water has the power to cleanse their sins.

We took the pilgrim trail from Gangotri to Gaumukh in June, joining groups of Indian pilgrims walking or riding ponies up to the wooded camp at Chirbasa. After Chirbasa the narrow trail winds along the edge of stone cliffs high above the river. The exposure is frightening, so only a few sadhus and trekkers go beyond this point. We had planned to walk for several days beyond Chirbasa, but at our camp at Bhojbasa, on the second night, our guides, Nitesh and Rajeev, received news that the trail beyond was washed out, due to a recent landslide. As a result, it was uncertain if we could hike safely to Gaumukh and on to Tapovan, the meadow several miles farther. The views from Bhojbasa were spectacular, with the summit of Mount Shivling (elevation 21,467 feet) rising above the lower peaks in the foreground and three of the peaks of the Bhagirathi Massif—Bhagirathi I (elevation 22,493 feet), Bhagirathi II (elevation 21,365 feet), and Bhagirathi III (elevation 21,175)—straight ahead in the distance. But we wanted to reach the holy glacier and see more of Shivling, a sacred peak regarded as the divine phallus of Shiva.

We started out the next day on a trail riddled with car-sized boulders that looked as if they had just fallen from the unstable ridges above us. The only other people on the trail were sadhus, the religious ascetics who take a vow of poverty and renounce worldly life. Like elsewhere in the Himalaya, they were recognizable by their appearance—penetrating eyes, strands of long, matted hair, painted foreheads, and

Shivling Peak, Garhwal Himalaya, Uttarakhand, India, 2018

loose clothing of yellow or saffron color. We have encountered wandering sadhus many times, especially near Hindu holy places such as Gaumukh. Although unpredictable and sometimes menacing, they seem to have an affinity for Bill and will cheerfully pose for photographs, with only a few exceptions. At the Kumbh Mela celebration we attended at the town of Hardiwar, India, in 1986, just the sight of cameras around our necks elicited scowls and angry gestures from a procession of the hard-core naked and ash-covered Naga sadhus. The Kumbh Mela, which takes place on a staggered basis every twelve years at four places in India, draws millions of Hindu pilgrims and holy men who come to celebrate together.

We joined a group of friendly sadhus on the trail and walked for several hours toward Gaumukh until the trail disappeared under piles of newly fallen rocks and boulders. In the distance we could see the opening of the legendary glacier, but, with only loose rock and no trail, we could safely get no closer. We watched the sadhus turn around and then scramble down the hillside to the edge of the river and splash themselves with the holy water, so we did the same before turning back to our camp.

It was not the first time we were unable to reach a planned destination in the mountains. Many different factors, from natural barriers such as rockslides or swollen rivers to inaccurate maps or government restrictions, had stopped us before. As we hiked back to our camp we thought wistfully of all the places in the Himalaya we had yet to see or wished to return to. At this point, it might be the lack of time that would stop us, but we pledged to each other to keep trying.

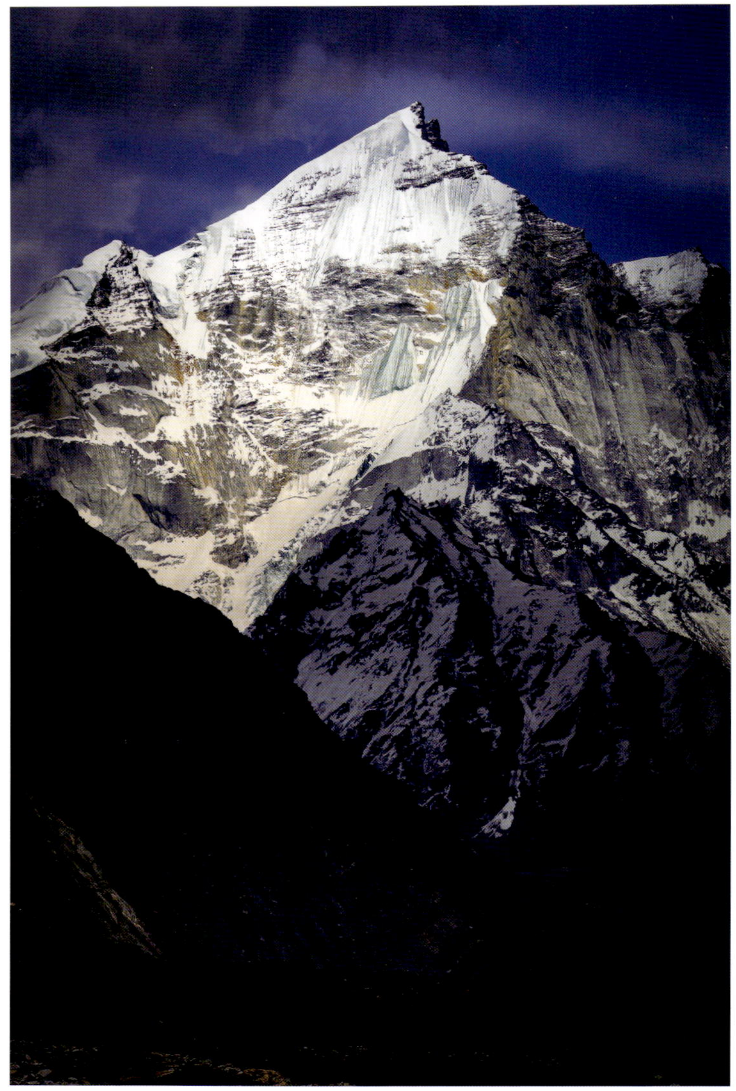

Bhagirathi I, Gangotri Glacier, Uttarakhand, India

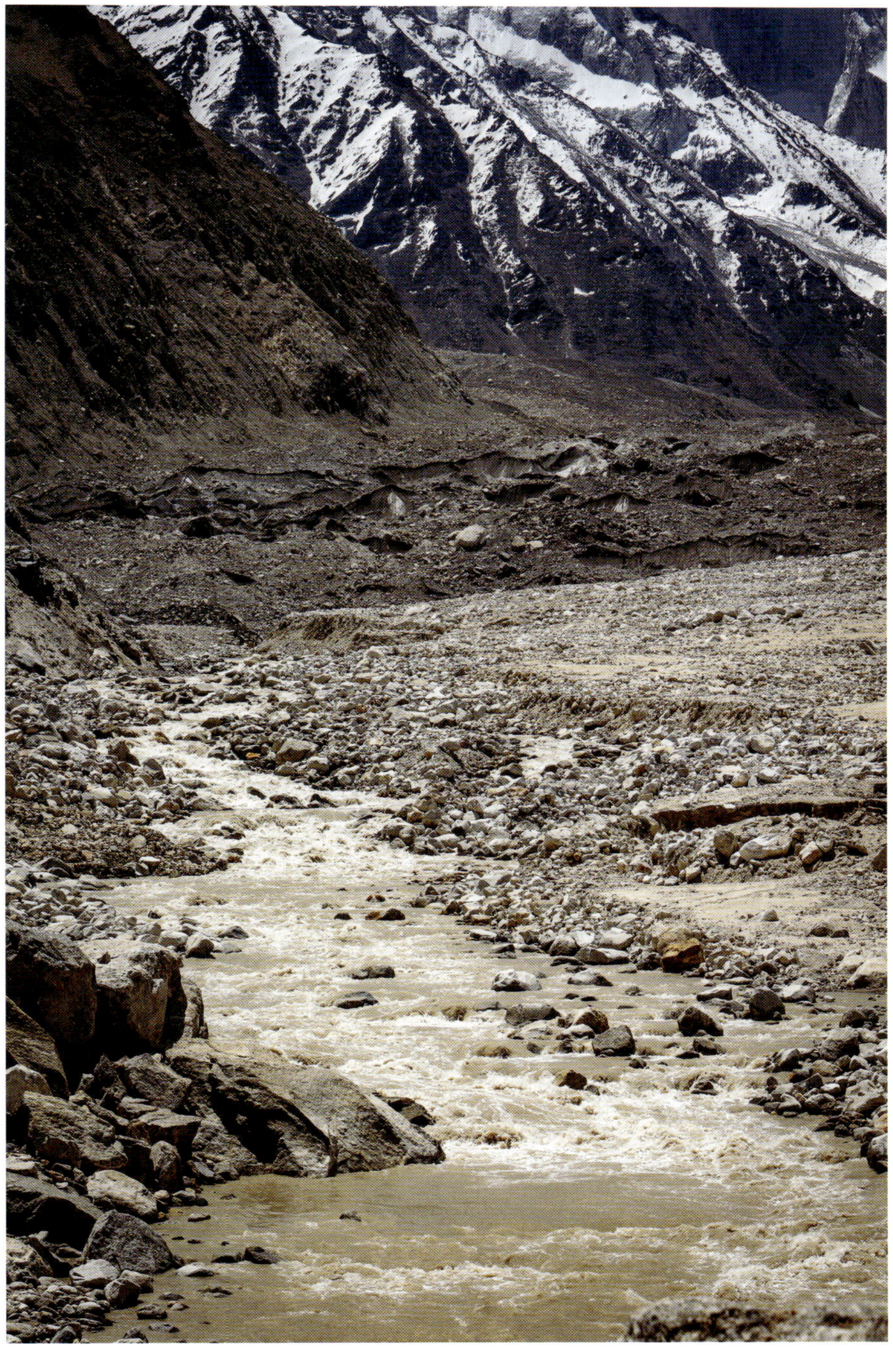

Gangotri Glacier and the Bhagirathi River, the source of the sacred Ganges, below Bhagirath I Peak, Uttarakhand, India, 2018

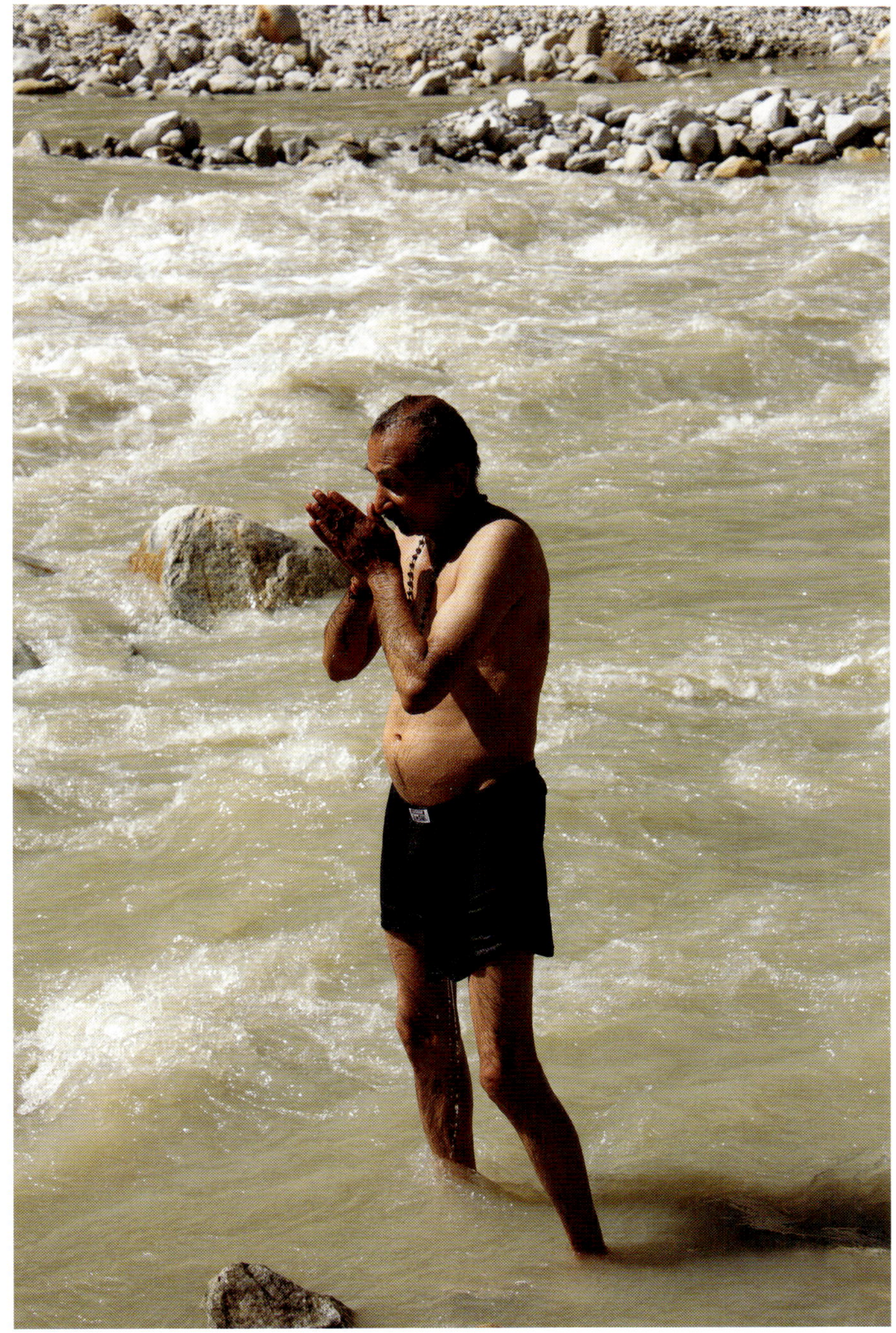

Prayers in the Bhagirathi River, a principal tributary of the Ganges River, Gangotri, India, 2018

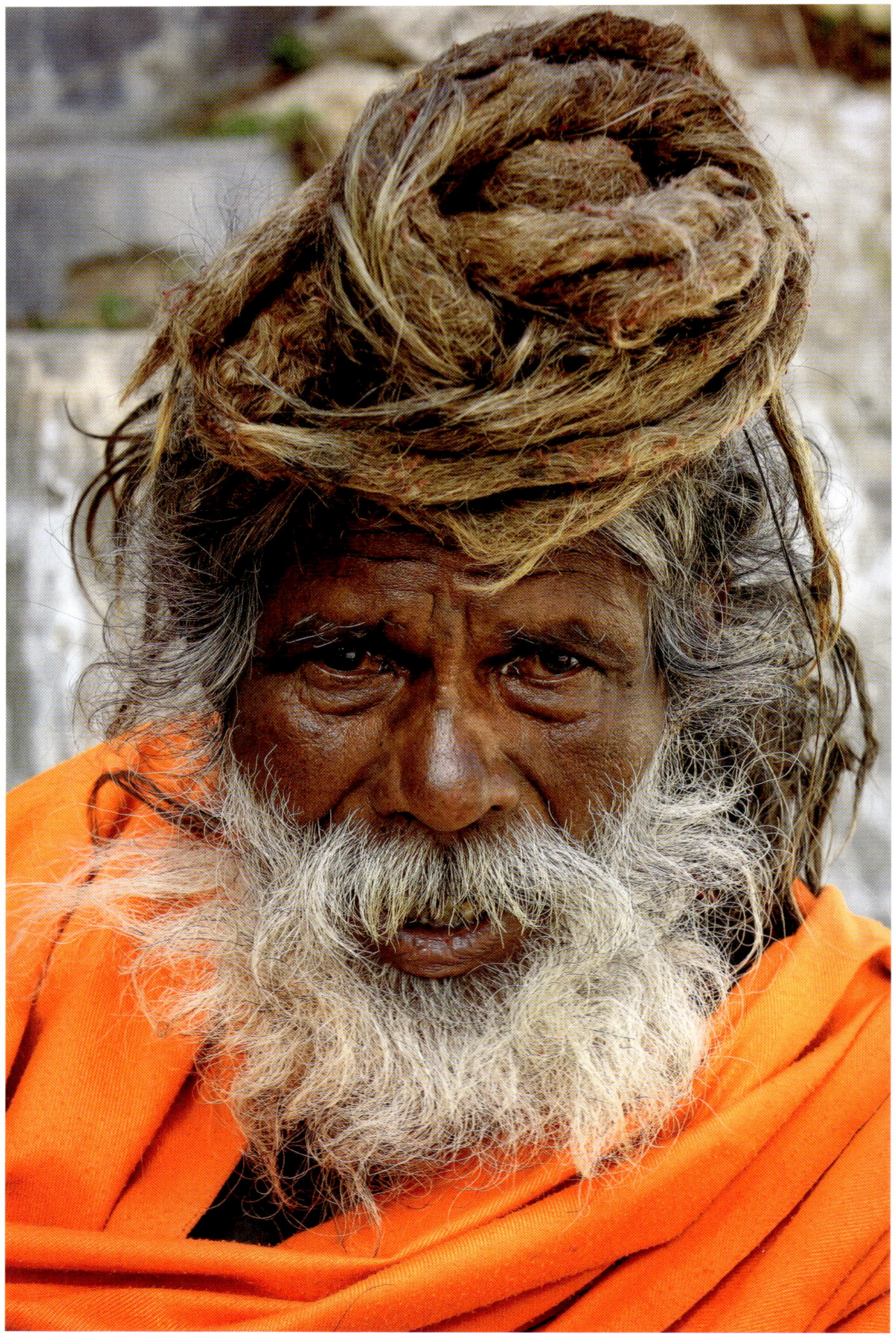

An Indian sadhu returning from a pilgrimage to the source of the Ganges, Uttarakhand, India, 2018

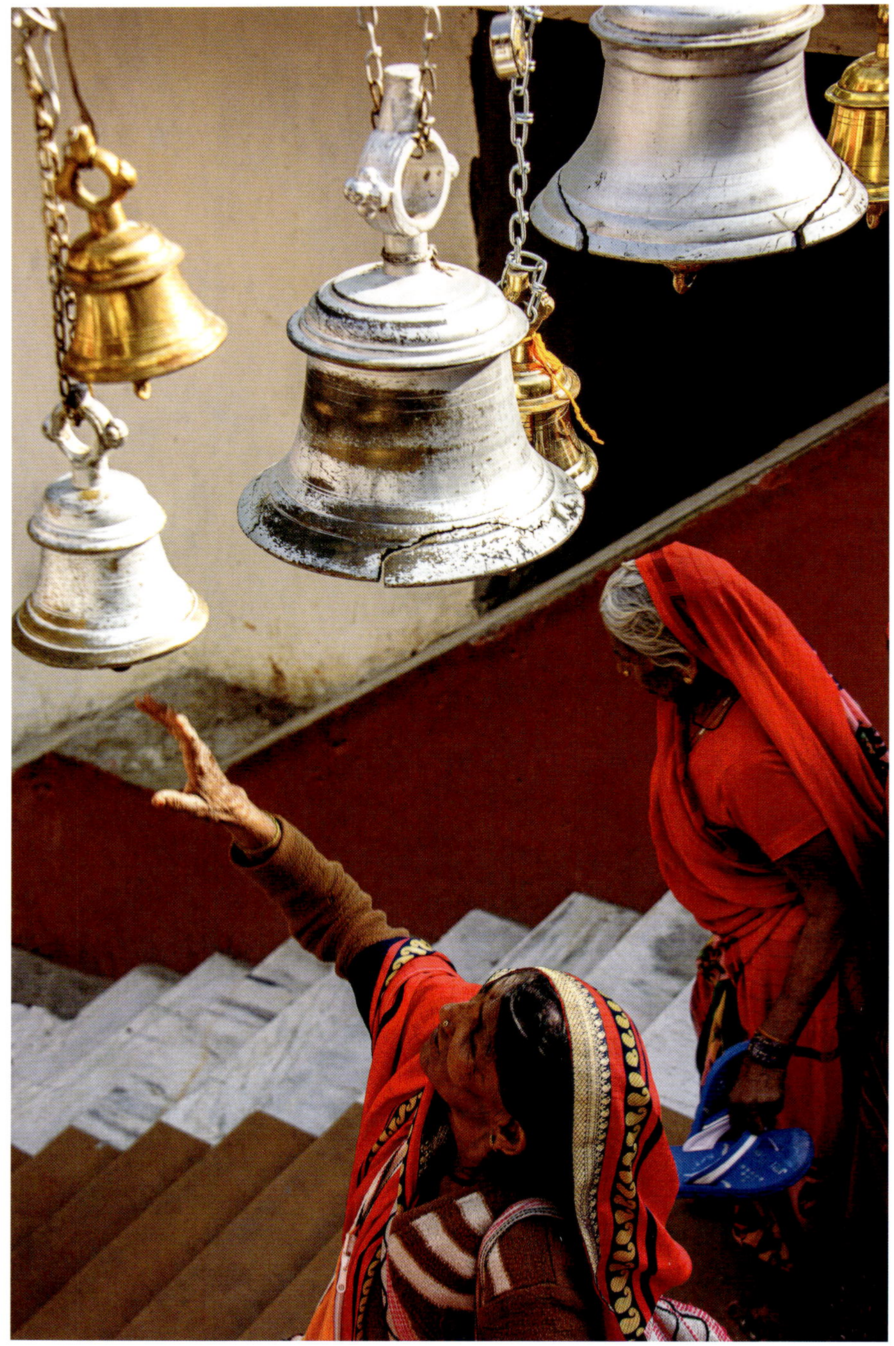

The sacred site of Gangotri, where Hindu pilgrims bath in the Bhagirathi River, Uttarakhand, India, 2018

Opposite: On the trail to Gaumukh, the source of the sacred Ganges River, Uttarakhand, India, 2018
Following spread: The trail to Gaumukh, Uttarakhand, India, 2018

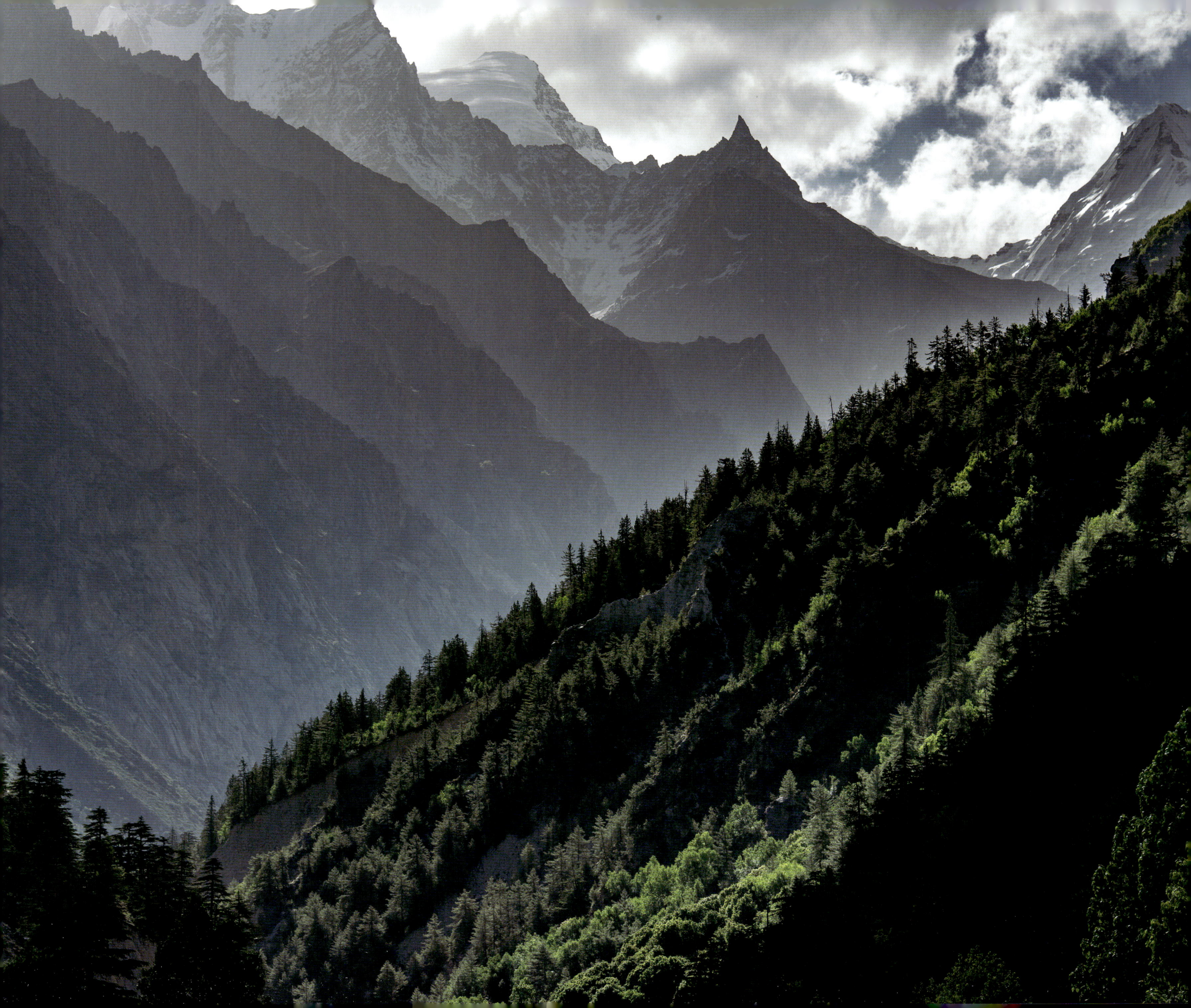

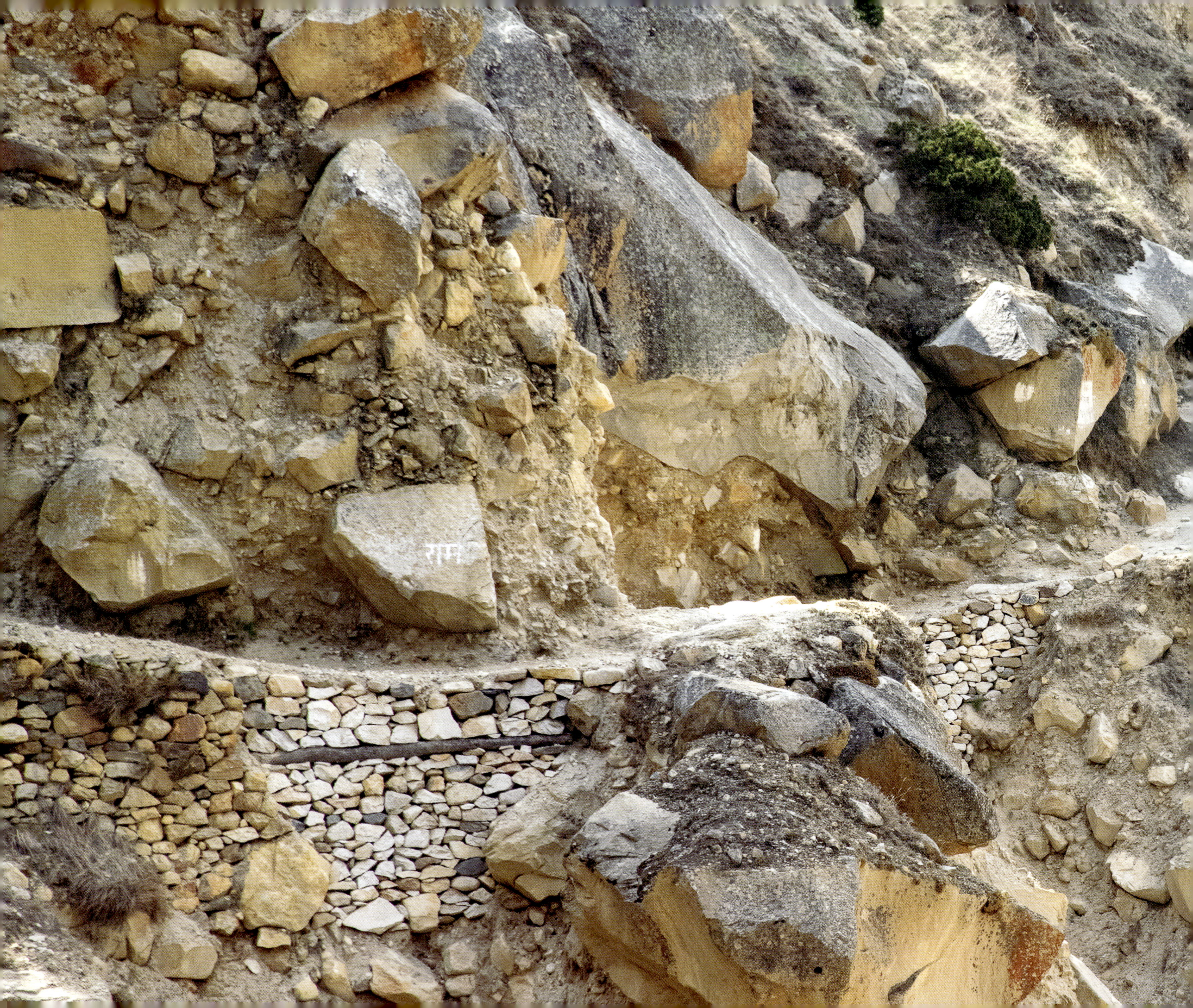

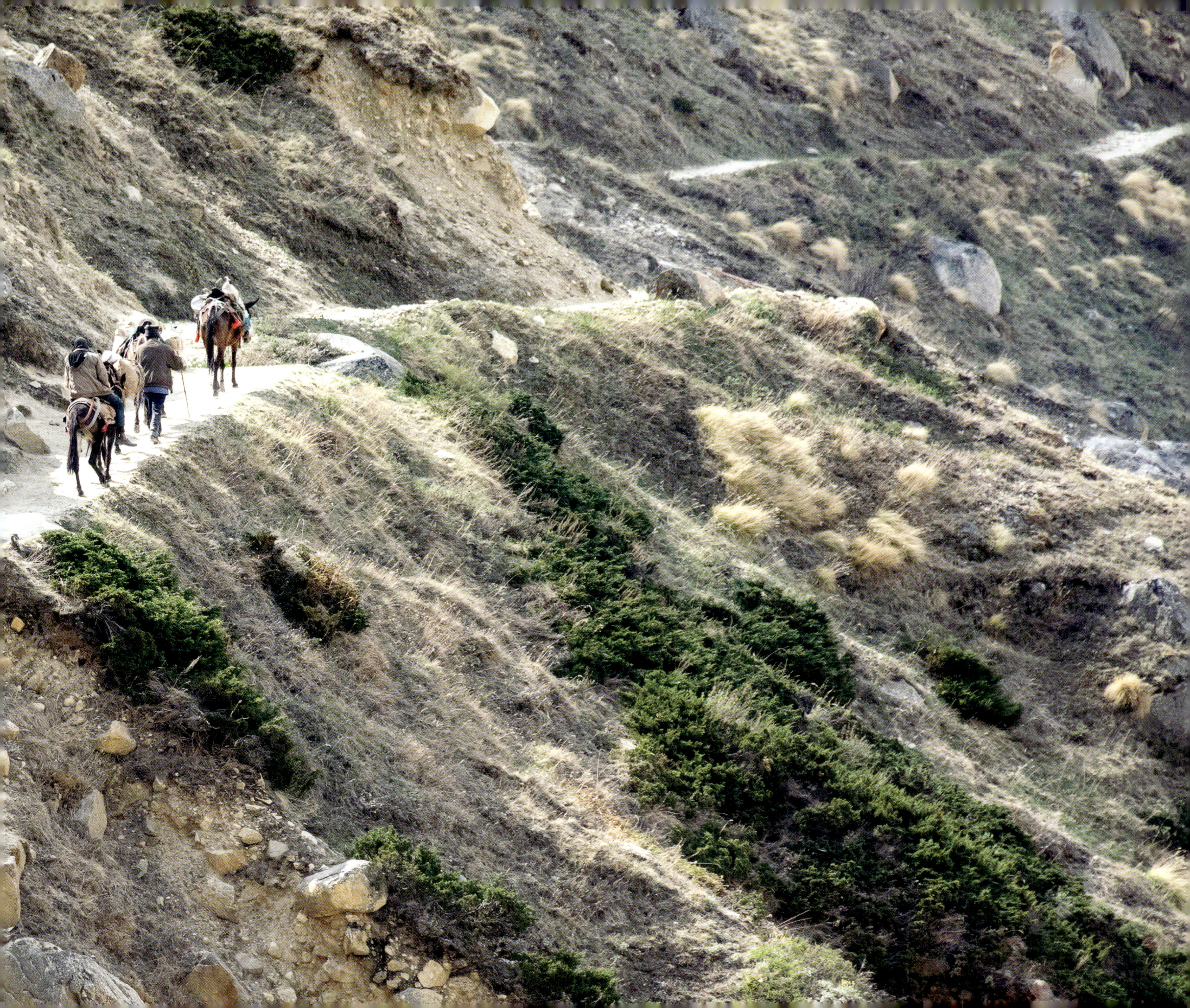

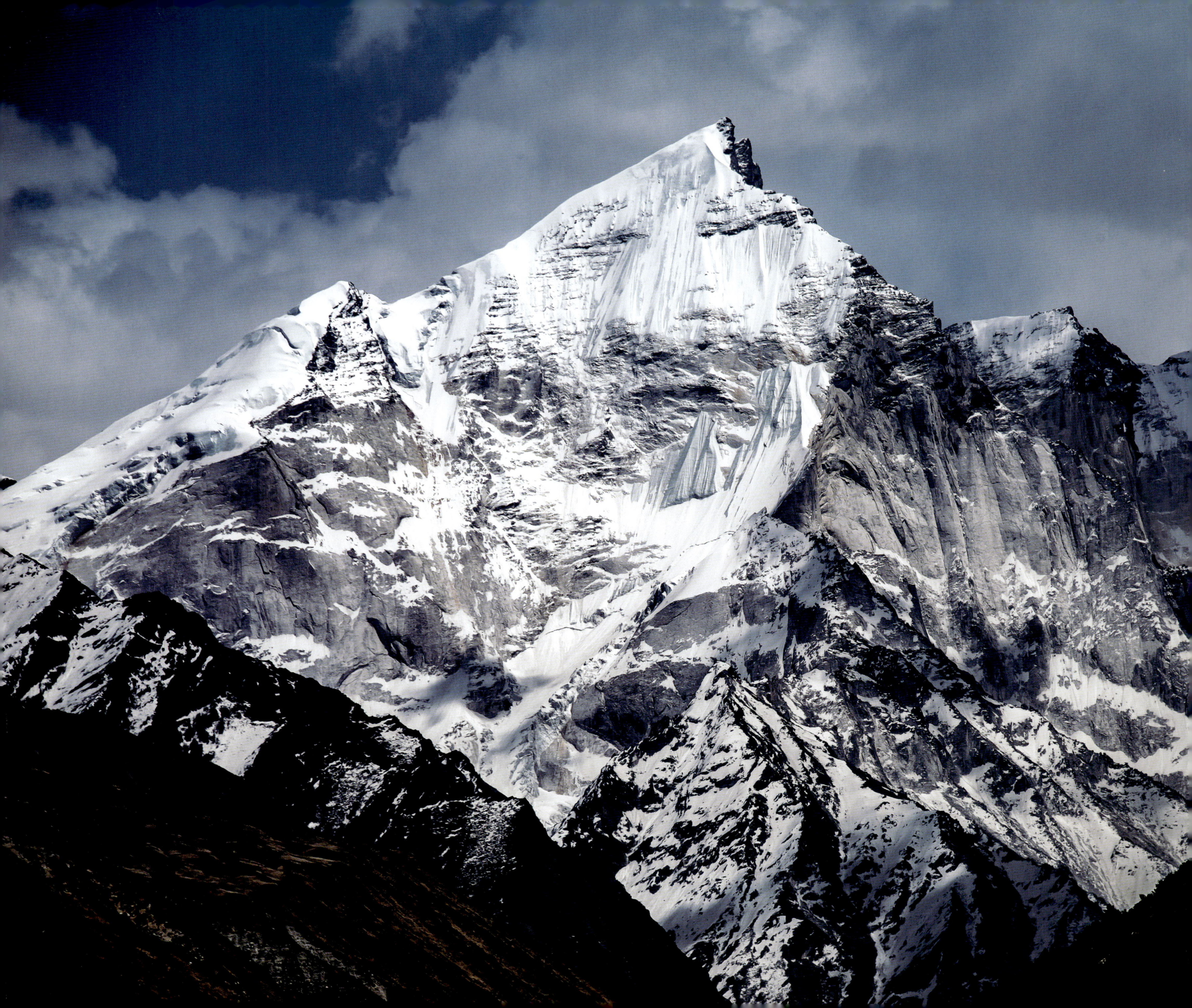

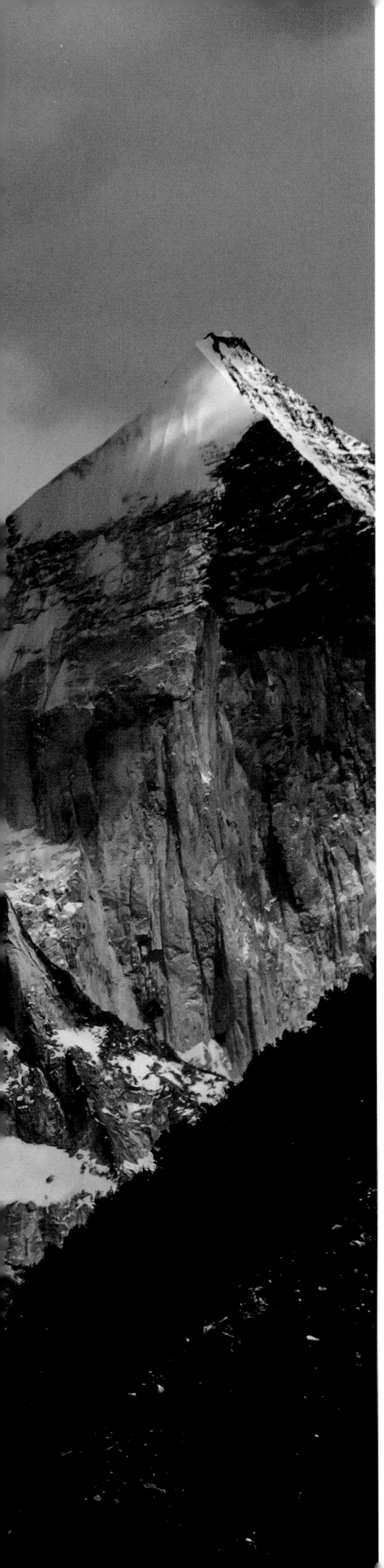

Epilogue

Many things have changed in the Himalaya since we started visiting there in 1981. Like everywhere else in the world, population growth and urbanization have strained resources both in cities and remote mountain areas. Much of the Himalaya is still rural, but the percentage of urban dwellers is growing. Kathmandu was a sleepy, small town when we first visited it in 1981, but today it is one of the fastest-growing cities in South Asia. Even Bhutan, which had been overwhelmingly rural, is seeing an increase in urbanization. In Tibet the Chinese government has encouraged the in-migration of ethnically Chinese citizens to the region, a clearly political policy that threatens to dilute Tibetan culture and identity.

There has also been explosive growth in the number of tourists to the cities and popular mountain regions of the Himalaya in recent decades. Nearly 1,200,000 foreign tourists visited Nepal in 2019. The number of tourists to the Khumbu region below Mount Everest has also grown considerably from fewer than 10,000 per year during the late 1980s to more than 60,000 in 2019. At Mount Everest, the highest place on Earth, long lines of climbers waited their turn to scale the last section of the peak.

Tourists and trekkers infuse cash into local economies by staying at inns, buying food, and employing inhabitants for trekking services. The increased demand for goods and services in undeveloped mountain communities by tourists can be overwhelming and poses challenges to the supply chain of goods for locals. For example, along popular trekking routes tourists add considerably to the demand for firewood, which has been traditionally a vital source of fuel for cooking and heating for people living in the foothills of the Himalaya. Programs promoting alternative fuel sources such as kerosene and solar as well as sustainable, community forestry programs have helped alleviate the strain in some locations, as has greater control of commercial timber operations. In Bhutan, a country that reveres its natural environment, its constitution requires a minimum of sixty percent of forest cover and fifty percent of the landscape to be under protected status.

During the past forty years, the improvement of infrastructure, especially roads, throughout the Himalaya has opened up opportunities for once-remote communities, although often at a cost. In Ladakh, one of the highest inhabited regions in the world, roads constructed by the military in this sensitive area adjacent to China have paved the way for tourists like us into once-closed areas such as the Nubra Valley. In Nepal, an emphasis on decentralized economic development means the provision of access to every town and sizeable settlement, so road systems are being constructed in places that once had only narrow trails.

Bhagirathi I above Gangotri Glacier, Uttarakhand, India, 2018

We saw firsthand the impact of improved roads when we visited Mustang in 2013. At that time, a right-of-way had been carved out of the steep hillsides, but the road was still rough and washed out in most places. In spite of this, on the good sections we saw kids in jeeps traveling to a school in a neighboring village and farmers hauling goods to more distant markets. The same is true in northern Pakistan where narrow, unpaved roads are a lifeline for some communities. The downside is the impact on fragile ecosystems. Removing vegetation and cutting away hillsides for roads makes the slope unstable and prone to erosion and landslides. Often roads in mountain regions are built along a river, so the debris from construction and landslides inevitably ends up in the river. Landslides are so common in some places that small teahouses have sprung up at these sites, and often jeeps and buses are available on both sides of the slide to carry passengers after they have scurried over the dirt and rocks on the road to get to the other side.

All of these issues are exacerbated by warming atmospheric temperatures. In recent years, glaciers in the Himalaya have reduced in size and retreated at an alarming rate. This affects the amount of seasonal water flow to millions of Indians living downstream who depend on it and increases the risk of catastrophic floods. Interestingly, glaciers in the Karakoram have remained somewhat stable, probably due to a weather pattern that helps keep them cold and dry during the summer.

Wildlife is also affected by all the same issues that impact people. Population growth, deforestation, and the expansion of farming into previously wild areas threaten the lifestyle and very existence of some species. In the Himalaya, in 2022, 163 species are considered globally threatened, including the Himalayan black bear, snow leopard, Ibex, blue sheep, red panda, golden langur, and tiger. Organizations such as the non-profit Snow Leopard Conservancy are working to ensure the species' survival by promoting community stewardship in their habitat areas and by encouraging incentives that can minimize human-wildlife conflict.

In Nepal, many of the regions we visited on our way to the base camps of the highest mountains have been designated as national parks and conservation areas, including the zones around Annapurna, Everest, Kangchenjunga, Makalu, and Manaslu. People living in the buffer zones outside the boundaries of the parks and conservation areas are allowed to utilize the resources there in a sustainable manner, which has encouraged cooperation. Himalayan regions in northern India are also protected by their designation as national parks, sanctuaries, and ecological zones.

These programs and others to protect the natural resources of the Himalaya have been developed with a recognition that it is a sacred landscape with some of the most beautiful natural features on Earth. For visitors like us, it is a source of inspiration. As Edwin Bernbaum eloquently notes in his introductory essay:

> The sense of another reality within or beyond the one we know is essential. Without it we slowly die of spiritual thirst, becoming desiccated husks of ourselves. The mountains of the Himalaya stand as constant reminders of that reality and the deeper life that flows from it, like a river from its source.

Acknowledgments

So many people have supported us in our travels over the years, both at home and abroad. In addition to our families, we thank our friends Yuko Matsumoto and Al Fleischman, for encouraging us to quit our jobs and travel, as they had done earlier; to Rick Kattenburg, for checking our stacks of Kodak slides after processing; and to the late Bill Daoust and Christine Bohar, for managing our affairs while we were away for long periods.

We are indebted to the guides and trekking staff members who showed us the way on the trail, fed us, carried our duffel bags, and shared their good cheer at the end of each day. Jagat Man Lama, now with More than Mountain Treks, was the first of many capable sirdars who showed us the hidden corners of Nepal. Ram Rai, whom we first met in 1985, remains a good friend. Special thanks to Ram and to Jalwant Gurung, of Crystal Mountain Treks in Kathmandu, for their guidance and logistical support over many years.

Ikram Beg, of Adventure Center Pakistan, led our trek to K2 Base Camp in 1985 and is a respected senior member of the mountain guide community in Pakistan. We are happy to still share stories with Ikram.

In India, Rajeev Rai, Sanjeev Rai, and Nitesh Sati, of White Magic Adventure Travel, guided us to many sacred Hindu and Buddhist sites, always with insightful commentary. Pawan Bisht, co-founder of White Mountain, has always been a source of knowledge and inspiration. Thanks also to Ramdev, our horseman in Ladakh, and to Sarosh Inayat, our guide to Kuari Pass in the Garhwal. Tashi Jorpel, our guide in Bhutan, shared his deep knowledge and love for his beautiful country.

We also share our gratitude to the many people who kindly accepted our invitation to allow us to photograph them. Their acceptance, kindness, humility, and good karma enabled this book to become a reality.

In addition, we thank George F. Thompson, the publisher, for his belief in this book and his many insights and editorial skills. David Skolkin, the brilliant book designer, crafted a volume that we certainly are proud of. Joanna Hurley's years of experience were invaluable throughout the management and editorial processes. Morgan Pfaelzer's maps added a sense of reality to our many steps taken throughout the Himalaya over the past forty years.

We also would like to thank our good friends Michael Tobias and Jane Gray Morrison for their inspirational foreword. And finally we thank Edwin Bernbaum, for his brilliant and scholarly essay on sacred mountains, his long-standing support for the people of the Himalaya, and his friendship over the past four decades.

Bibliography

Allen, Charles, *A Mountain in Tibet: The Search for Mount Kailas and the Sources of the Great Rivers of India* (London, UK: Futura Publications, 1982).

_________, *The Search for Shangri-La: A Journey into Tibetan History* (London, UK: Abacus, 2001).

Bernbaum, Edwin, *Sacred Mountains of the World,* 2nd edition (London, UK & New York, NY: Cambridge University Press, 2022).

Blum, Arlene, *Annapurna: A Women's Place* (San Francisco, CA: Sierra Club Books, 1998).

Bonington, Chris, *Annapurna South Face* (New York, NY: McGraw-Hill, 1971).

_________, *Everest: The Hard Way* (New York, NY: Random House, 1976).

Bowman, W. E., *The Ascent of Rum Doodle* (London, UK: Max Parrish, 1956).

Curran, Jim, *K2, Triumph and Tragedy* (Seattle, WA: Mountaineers Books, 1987).

_________, *K2, The Story of the Savage Mountain* (Seattle, WA: Mountaineers Books, 1995).

Craig, Robert W., *Storm and Sorrow in the High Pamirs* (New York, NY: Simon and Schuster, 1977).

Dupree, Nancy Hatch, *The Valley of Bamiyan* (Kabul, Afghanistan: Afghan Tourist Organization, 1967).

Govinda, Lama Anagarika, *The Way of the White Clouds: A Buddhist Pilgrim in Tibet* (second Indian Edition: New Delhi, India: B.I. Publications, 1982).

Govinda, Li Gotami, *Tibet in Pictures*, Vols. 1 & 2 (Emeryville, CA: Dharma Press, 1979).

Harvard, Andrew, and Todd Thompson, *Mountain of Storms: The American Expeditions to Dhaulagiri, 1969 and 1973* (New York, NY: New York University Press, 1974).

Herzog, Maurice, *Annapurna: First Conquest of an 8,000-Meter Peak* (New York, NY: E. P. Dutton & Co., 1953).

Hillary, Sir Edmund, and Desmond Doig, *High in the Thin Cold Air* (Garden City, NY: Doubleday., 1962).

Hunt, Sir John, *The Conquest of Everest* (New York, NY: E.P. Dutton & Co, 1954).

Ishii, Hiroshi, and Pradyumna P. Karan, *Nepal: A Himalayan Kingdom in Transition* (New York, NY: United Nations University Press, 1996).

Isserman, Maurice, and Stewart Weaver, *Fallen Giants: A History of Himalayan Mountaineering from the Age of Empire to the Age of Extremes* (New Haven, CT: Yale University Press, 2008).

A Note on the Photography

I consider myself a documentary photographer, and my photographic work over four decades has focused on chronicling Indigenous cultures and their environments. My objective for this book was to capture in photographs the cultures of those who inhabit regions of the vast Himalayan range as well as the rugged environments that define their lives.

All of the photographs in this book were taken with Nikon camera equipment. On our first visits to the Himalaya in 1981 and 1982, I carried a Nikon F3. This camera, introduced in 1981, had both manual and semi-automatic exposure control. This rugged, professional camera set the standard for photographers embarking on treks, overland excursions, and mountaineering expeditions. On our return to the Himalayas in 1985, 1986, 1991, and 1993, my camera of choice was the Nikon F3HP. The primary difference between the F3 and the F3HP is the viewfinder, allowing the photographer to view entirely the viewfinder's full-frame image. The photographs in this volume from this early period (1981, 1982, 1985, 1986, 1991, and 1993) were taken with a Nikkor 24 mm lens, a Nikkor 50 mm lens, or a Nikkor P 105 mm lens and were digitally processed from scanned Kodak Kodachrome 25, 35 mm transparencies. My later images in this volume, taken from 2008–2018, were produced using five different Nikon Digital Single Lens Reflex (DSLR) cameras: D80, D600, D610, D810, and the D850. All of the lenses used with these cameras were Nikkor fixed focal length 24, 35, 50, and 105 mm and Nikkor 14–24, 24–70, and 28–300 mm zooms. I used a Nikon neutral color (NC) filter on all of the lenses, except the 14-24 mm zoom, to reduce glare and reflections. Most of my photographs were taken hand-held, without a tripod, although, at times, I used a tree or a rock to help steady the camera in low light. Camera flash and artificial light were never used.

Mountain photography certainly has its challenges. Bright light, high contrasts, intense whites and, at times, bitterly cold temperatures are all issues the high-altitude mountain photographer confronts on a daily basis. The weight of one's equipment is also a challenge. The weight of the early Nikons was ideal, unlike the much heavier digital equipment I use today. Luckily, Kodachrome 25 transparency film, with its long-term stability, retains its accurate color and density over many years. The Kodachrome transparencies scanned for this volume were as sharp as the day they were taken, in many cases, after 40 years.

Travels Across the Roof of the World: A Himalayan Memoir was brought to publication in an edition of 1,250 hardcover copies. The text was set in Perpetua and Akkurat, the paper is Condat Perigord matt (FSC), 150 gsm weight, and the book was professionally printed and bound by Pristone Printing Ltd. in Singapore.

Publisher: George F. Thompson
Project and Editorial Director: Joanna T. Hurley
Editorial and Research Assistant: Mikki Soroczak
Manuscript Editor: Purna Makaram
Book Design and Production: David Skolkin
Map Research: Anne Frej
Map Design: Morgan Pfaelzer

Published in 2022. First hardcover edition.
Printed in Singapore on acid-free paper.

George F. Thompson Publishing, L.L.C.
217 Oak Ridge Circle
Staunton, VA 24401–3511, U.S.A.
www.gftbooks.com

30 29 28 27 26 25 24 23 22   1 2 3 4 5

The Library of Congress Preassigned Control Number is 2021951025.

ISBN: 978–1–938086–93–9